2002

Human Evolution through Developmental Change

Human Evolution through Developmental Change

Edited by
Nancy Minugh-Purvis
and
Kenneth J. McNamara

The Johns Hopkins University Press
Baltimore and London

The Johns Hopkins University Press
2715 North Charles Street
Baltimore, Maryland 21218-4363
www.press.jhu.edu

Library of Congress Cataloging-in-Publication Data

Human evolution through developmental change / edited by Nancy
Minugh-Purvis and Kenneth J. McNamara.
    p.   cm.
Includes bibliographical references (p.   ) and index.
ISBN 0-8018-6732-0 (hardcover)
1. Human evolution.   2. Heterochrony (Biology).   3. Fossil
hominids.   I. Minugh-Purvis, Nancy, 1953–   .   II. McNamara, Ken.
GN281.4.H85   2001
599.9398—dc21   00-012970

A catalog record for this book is available from the British Library.

Contents

Part II   The Evolution of Hominid Life-History Patterns

Part III   The Evolution of Hominid Development

Foreword

> Throughout most of the twentieth century the obvious connection be-
> tween phylogenetic transformations in shape of organisms and the un-
> derlying modifications of the genetic systems controlling ontogeny has
> been largely ignored except by a few outsiders to the neo-Darwinian
> synthesis, which was built of other elements. That synthesis was in-
> complete.
>
> —R. A. Raff and T. C. Kaufman, *Embryos, Genes, and Evolution: The*
> *Developmental-Genetic Basis of Evolutionary Change*

The few, outstanding zoologists to whom Raff and Kaufman were referring probably number no more than a half-dozen and are principally (major publication dates in parentheses) St. G. Mivart (1871), William Bate-
son (1894), Walter Garstang (1922), Gavin de Beer (1930, 1958), Julian S.
Huxley (1932, 1942), and Richard Goldschmidt (1933, 1940). All were, of
course, heard but scarcely heeded in regard to the role and significance of
developmental mechanisms in phylogenesis, at least within the context of
the modern synthesis.

Developmental biology has ultimately and rightfully gained a central
position in evolutionary studies. There is scarcely an aspect of modern evolutionary biology that does not rely on or profit substantially from its perspectives and findings. How, then, has this patent if lamentable absence of
any developmental perspective within the framework of the modern evolutionary synthesis come to be? Some important historical perspectives, influential in my own thinking, have been especially Jane M. Oppenheimer's
Essays in the History of Embryology and Biology (1967), *A History of Embryology* (T. J. Horder, J. A. Witkowski, and C. C. Wylie, eds., 1986), and Scott
F. Gilbert's edited volume, *A Conceptual History of Modern Embryology*
(1991).

Fifty years ago, in my own graduate studies, embryology seemed largely
devoid of any evolutionary and genetic reference. Generally speaking, genetics was focused largely on its transmissional, physiological, and popula-

tional aspects, while human genetics, specifically, was essentially medical genetics. *Factors of Evolution* (1949), by I. I. Schmalhausen, has been thought by Theodore Dobzhansky to supply "as it were, an important missing link in the modern view of evolution." However, at the time, this surely did not seem to be so. Similarly, Richard Goldschmidt's earlier *The Material Basis of Evolution* (1940), by a practicing developmental biologist and physiological geneticist, utterly failed to capitalize on this opportunity to forge this missing link in the emergent synthetic chain.

On the contrary, the first successful effort in my view was that of C. H. Waddington's *The Strategy of the Genes* (1957) and a successor volume, *New Patterns in Genetics and Development* (1962). Most of the concepts and guiding perspectives were set out in his earlier volumes, devoted either to genetics (*An Introduction to Modern Genetics,* 1939) or to development (*Principles of Embryology,* 1956). Because of other preoccupations at the time, I did not sufficiently appreciate this fact until we met on the occasion of the well-known Darwin Centennial held at the University of Chicago in November 1959. In fact, developmental biology was explicitly excluded from the concerns of that occasion. Nonetheless, the overall effect on evolutionary studies was muted, and it was to be another decade before there were fruitful consequences. Incidentally, that particular term J. S. Huxley and I shared adjacent offices. His groundbreaking volume, *Problems of Relative Growth* (1932), came under discussion on many occasions, particularly with reference to the growing hominid fossil record and the nature and source of increasingly purported taxic distinctions. Huxley had also recognized in *Evolution: The Modern Synthesis* (1942)—my first inclusive introduction to the subject during the war years—the relevance of development in evolutionary transformation, devoting a section of a chapter to "the consequences of differential development." He remarked repeatedly on the necessity for elucidation of developmental patterns (and differences in them) in the hominization process, as requisite sources for insights into selective factors attendant on morphological change(s) and consequent adaptations.

In the area of primatology, A. H. Schultz was extensively concerned with aspects of growth and development in nonhuman primates. His explications of growth and development of chimpanzee and orangutan (in *Contributions to Embryology,* Carnegie Institution, 1940 and 1941, respectively), largely from his own collections, were unique for their time. This interest is epitomized in his *Life of Primates* (1969), the longest chapter of which is devoted to growth and development. These efforts both stimulated and were completed by the subsequent cross-sectional study of growth of gorillas (by Francis E. Randall) and, more extensively, longitudinal growth of the chimpanzees (by James A. Gavan), during and after the war years (published in *Human Biology*, 1943–44 and 1953, respectively). Subsequently, other investigators were to provide longitudinal studies of postnatal growth in two species of *Macaca*, the only taxon of Cercopithecoidea so studied thus far. Human growth studies were long a flourishing part of biological anthropology but then largely lacked an evolutionary linkage or perspective.

Surely, the subsequent modern and major stimulus for this perspective was S. J. Gould's *Ontogeny and Phylogeny* (1977, with roots to 1966 in a major essay in *Biological Reviews*). Its effect was immediate, substantive, and far-reaching; J. S. Huxley would have lauded it mightily. However, the first significant inclusions of this perspective, in a text, were by M. L. McKinney in *Heterochrony in Evolution* (1988) and by J. S. Levinton in *Genetics, Paleontology, and Macroevolution* (1988).

This present volume, appearing at the onset of the millennium, is a salutary and welcome documentation of the enhanced application of evolutionary developmental biology to higher primates, and particularly to hominins, extant and extinct. It is certainly the most inclusive treatment afforded thus far, and it fills a need arising from previous wide-ranging overviews of this general subject by M. L. McKinney and K. J. McNamara (*Heterochrony: The Evolution of Ontogeny*, 1991) and by K. J. McNamara (*Shapes of Time*, 1997).

Clearly, there are various aspects to developmental change. Allometry reflects shape change correlated with decrease or increase in size. The appearance of evolutionary novelty (apomorphy) is a central concern, and delineation of the process(es) behind it is a major issue. Heterochrony, reflecting change in developmental timing, explicitly "displacement of a feature relative to the time that this same feature appeared in an ancestral form" (de Beer's initial formulation), concerns such shifts in regard to organs, structures, features, traits, and shapes (but not rates or timings as processes). Explicitly, following Gould (1977), heterochrony is "phyletic change in the onset or timing of development, so that the appearance or rate of development of a feature in a descendant ontogeny is either accelerated or retarded relative to the appearance or rate of development of the same feature in an ancestor's ontogeny."

All the contributions in the initial part of this volume are concerned with aspects of differential growth (onsets and offsets) rates, heterochronic effects in general and in respect to particular organs, systems, character complexes. The extent and nature of progress in such studies are clearly demonstrated. However, distinctions in usage of the heterochonic perspective, of comprehension and application of particular patterns, reveal the necessity, still, for further elucidation at both the conceptual level and that of praxis.

The extant hominins, humanity included, comprise a paucity of their past diversity, as is increasingly well documented by an enhanced fossil record. Varied patterns of growth, development, and maturation are displayed among extant taxa, and more intensive investigation of these phenomena, both longitudinally but usually cross-sectionally, has very substantially enhanced the sparse knowledge of decades past. Nonetheless, many aspects of comparative biology and its developmental genesis of extant taxa remain poorly documented and not well understood. Extinct taxa will ultimately afford documentation of alternative patterns and antecedents to conditions manifest in descendant and extant forms. The accelerative pace of study focused on life-history patterns and their variations promises to revolutionize some ill-formulated conceptions of evolutionary biology now still prevalent.

Last, the vastly enhanced documentation of the hominin fossil record now reveals the remarkable diversity of past taxa and adaptations. Even a minimalist perspective cannot subsume the spatiotemporal diversity in any simplistic, linear scheme. The developmental biology of such past hominin representatives has increasingly become the subject of new and innovative study, such that substantial and important contrasts have been quite well documented in some systems between purported taxa of specific or lesser rank. The potential of such studies, often employing greatly enhanced technological resources, is demonstrably great. Ultimately, developmental biology must come to play a significant role in character recognition, evaluation, and integration, a seriously persistent problem in view of the widespread application of cladistic procedures for assessment of (sister) group relations. Correlative evidence obtained from different paths of investigation promises to enlighten problems currently controversial or otherwise seemingly intractable. Overall this volume demonstrates uncommonly well both the current status of such knowledge and the expanding scope of research leading to future enlightenment and resolution.

F. Clark Howell
Laboratory for Human Evolutionary Studies
Museum of Vertebrate Zoology
University of California-Berkeley

Preface

In recent years, biologists, paleontologists, and paleoanthropologists, separately and together, have come to recognize development as an indispensable ingredient of evolutionary change. This would have been no surprise to our colleagues of the nineteenth century, many of whom merged the phenomena of ontogeny and phylogeny into a single, intricate continuum. But as the new synthesis began to reshape views of evolution in the 1920s and 1950s, genetic mechanisms were quickly favored over developmental phenomena as the driving forces of evolution. Not until the close of the twentieth century did an awareness resurface that the link between phenotype and genotype is development, that the ontogenetic perspective is the new synthesis through which twenty-first century science will understand evolutionary change.

During the 1970s paleoanthropologists began to excavate museum archives as well as geological strata as interest grew in the large number of long-ignored immature hominid remains languishing in museum storage. Typically more fragmentary than the remains of adults, infant, juvenile, and adolescent hominid fossils posed the additional annoyance of not always exhibiting the morphology seen in adults and so crucially needed for the typological studies traditionally characterizing paleoanthropology. By the 1980s, however, as controversy over the nature of modern human origins fermented and the necessity for studies beyond morphological analysis of adult phenotype became self-evident, interest in developmental studies in paleoanthropology intensified.

During this same period, remarkable strides in molecular biology led to the emergence of the discipline of evolutionary developmental biology. For the first time, evolutionary relationships between taxa, previously suspected on the basis of paleontological research, could be tested in the laboratory and the ontogenetic links leading to the appearance of new phenotypes could be demonstrated in the lab as well. Moreover, it became apparent that this generation of new phenotypes—this evolutionary patterning—was due to variations in developmental rate and timing, known as *heterochrony*, operating at levels from molecules to organ systems.

In 1996, encouraged by Solomon Katz, former head of the Anthropology Section for the American Association for the Advancement of Science (AAAS), Nancy Minugh-Purvis and Michael McKinney organized a symposium, "Heterochrony: Merging Evolutionary Perspectives in Biology, Paleontology, and Paleoanthropology," attempting further interdisciplinary dialogue on heterochrony between evolutionary biologists from various fields. Twelve papers presented at the Baltimore AAAS meeting discussed topics ranging from the effect of heterochronic patterning on the evolution of behavior to morphogenesis. Even in the planning stages, the intensity and enthusiasm of exchanges between the participants indicated the need for a book on the subject in which students of evolution from various subdisciplines could share their different approaches to common problems.

When, in late 1997, other commitments prevented Mike McKinney from continuing as a coeditor, Ken McNamara enthusiastically jumped into coeditorship. Several of the original participants were unable to contribute to this volume, but we are pleased to have included some chapters by authors who were unable to attend the original meeting. This, we believe, has resulted in a well-balanced examination of heterochronic theory and its application to current issues in hominid evolution.

The book is organized into three parts focusing on, first, more theoretical applications of heterochronic theory to the hominid fossil record; second, the relationship of developmental change to aspects of hominid life history; and, third, the role of heterochronic change in the evolution of Pliocene to late Pleistocene hominids. In Part I, "Evolution and Development," after a brief introduction to heterochrony, Brian Hall introduces the reader to the biological basis of heterochronic change and David Alba, Gunther Eble, Brian Shea, and Ken McNamara discuss theoretical frameworks for heterochronic analysis. Part I ends with the contributions of Susan Crockford and Sean Rice, which look at biological complexes that influence the evolution of behavior. Crockford provides a provocative, testable hypothesis on the possible role of the thyroid in heterochronic patterning, and Rice looks at heterochronic modeling as it applies to the most distinctively complex human feature—the brain.

Part II, "The Evolution of Hominid Life History Patterns," picks up where Part I left off. It begins with Mike McKinney's engaging examination of human brain evolution, which is followed by Nina Jablonski, George Chaplin, and Ken McNamara's chapter investigating the evolution of unique signal characteristics of the earliest hominids, such as bipedalism, increased body size, and increased brain size. Rebecca German and Scott Stewart examine important, but often neglected, questions concerning the ontogeny of sexual dimorphism and its role in hominid evolution. Exploiting the rich dental fossil record, Jay Kelley examines the selective forces behind life-history evolution in Miocene apes in comparison with living apes, including humans. Robert Anemone examines dental development as an indicator of life-history patterns in early hominids, and Kevin Kuykendall discusses standards for assessing the dental evidence for life history in living African apes. Sue Parker's chapter completes this section with an interesting model for the examination of comparative cognitive development in primates, including hominids, based on extrapolations of neurological maturity from the evidence for dental development.

Part III, "The Evolution of Hominid Development," provides a chronological survey of heterochronic change in the hominid fossil record from the Pliocene to the late Pleistocene. Fernando Ramirez Rozzi's examination of differences in dental development provides thought-provoking evidence for heterochronic diversity early in the hominid fossil record. Christine Berge's chapter, as well as that of Andrew Nelson and Jennifer Thompson, provide rare glimpses of heterochrony in the hominid postcranial complex—a difficult area of study, given the extremely rare and incomplete nature of immature postcrania, but a record that they demonstrate is, nevertheless, a valuable source of developmental data. A developmental study of the *Homo erectus* cranium by Susan Antón challenges some traditional interpretations of the Chinese and Indonesian samples. Maureille and Braga provide an interesting study tracing developmental changes in the premaxillary suture from early to later hominids. Frank Williams, Laurie Godfrey, and Michael Sutherland find considerable differences in the growth patterning of the craniofacial complex of Neandertals and modern humans, providing an interesting contrast to the final chapter of the book, in which Nancy Minugh-Purvis finds continuity of heterochronic change in the neurocranium from Neandertals to modern humans in the Upper Pleistocene.

A few comments are needed regarding editorial practices in this volume. We have left the ever-contentious spelling of the term for late Pleistocene archaic Europeans to the discretion of our individual authors. As a result, readers will encounter *Neandertal* where authors have followed the convention of omitting the *h*, which was removed from the word *thal* by modern German orthography; others, particularly English colleagues, prefer the term *Neanderthal*, which retains the *h* persisting in the taxonomic nomen *Homo sapiens neanderthalensis*. Similarly, we have not been dogmatic editors in the formal naming of the Kenyan hominid specimen KNM-WT 15000, formerly known colloquially as *Turkana Boy*, but now called *Nariokotome Boy*. Some authors prefer to assign him to *H. erectus*, whereas others call him *H. ergaster*. There are also some differences of opinion concerning heterochronic terminology (as there have been for over a hundred years). In Chapter 2 David Alba discusses the variants in terminology in use at present. While we have our own predilections, we refrain from wielding the editorial big stick but have tried to ensure that each author clearly explains the meaning of the term that he or she is using. Our aim, of course, is to enlighten, not create confusion.

Finally, we sincerely thank all of our contributors, but especially F. Clark Howell for his thoughtful foreword and Mike McKinney, who was instrumental in organizing the original AAAS conference and this book and whose title this volume bears. We also thank the many outside reviewers who took the time to read and provide useful comment on the chapters contained herein and the enthusiastic and capable editorial staff at Johns Hopkins University Press: Ginger Berman, early in this project, and Wendy Harris, in the later stages. Ken McNamara also thanks Danielle West for her help with some of the illustrations, and Nancy Minugh-Purvis thanks Doug Purvis for his love, support, and patience throughout this project.

Contributors

David M. Alba, Institut de Paleontologia M. Crusafont, Barcelona, Spain

Robert L. Anemone, Department of Anthropology, Western Michigan University, Kalamazoo, Michigan

Susan C. Antón, Department of Anthropology, Rutgers University, New Brunswick, New Jersey

Christine Berge, Laboratoire d'Anatomie Comparée, Musée National d'Histoire Naturelle, Paris, France

José Braga, Laboratoire d'Anthropologie des Populations du Passé, Université Bordeaux I, France

George Chaplin, Department of Anthropology, California Academy of Sciences, San Francisco, California

Susan J. Crockford, Pacific Identifications, Victoria, British Columbia, Canada

Gunther J. Eble, National Museum of Natural History, Smithsonian Institution, Washington, D.C., and Santa Fe Institute, Santa Fe, New Mexico

Rebecca Z. German, Department of Biological Sciences, University of Cincinnati, Cincinnati, Ohio

Laurie R. Godfrey, Department of Anthropology, University of Massachusetts, Amherst, Massachusetts

Brian K. Hall, Department of Biology, Dalhousie University, Halifax, Nova Scotia, Canada

Nina G. Jablonski, Department of Anthropology, California Academy of Sciences, San Francisco, California

Jay Kelley, Department of Oral Biology, College of Dentistry, University of Illinois, Chicago, Illinois

Kevin L. Kuykendall, Department of Anatomical Sciences, Medical School,

University of the Witswatersrand, Parktown, Republic of South Africa

Bruno Maureille, Laboratoire d'Anthropologie des Populations du Passé, Université Bordeaux I, France

Michael L. McKinney, Department of Geological Sciences, University of Tennessee, Knoxville, Tennessee

Kenneth J. McNamara, Department of Earth and Planetary Sciences, Western Australian Museum, Perth, Western Australia, Australia

Nancy Minugh-Purvis, Department of Neurobiology and Anatomy, MCP Hahnemann University, Philadelphia, Pennsylvania

Andrew J. Nelson, Department of Anthropology, University of Western Ontario, London, Ontario, Canada

Sue Taylor Parker, Department of Anthropology, Sonoma State University, Rohnert Park, California

Fernando Ramirez Rozzi, CNRS-EP1781, Station Marcellin Berthelot, Meudon la Forêt, France

Sean H. Rice, Department of Biology, Yale University, New Haven, Connecticut

Brian T. Shea, Department of Cell and Molecular Biology, Northwestern University, Chicago, Illinois

Scott A. Stewart, Department of Biological Sciences, University of Cincinnati, Cincinnati, Ohio

Michael R. Sutherland, Statistical Consulting Center, University of Massachusetts, Amherst, Massachusetts

Jennifer L. Thompson, Department of Anthropology, University of Nevada, Las Vegas, Nevada

Frank L'Engle Williams, Department of Anthropology and Geography, Georgia State University, Atlanta, Georgia

Human Evolution
through
Developmental Change

What Is Heterochrony?

KENNETH J. MCNAMARA

Heterochrony plays a central role in evolution. It provides the raw material for natural selection to work on and operates from the intraspecific to the highest taxonomic levels. In this book we explore the extent to which heterochrony has molded and directed the course of human evolution.

Essentially, heterochrony can be defined as *change to the timing and rate of development.* Each individual organism has an ontogeny—its life history, from conception until death. As organisms develop they not only increase in size but also undergo changes in shape. Furthermore, most organisms have a finite period of growth, often terminated by the onset of sexual maturity. At this time, growth slows down appreciably or stops. Different species grow at different rates, and a certain amount of variation is present among the individuals making up a given species. Furthermore, different parts of each individual may grow for different lengths of time and at different relative rates. Heterochrony, then, is the concept of changes to these rates and durations of growth of all, or part, of an organism, compared with its ancestor.

Heterochrony produces two major effects. Compared with an ancestor, a descendant (either a descendant species or an individual within a population) can show either less growth of particular morphological features or more growth. If there is less growth during ontogeny, the descendant adult will resemble the juvenile condition of the ancestor. This is known as *paedomorphosis.* Conversely, if the descendant undergoes greater development, it is said to show *peramorphosis.* These descriptive terms portray the descendant morphological pattern. Each of these two phenomena can be generated by three different processes. Paedomorphosis occurs if the duration of growth of the descendant form is prematurely truncated (*progenesis*), or if the growth rate is less in the descendant than in the ancestor (*neoteny*), or if onset of growth is delayed (*postdisplacement*) (see Figs. 2.1–2.3). Progen-

1

esis can often affect the whole organism if onset of sexual maturity occurs prematurely, or it may involve only certain traits. Neoteny and postdisplacement generally affect only certain traits, not the entire organism.

Conversely, peramorphosis occurs if duration of growth in the descendant is extended (*hypermorphosis*), or if the actual growth rate is increased in the descendant (*acceleration*), or if onset of growth is earlier in the descendant (*predisplacement*) (see Figs. 2.1–2.3). Hypermorphosis can affect the whole organism if onset of sexual maturity is delayed, or it can target just certain traits. Acceleration and predisplacement affect only certain traits, not the entire organism (see Chap. 2 for a more detailed discussion of the heterochronic processes and other synonymous terminology).

Heterochronic terminology can be applied both to the appearance during ontogeny of meristic characters (i.e., discrete structures formed during ontogeny, such as the number of skeletal elements) and to the subsequent changes in shape of these traits. These are called *mitotic heterochrony* and *growth heterochrony*, respectively. In many organisms, mitotic heterochrony, induced especially by pre- and postdisplacement, can be important very early in development because variations can occur between ancestors and descendants in the timing of onset of development of major morphological features. Neoteny and acceleration will be particularly influential during subsequent ontogenetic development. Progenesis and hypermorphosis frequently come into play at a late stage in ontogeny, as they reflect variations in the time of offset of growth.

Most organisms not only increase in size as they grow but also change shape. This relationship between size and shape is known as *allometry*. If, during ontogeny, the size and shape of a structure remain the same relative to overall body size, growth is isometric. In reality, few, if any, organisms are known to grow isometrically. Generally, during ontogeny a particular structure will change shape and size relative to the size and shape of the entire organism. If there is an increase in size, growth is said to occur by *positive allometry*. Conversely, if there is a relative decrease in size, growth is said to show *negative allometry*. Heterochrony involves changes not only in time, but also in shape and size. Consequently, there is a close relationship between allometry and heterochrony. The effect of changes to growth rates is to change allometries. Increase in allometry is said to be expressed phylogenetically as peramorphosis, whereas reduction of allometry produces paedomorphosis (see Chap. 4 for a discussion of problems with this interpretation). Extensions or contractions of the period of growth (i.e., hypermorphosis or progenesis) exacerbate or reduce the effects of allometric changes. Consequently, those organisms that undergo pronounced allometric change during growth are more likely to generate very different descendant adult morphologies if rates or durations of growth are changed.

Peramorphosis is considered to be important in generating increases in body size by either hypermorphosis or acceleration. Consequently, peramorphosis is often associated with paedomorphic trends in some traits. This is known as *dissociated heterochrony*, where some morphological traits are peramorphic while others are paedomorphic. Organisms are rarely all peramorphic or all paedomorphic. Peramorphic and paedomorphic traits may possibly be interrelated, in that paedomorphic traits may be developmental trade-offs for peramorphic features. Trends toward increased body size mean that there must be a greater input of energy to enable the organism to attain a larger body size, and often more complex morphological features, than in its ancestor.

Classic examples of developmental trade-offs are found in dinosaurs, ratite birds, and hominids. In hominid evolution the trend has been for an increase in body size and peramorphic evolution of the brain and lower limbs, both increasing in size and complexity (see Chaps. 7–9, in particular). Developmental trade-offs, in particular for the larger brain, include a reduction in size of the gut and other traits, such as a paedomorphic reduction in jaw and tooth size (see Chap. 5 for a more detailed discussion).

A close relationship is known to exist between heterochrony and life-history strategies. These include factors such as size at birth; growth rates; age at maturation; body size at maturity; the number, size, and sex of offspring; and length of life. Attempts to categorize life-history traits have been of limited success. The most widely known is the *r-K* continuum, a descriptor of environments and the life-history traits of their inhabitants. While the *r-K* continuum is an oversimplification of life-history strategies in general, it does seem to work at higher taxonomic levels. The *r*-selected extreme categorizes unpredictable, often ephemeral environments. Selection pressure for organisms inhabiting such environments targets those that mature rapidly, have short life spans, and are small in size. These are traits produced by progenesis. Such organisms typically produce large numbers of offspring. At the other extreme are *K*-selected populations, inhabiting constant, predictable environments. Characteristic features of organisms inhabiting such environments are delayed onset of reproduction, long life span, and large body size, all traits typical of hypermorphosis. These organisms produce few, large offspring. As well as being produced by progenesis, *r*-selected characteristics may also be produced by acceleration. Conversely, many *K*-selected organisms seem to show some traits that evolved by neoteny, as well as hypermorphosis. It is probably for this reason that these two pairs of processes often go together. The relationship between heterochrony and life histories is explored in Part II.

Heterochrony can affect life-history traits in other ways. Many organisms undergo pronounced morphological, behavioral, and ecological changes during ontogeny. Changes in the timing of transition from one phase to another (called *sequential heterochrony*) can cause significant morphological and behavioral changes to the descendant adult (see Figs 5.1 and 5.2). Some contributors to this book argue that this has played a significant role in human evolution (see Chaps. 5, 8, and 9).

For a general text on heterochrony, the reader is referred to *Shapes of Time: The Evolution of Growth and Development*, by Kenneth J. McNamara (1997). For a more detailed study, see *Heterochrony: The Evolution of Ontogeny*, by Michael L. McKinney and Kenneth J. McNamara (1991).

Part I

Evolution and Development

Chapter 1

Evolutionary Developmental Biology

Where Embryos and Fossils Meet

Brian K. Hall

The title of this volume, *Human Evolution through Developmental Change*, is based on the premise that evolutionary change is reflected in altered development (ontogeny) and that alteration in development can bring about evolutionary change. Such thinking has not always been standard, although notions of some connection between evolution and development (phylogeny and ontogeny) have long occupied evolutionary biologists, anthropologists, and paleontologists (Osborn 1894; de Beer 1940; Simpson 1944; Gould 1977; Mayr 1982; Moore 1986). The parallelism between ontogeny and phylogeny that preoccupied naturalists in the nineteenth century—symbolized and actualized in the work of Karl Ernst von Baer, Ernest Haeckel, the Hertwig brothers, Anton Dohrn, Frank Balfour, and many others—saw the connection as driven by a series and sequence of phylogenetic events that were repeated in ontogeny or individual development. It would be the 1920s before the tables were turned and ontogeny was seen, not to repeat phylogeny, but to create it (Garstang 1922; see the papers in Hall and Wake 1999 for evaluations of Garstang's legacy). In 1919, Otis Whitman, the first director of the Marine Biological Laboratory at Wood's Hole, was espousing that "all that we call phylogeny is to-day, and ever has been, ontogeny itself. Ontogeny is, then, the primary, the secondary, the universal fact. It is ontogeny from which we depart and ontogeny to which we return" (Whitman 1919, 178). Bolk's theories of fetalization (e.g., Bolk 1926) typified, for the transformation of nonhuman primates to humans,

7

the ontogenetic approach applied by others to the remainder of the animal kingdom.

Evolutionary Developmental Biology

Nowadays, we have evolutionary developmental biology as a grand-sounding title for the study of the relationships between development and evolution, and we are convinced that, along with the name, comes enhanced understanding of how evolution is effected through developmental change (Hall 1996a, 1998a; Gilbert et al. 1996; Raff 1996). But evolutionary developmental biology is more than a name for an emerging subfield of biology. It is a reflection of a level of analysis, synthesis, and understanding not possible through the study of evolution or development alone. Purugganan summarized the status quo: "It is clear that a new field—evolutionary developmental biology—is emerging, integrating concepts and techniques in classical developmental biology, molecular genetics, phylogenetic systematics and population biology. Moreover, this fledgling field is ready to go beyond model systems and explore the development of evolutionary important non-model species" (Purugganan 1996, 7).

That we need this third level of analysis is illustrated in the following statements about evolution, development, and evolutionary developmental biology. Evolution acts at at least three levels:

1. the change in gene frequency, where the unit of study is the allele, gene, or population;
2. the appearance of new characters, where the unit is any recognizable organismal character—cell, tissue, organ, physiology, behavior, cultural activity; and
3. the appearance, adaptation, and/or spread of species, where the unit is the species, species assembly, population, or community.

The common denominator at all levels of evolutionary change is genetic change through time. The level of biological organization through which change is effected is ontogeny. The discipline through which we study and understand how development relates to evolution is evolutionary developmental biology. The fundamental units through which change is effected are the gene, species, or population for evolution, the embryo for development, and the cell or fields of cells for evolutionary developmental biology.

Hierarchical Control

It is now a truism to state that embryos and embryonic development are organized hierarchically; see Valentine and May (1996) for an evaluation of hierarchies in biology and paleontology. Because of homology, all comparative biology is intrinsically hierarchical (Hall 1994). The potential for interaction and communication increases as the embryos of multicellular animals pass from the single-celled egg or zygote to the multicellular blastula, to the reorganized and more morphologically complex and multilayered gastrula, to the emerging organ systems of the neurula, and so on. Further distinguishing evolutionary developmental biology as a separate level of analysis with its own assumptions and mechanisms and therefore its own unique contribution is recognition and analysis of the interactive and hierarchical organization of both developmental and evolutionary change (Hall 1983, 1998a).

Of course, evolutionary developmental biology is not the only aspect of biology to take a hierarchical approach. A hierarchical approach is meaningful wherever emergent properties appear in crossing from one level to another (Templeton 1982). In other areas, perhaps that of greatest relevance to human evolution through developmental change is the hierarchical analysis required in constructional morphology (Seilacher 1988; Vogel 1991) and in systematics, ecology, and behavior (Brooks et al. 1995).

Maternal and Zygotic Control

We can distinguish informational hierarchies in development. In the egg, zygote, and blastula, we see spatial heterogeneity, much of it derived from maternal gene products or organelles deposited during oogenesis. Maternal cytoplasmic control is limited to very early development, in many species extending only for several cell divisions. Cohen (1979) defined the stage when maternal control is replaced by zygotic as the phyletic (now phylotypic) stage. Cohen and Massey (1983), Buss (1987), and Davidson et al. (1995) emphasized that the basic body plan is established under maternal cytoplasmic control before the zygotic genome begins to control development. We owe to Lewis (1978), Nüsslein-Volhard and Wieschaus (1980), and the many who followed the knowledge that homeobox genes provide the genetic basis of such patterning.

The utility of *Hox* genes for investigating origins of *Baupläne* is amply illustrated by a study in which expression boundaries of an amphioxus *Hox* gene (*AmphiHox* 3) in the developing dorsal nerve cord were compared with the expression pattern of the mammalian homologue, *Hox*-2.7. Holland et al. (1992) concluded that the vertebrate brain is homologous to an extensive region of the amphioxus nerve cord, a finding consistent with the vertebrate head as an elaboration of a preexisting body region. These findings are not consistent with the vertebrate head as a novelty that lies entirely anterior to the prevertebrate head.

Further examples of shifts in positions of *Hox* genes and evolutionary change in basic body plan—in this case, axial structures (numbers of vertebrate, positions of limbs along the body axis) associated with the vertebrate body plan—are the expression studies of *Hox* genes in chicks and mice (Burke et al. 1995). Axial structures track *Hox* gene expression patterns because *Hox* genes play important roles in determining the positions of those axial structures. Furthermore, vertebrae take on characters of more anterior or posterior vertebrae after expression boundaries of *Hox* genes are shifted in transgenic mice or by the administration of retinoic acid, an upstream regulator of *Hox* genes (Valentine 1990; Gruss and Kessel 1991). *Hox* genes specify vertebrae to specific body regions as part of establishing the Baupläne. Slack et al. (1993) extended such pattern specification to the evolutionary stability of conserved developmental (phylotypic) stages. Given such evolutionary stability, it is perhaps not surprising that there are so few basic types of animals (Hall 1996b, 1998a).

Interactions

Spatial heterogeneity in the gastrula is initiated following complex morphogenetic movements that rearrange germ layers and permit waves of inductive interactions between developing parts. Temporal and spatial heterogeneity characterizes the neurula as secondary and tertiary interactions subdivide and pattern developing organs and so regionalize the organs and the embryo itself. Cascades of interactions—whether gene to gene, cell to cell, organ to organ, or even organism to organism—are the visible manifestation of this epigenetic organization of embryonic development (Hall 1983, 1987, 1998a).

Epigenetics and Emerging Properties

Epigenetics is the sum of the genetic and nongenetic factors acting upon cells to control gene expression selectively to produce increasing phenotypic complexity during development and evolution (Hall 1990, 1998b). Epigenetics is therefore to evolutionary developmental biology as natural selection is to evolution or as differentiation, morphogenesis, and growth are to development. A major contribution made by evolutionary developmental biology to evolutionary biology over the past decade has been the knowledge that developmental and evolutionary change must be viewed as hierarchical, interactive, and characterized by emergent properties not apparent from analysis at one level alone (Hall 1998a–c; Raff 1996; Goodwin et al. 1993).

An example will illustrate this point. Teeth are composed of enamel and dentine. To know that enamel is formed from epithelium and dentine from mesenchyme does not reveal or even hint that neither tissue type (nor teeth) can form unless epithelium and mesenchyme interact in a complex set of inductive interactions (Lumsden 1988; Thesleff and Sahlberg 1996). Extend this understanding to knowledge that teeth arose during evolution from denticles associated with the dermal armor of jawless fishes more than 400 million years ago, and we begin to glimpse how those dermal denticles must have appeared as a consequence of the origin and elaboration of epigenetic tissue interactions (Hall 1987; Smith and Hall 1993; Graveson et al. 1997). A unit of analysis (the cell) and a mechanism of integration (epigenetics) not seen in either development or evolution are necessary to provide a full understanding of both.

Evolutionary developmental biology is also integrative and hierarchical because it takes account of parental and zygotic genotypes, life-history evolution, ecology, and environmental influences to explain how the phenotype is more than the outward manifestation of the genotype—the problem of the genotype-phenotype map and its evolvability (Wagner and Altenberg 1996). This hierarchical approach is evident in much recent writing, of which Hall (1998a), Liem and Wake (1985), and Valentine (1990) are representative studies. Cheverud (1996), Cheverud et al. (1991), and Atchley and Hall (1991) are some of the few attempts to integrate epigenetic elements into quantitative genetics models.

Neither epigenetics nor evolutionary developmental biology would be necessary if there were a one-to-one correspondence between genotype and phenotype. In that case, the concern of this volume would not be with human evolution through developmental change but rather with human evolution through genetic change. Overviews such as that by B. H. Smith (1992) on life history and the evolution of human maturation or Shea (1992) and Macho and Wood (1995) on allometry/heterochrony and the evolution of human maturation would be cast in a vastly different light. All paleoanthropologists would be geneticists and not merely aware of genetics (Minugh-Purvis 1996).

The Evolution of Morphology

The most obvious aspect of the biological world for which evolutionary developmental biology can enhance our understanding is evolutionary alteration in morphology, the evolution of the phenotype. No morphological change, whether it appears during an individual lifetime or across successive generations, can occur without altering some, often many, aspects of development. This is as true for a subtle change in the shape of the condylar process of the mammalian dentary as it is for modification of the dentition during ontogeny or speciation of cichlid fishes, transformation of the fin to a limb at the origin of the tetrapods, or modification of mandibular or skull morphology in the transitions from anthropoid primates to hominoids. No matter how great or small the phylogenetic change, some alteration in development is at its base.

D'Arcy Thompson (1942) tried to approach such transitions through transformations among adult shapes using a dynamic technique rendered sterile by its application to adults. Huxley (1932) approached the problem from a more ontogenetic perspective, seeing transformation as essentially a problem in correlated growth between parts. Again, because the approach did not relate embryos to adults but rather compared parts within an organism at an instant during the life cycle, a wonderfully intuitive approach failed to link ontogeny and phylogeny. Heterochrony shows greater promise, although it certainly is not the only mechanism responsible for the evolution of morphology.

Heterochrony

De Beer (1940) proposed a more developmental method when he espoused the analysis of timing of development, not just within a single life cycle but in a descendant in comparison to timing of the same process in an ancestor. This phylogenetic concept with a developmental mechanism was introduced by Haeckel (1866) under the name *heterochrony*. De Beer's analysis, although pioneering, was theoretical. It failed to ignite a fusion of development and evolution.

Gould (1977) simplified de Beer's eight heterochronic processes into two—acceleration and retardation. Alberch et al. (1979) refined Gould's approach into the now-familiar ontogenetic trajectory, and heterochrony raced ahead of its time. It seemed that everyone could find evidence for heterochrony or, at least, justify use of the term to explain phenotypic changes in their favorite organism. Other mechanisms linking development and evolution were ignored or not sought. Even Haeckel's second process linking ontogeny and phylogeny, heterotopy, was passed over (but see below). Heterochrony was too seductive, even if only rarely was it asked, "Heterochrony of what?" or "Do we really know the timing of the process in the ancestor?"

I want to use heterochrony as the springboard for much of the balance of this chapter by asking such basic questions as whether embryos can measure the passage of time in a way that would render the process phylogenetic; see Hall and Miyake (1995a) for a discussion of this issue. I also want to ask whether timing of developmental events can occur at any stage in the life cycle. Von Baer and Haeckel, and I suspect many current biologists and anthropologists, believed or believe that early development either is immune to change or, if altered, would so drastically deflect subsequent development that early changes would be lethal or actively selected against; but see Richardson (1995) for an opposing view. If developmental processes can change throughout the life cycle, how are we able to recognize common stages in the ontogeny of related organisms? Then I will take up the cell (or rather groups of similar cells) as the fundamental unit that evolutionary developmental biology should be addressing and discuss respecification of cell lineages and modifications of cell-to-cell interactions as mechanisms that can bring about evolutionary change.

From cell populations it is a short step to pattern specification and the essential role that cell lineages play in specification of phenotypic patterns during development and in respecification of patterns during evolution. Several macroevolutionary examples—the origin of mammalian middle ear ossicles from elements of the reptilian lower jaw, the origin of external and internal cheek pouches, the progressive replacement of the dermal exoskeleton by a cartilaginous endoskeletal during vertebrate evolution— illustrate the utility of an epigenetic approach to cells as the fundamental unit and epigenetics as the fundamental process in the arena where embryos and fossils meet. Such alterations can be based in heterochrony as proposed for modification of body plans in the evolution of the snake skull, turtle shell, and other organ systems (Hall 1998a). Duboule (1994) proposed a mechanism to couple heterochrony to altered expression patterns of *Hox* genes in specific populations of cells with high mitotic rates (i.e., a means to couple heterochrony to Baupläne). Such alterations can also be based in heterotopy, the second major mechanism proposed by Haeckel to explain evolutionary change in morphology.

Measuring Time

Gould (1977) simplified heterochrony into acceleration or retardation. He emphasized that either process required a standardized baseline against which to measure whether development had been advanced or slowed. Alberch et al. (1979) specified three temporal standards: onset, offset, and rate. Several authors have explored the embryological and cellular mechanisms that drive such changes; see, for example, Hall (1984, 1990, 1998a) and Duboule (1994).

In practice, only three criteria can be used to standardize comparisons of shifts in the timing of development of the same organ in different individuals (species) or to allow rates of development of different organs to be compared in the same individual:

1. some measure of maturation, such as life cycle transitions (embryo to larva, hatching, sexual maturity), or attainment of a particular degree of morphological complexity—usually measured as embryonic stages;
2. size (absolute or relative, final or at specific landmark stages), or
3. age (absolute or relative) or age at a particular stage of maturity, size, or morphological development.

Provided that the metric selected allows timing to be treated as a relative concept, then the measure may be developmental (morphological stage, developmental events, numbers of cell cycles), physiological (energy metabolized/unit body [organ] weight), or growth driven. Hall and Miyake (1995a) discussed these criteria and their application in different organisms, stages of development, or approaches (morphological, physiological, maturational, or developmental time). Miyake et al. (1996a) developed a staging table for inbred C57BL/6 mice that is typical of the detail required and with which we have analyzed variability in the timing of equivalent developmental processes in three inbred strains of mice (Miyake et al. 1997).

Bates (1994) and Hart (1996) discussed these criteria in the context of alteration or loss of the larval stage from the life cycle, B. H. Smith (1992) for their application to human maturation and evolution, and K. K. Smith (1996) for their application to four marsupial and five placental mammalian taxa.

Heterotopy

Heterotopy describes for position what heterochrony does for time but involves altering the timing of development. When homologous organs can be shown to develop from different embryonic regions—classically from different germ layers—Haeckel's concept of heterotopy is operating. Haeckel's concern was with the origin of germ cells and gonads. Mesoderm arose later in animal evolution than did either ectoderm or endoderm. Haeckel argued that reproductive systems that were mesodermally derived in triploblastic animals must have had endo- or ectodermal origins in their diploblastic ancestors.

A heterotopic change that both includes elements of heterochrony and provides a valuable illustration of the origin of an apparently macroevolutionary change by microevolutionary changes is the origination of external cheek pouches in geomyoid rodents (pocket gophers, kangaroo rats, and their allies). Unlike internal pouches that open inside the mouth and are lined with mucous-secreting epithelium, external pouches open outside the mouth and are lined with hair that forms following cellular interaction between the pouch epithelium and superficial mesenchyme.

Transformation to an external pouch, although long mysterious, is readily explained from an analysis of pouch development (Brylski and Hall 1988a, 1988b). Because of differential growth between lip, pouch, and facial epithelia, the pouch opening is coupled to the epithelium of the corners of the mouth and so develops outside the mouth. Growth of a cord of cells into superficial, hair-forming mesenchyme and the ability of the epithelial cells to interact with the mesenchyme in this ectopic location allow hair rather than mucous glands to form. (The ability of epithelia to modulate between a mucous cell and a keratinized or haired phenotype is well documented; see Hardy 1992.) Differential growth (heterochrony), leading to a shift in position (heterotopy), coupled with the ability to respond to more than one inductive environment (developmental plasticity) derives an external pouch from an initially internal rudiment.

The induction of Meckel's cartilage in different vertebrate classes at different times during development and by spatially distinct epithelia is a further example of phenotypic change based on heterochrony and heterotopy (Hall 1983, 1984, 2000). Delayed induction in mammals (more strictly in mammal-like reptiles) was a necessary prerequisite to freeing portions of the skeleton of the reptilian lower jaw for transformation into mammalian middle ear ossicles. As Rowe (1996) has shown, middle ear ossicles and brain co-evolved in mammals. A heterochronic shift in brain growth is proposed as the step that provided the conditions required for formation of the middle ear ossicles.

Zelditch and colleagues at the University of Michigan have begun an analysis of the evolution of form in which they see heterotopy as a means to produce new ontogenetic trajectories that provide an escape from ancestral ontogenies (Zelditch et al. 1992; Zelditch and Fink 1996). The epigenetic approach they use allows developmentally individualized parts (regions of the skeleton) to be analyzed both regionally and temporally. This insightful proposal and its application to morphological change within fishes and mammals represent an exciting resurrection of heterotopy as an important but vastly understudied mechanism for evolution through developmental change. As they cogently remind us: "From the perspective of the developmental biology of pattern formation, spatial phenomena are just as fundamental as temporal phenomena. *In development, time is no more primary or basic than space*" (Zelditch and Fink 1996, 252–3, italics mine).

The Cell

An enormous amount of work on cell lineages was undertaken in the latter part of the nineteenth century and the first decades of the twentieth century; see Wilson (1896) and Lankester (1911) for contemporary analyses. Haeckel's concept of heterotopy and the entrenchment of the germ layer theory were in considerable jeopardy from lineage studies demonstrating that germ layers arose from different cell lineages in different organisms. Which took precedence, germ layers or cell lineages?

A cell lineage approach has shown that early developmental events can undergo dramatic changes over the phylogenetic history of a group. Despite von Baer's "law" of the uniformity of early embryos, development before the phylotypic stage can be quite varied (Hall 1997, 1998a; Richardson 1995). Examples include divergent forms of cleavage and blastula formation. The presence of the type of flattened "blastodisk" found in avian embryos rather than a spherical blastula with blastopore in directly developing amphibians such as *Gastrotheca riobambae* and *G. plumbea,* egg-brooding hylid frogs, is a dramatic example (Elinson 1987). The divergent patterns of egg size, cleavage, and cell lineage seen in directly developing echinoderms that have lost the feeding-larval stage or the loss of primary mesenchyme in primitive sea urchins such as *Eucidaris tribuloides* (Raff 1996; Hall and Wake 1999) represents a further set of examples elaborated below.

Our general understanding of echinoderm development is that a larval stage is interposed between embryo and adult (i.e., that development is indirect). Some lineages of cells in indirectly developing sea urchins specify only larval cell types, such as the larval skeleton. These cells make no contribution to the adult. However, many species of sea urchins have lost or greatly reduced the larval stage. Larval lineages are not needed in such species, which display direct development. We might expect these larval lineages to have been eliminated. Unexpectedly, directly developing species have retained these cell lineages. Even more unexpectedly, the lineages have been respecified to form adult structures that normally form from different germ layers. So, in a classic example of heterotopy, cells that form larval mesenchyme (a mesodermal derivative) in indirect developers form part of the gut (an endodermal derivative) in direct developers.

A Model

Atchley and Hall (1991) proposed a model for the development and evolution of complex morphological structures in which the parameters of groups of similar cells were fundamental. Gilbert et al. (1996) called for renewed emphasis on cells and morphogenetic fields as major units of ontogeny bringing about changes in evolution. Our model was applied to the mammalian dentary, which, although a single bone, has distinct structural components, such as coronoid, condylar, and angular processes, alveolar units associated with the molar, and incisor teeth (Fig. 1.1). We argued that each structural component was the physical manifestation of underlying morphological components, each based on a distinct lineage of cells.

Three cell populations were distinguished for the rodent dentary:

1. a skeletogenic population from which the bone of the body (ramus) of the dentary arises;

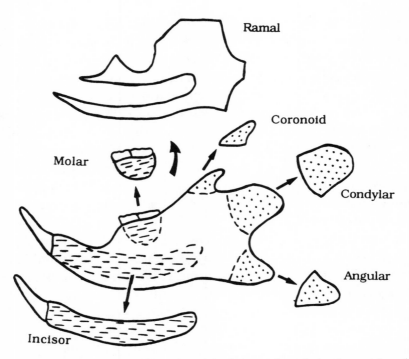

Fig. 1.1. A diagrammatic representation of the dentary of a typical rodent, such as the mouse, shown intact (*center*) and with the individual structural components shown in isolation. Each structural component is based on a separate lineage of cells: a population of osteogenic cells that forms the ramus, an odontogenic/osteogenic population that forms the alveolar units (molar and incisor), and a chondrogenic/osteogenic population from which the coronoid, condylar, and angular processes arise. See text for details.

2. an odontogenic population from which the dentine of the teeth and the alveolar bone arise, and
3. a third population that is osteo- and chondrogenic but not odontogenic (Table 1.1) and from which the major processes at the posterior of the dentary arise (Fig. 1.1).

The significance of such populations, or *condensations* as they are often termed, lies in their cohesion as fields of cells, that selective gene activation initiating cell differentiation is initiated within such condensations, and that deviations in condensation size, location, or time of appearance are critical events in initiation of the abnormal morphology associated with mutations, syndromes, selection, and by extension, evolutionary change; see Hall and Miyake (1992, 1995b) for overviews and Miyake et al. (1996b) for a stage-

Table 1.1   The Major Elements in the Hierarchy Leading
to Formation of the Mammalian Dentary

Morphological Unit	*Dentary*
Cell lineages	Odontogenic of alveolar units producing dentine and bone
	Osteogenic of body of dentary producing bone
	Skeletogenic of dentary processes producing cartilage and bone
Levels of control	*Fetal genotype*, including selective action of genes in cell lineages and in tissues directly associated with dentary (muscles, teeth); fetal hormones; growth factors
	Maternal genotype
	Maternal environment including uterine environment, litter size, hormones and growth factors crossing placental barrier
	Epigenetic factors, including inductive cell interactions, tissue interactions, action of muscles, nerves, vascular system, teeth
	Environment, including such physical factors as temperature, weaning environment, postnatal nutrition

by-stage analysis of the onset of condensation during first arch development. Experimental evidence accumulating in other systems demonstrates that morphology can be altered by the action of such small populations of cells and that such populations can be selected for (Weber 1992).

Independent evidence for discrete genetic control over these lineages is accumulating from studies with transgenic mice. Thus, $Msx1^-$ homozygotes have major deficiencies in the alveolar portions of the dentary, while the ramus and posterior processes are morphologically normal. *MSX1*, active in the odontogenic but not in the other cell lineages, appears to control initial proliferation and accumulation of the cell condensation (Satokata and Maas 1994). Goosecoid-null mice have normal alveolar components and odontogenic cell lineages but are deficient in the coronoid and angular processes. The condylar process and ramus are normal (Rivera-Pérez et al. 1995). In $MHox^{-/-}$ mice, the coronoid and condylar processes are reduced and the angular process is nonexistent (Martin et al. 1995). Different cell lineages are under different genetic control.

Each structural component of the dentary, as well as arising from a discrete cell population, is subject to extrinsic control from adjacent developing organs. There is, to use the terms of Richman and Mitchell (1996), both cell-autonomous and non-cell-autonomous control. The condylar process is dependent for its initiation and subsequent growth on the normal functioning of the lateral pterygoid muscle, while alveolar bone depends for normal development on the presence of the teeth but develops independent of muscle action. From such analyses Atchley and Hall (1991) constructed a hierarchical approach to morphology (Table 1.1). Development or evolution of morphology could be tackled either by working down from the structural components toward the controlling factors or by building up from the controlling factors to the final morphology. This hierarchical approach is paradigmatic for how one must approach the evolution of morphology through developmental change if one is to understand the proximate mechanisms responsible for altered morphology.

Cell Lineages and the Evolution of the Brain

Another example, from many, of the importance of populations or lineages of cells in patterning the phenotype during ontogeny and therefore of understanding a fundamental unit for evolutionary change in morphology comes from studies of the patterning of the central nervous system, especially the vertebrate hindbrain; see Lumsden and Graham (1996) for an up-to-date analysis.

The nineteenth-century notion that the brain is segmented into units or rhombomeres (r) has been confirmed with labeling studies. Expression boundaries of *Hox* genes establish the boundaries between adjacent rhombomeres. Neural crest cells arise from the neural tube as the neural folds close after rhombomere boundaries have been established. Initially, crest cells appear along the length of the hindbrain. Crest cells emerge from r1, 2, 4, 6, and 7, but not from r3 or r5, in a pattern that alternates rhombomeres except for r1. Crest cells are present adjacent to r3 and r5 but are eliminated by programmed cell death or apoptosis. Here we see exquisite hierarchical patterns of specification of cell lineages. Initially, all rhombomeres are specified to produce neural crest cells. Subsequently, only cells from alternate rhombomeres are permitted to migrate to form the mesenchyme that populates the developing head; see Hall (1999) for further details of the neural crest.

This hierarchical patterning is augmented by interaction between adjacent rhombomeres. Death of neural crest cells in r3 and r5 requires that those rhombomeres interact with their even-numbered neighbors. Molecular evidence suggests that r3 and r5 produce a bone morphogenetic protein-4 (BMP-4) and a homeobox gene, *Msx*-2, only when associated with even-numbered rhombomeres and that BMP-4 and its downstream product *Msx*-2 provide sufficient molecular signals to trigger apoptosis. Any further analysis that attempts to understand the mechanistic basis of patterning of the hindbrain, patterning of derivatives of hindbrain neural crest cells, or how patterning of the vertebrate brain has changed during evolution will have to treat rhombomeres as discrete, independent, yet interacting cell lineages.

The Dermal Exoskeleton

A final example comes from analysis of the developmental and evolutionary origins of the vertebrate dermal exoskeleton of sharks, armored fishes, and the like and of the teeth, the only remnant of the cranial dermal skeleton remaining in many vertebrates. An abundance of evidence summarized by Smith and Hall (1993) documents that the dermal exoskeleton and its derivatives, the teeth, are fundamentally constructed from dentine and bone, tissues that are of neural crest origin. The dentine is overlain with a tissue, enamel, that is of epithelial origin. As already noted, both tissues develop only after undergoing a complex series of epithelial-mesenchymal interactions. In early jawless fishes, dermal ossicles or a solid dermal armor often extended over much of the body. The presumption is that neural crest cells along the entire length of the neural tube were producing dermal denticles. However, this presumption is not compatible with studies on a modern, admittedly toothless vertebrate, the chick, in which the skeleton-forming neural crest is confined to the cranial region. Studies in other vertebrates, including those with teeth, confirm this general patterning of the neural crest into a skeleto- and odontogenic cranial neural crest and a nonskeletogenic and non-tooth-forming trunk neural crest; see Figure 1 in Hall and Hörstadius (1988).

How do we reconcile the embryological and fossil data? Are our ideas of parsimony of developmental mechanisms through evolutionary time incorrect? Did the dermal armor of ancient vertebrates all arise either from cranial neural crest or from mesoderm rather than from cranial neural crest? Alternatively, have the experimental studies on embryos from extant vertebrates misled us? The species studied were not chosen for their ability to resolve evolutionary problems and certainly not for their extensive dermal exoskeleton. They were chosen for convenience as model organisms from which patterns of development of the neural crest could be extrapolated to other organisms. Shortcomings of the use of model organisms when approaching evolution through development are discussed by Hanken (1993) and Bolker (1995). Evolutionary developmental biology, "this fledgling field, is ready to go beyond model systems and explore the development of evolutionary important non-model species" (Purugganan 1996, 7).

We set out to resolve the apparent paradox by examining embryos of a species with teeth, the Mexican axolotl, *Ambystoma mexicanum.* The aim was to determine whether trunk neural crest might have tooth-forming potential not used during normal development. Lumsden (1988) had already shown this to be so for the most rostral trunk neural crest of the mouse. Cranial neural crest of the Mexican axolotl was challenged in vitro with cranial ectoderm known normally to induce tooth formation, and formation of teeth was confirmed. Cranial neural crest also formed cartilage, as expected, from a population of skeletogenic cranial neural crest cells. Then trunk neural crest was similarly challenged with cranial ectoderm with which it would normally never come into contact but which can, and does, induce cranial neural crest to form teeth. The most rostral region of that trunk neural crest produced teeth but never cartilage (Graveson et al. 1997).

Tooth-forming capability does indeed extend beyond the cranial chondrogenic neural crest. Initially, the latter result was puzzling—we expected the boundaries of tooth- and cartilage-forming crest to coincide—until we realized the obvious. Cartilage and teeth represent two entirely separate vertebrate skeletons, the endoskeleton, based in cartilage, and the exoskeleton, based in dentine and bone. Given that endo- and exoskeletons have been separate since the origin of the vertebrates, we would not expect the cells producing them to display regional patterns similar to those of the neural crest (Graveson et al. 1997; Hall 1999, 2000).

That the neural crest as an embryonic region houses diverse populations of cells highlights the importance of a focus on lineages of cells. To map the neural crest as if it were a homogenous population of cells is to deny a half-billion years of evolutionary history. Similarly, to map the rhombomeres of the hindbrain without acknowledging the distinct cell populations in alternate rhombomeres would rob the analysis of the richness brought by a developmental approach to evolution. The merging of fossils and embryos in studies on the neural crest brings cell populations to the forefront as fundamental units of evolutionary developmental biology. Understanding how such cell populations function during development and have changed during evolution is an important challenge for those who wish to understand human evolution through developmental change.

Acknowledgments

Financial support for my research has been provided by the Natural Sciences and Engineering Research Council (NSERC) of Canada, the National Institutes of Health of the United States, and the Killam Foundation at Dalhousie University. For discussions on the topics covered in this chapter, I especially thank Bill Atchley, Ann Graveson, Tom Miyake, and Moya Smith.

References

Alberch, P., Gould, S.J., Oster, G.F., and Wake, D.B. 1979. Size and shape in ontogeny and phylogeny. *Paleobiology* 5:296–317.

Atchley, W.R., and Hall, B.K. 1991. A model for development and evolution of complex morphological structures and its application to the mammalian mandible. *Biological Reviews of the Cambridge Philosophical Society* 66:101–157.

Bates, W.R. 1994. Ecological consequences of altering the timing mechanism for metamorphosis in anural ascidians. *American Zoologist* 34:333–342.

Bolk, L. 1926. La récapitulation ontogenetique comme phénomène harmonique. *Archives of Natural History and Embryology* 5:85–98.

Bolker, J.A. 1995. Model systems in developmental biology. *BioEssays* 17:451–455.

Brooks, D.R., McLennan, D.A., Carpenter, J.M., Weller, S.G., and Coddington, J.A. 1995. Systematics, ecology and behavior. *BioScience* 45:687–695.

Brylski, P., and Hall, B.K. 1988a. Epithelial behaviors and threshold effects in the development and evolution of internal and external cheek pouches in rodents. *Zeitschrift für zoologische Systematik und Evolutionsforschung* 26:144–154.

———. 1988b. Ontogeny of a macroevolutionary phenotype: the external cheek pouches of Geomyoid rodents. *Evolution* 42:391–395.

Burke, A.C., Nelson, C.E., Morgan, B.A., and Tabin, C. 1995. *Hox* genes and the evolution of vertebrate axial morphology. *Development* 121:333–346.

Buss, L.W. 1987. *The Evolution of Individuality.* Princeton: Princeton University Press.

Cheverud, J.M. 1996. Developmental integration and the evolution of pleiotropy. *American Zoologist* 36:44–50.

Cheverud, J.M., Hartman, S.E., Richtsmeier, J.T., and Atchley, W.R. 1991. A quantitative genetic analysis of localized morphology in mandibles of inbred mice using finite element scaling analysis. *Journal of Craniofacial Genetics and Developmental Biology* 11:122–137.

Cohen, J. 1979. Maternal constraints on development. In D.R. Newth and M. Balls (eds.), *Maternal Effects in Development,* 1–29. Cambridge: Cambridge University Press.

Cohen, J., and Massey, B.D. 1983. Larvae and the origins of major phyla. *Biological Journal of the Linnean Society* 19:321–328.

Davidson, E.H., Peterson, K.J., and Cameron, R.A. 1995. Origin of Bilaterian body plans: evolution of developmental regulatory mechanisms. *Science* 270: 1319–1325.

de Beer, G.R. 1940. *Embryos and Ancestors*. Oxford: Clarendon Press.

Duboule, D. 1994. Temporal colinearity and the phylotypic progression: a basis for the stability of a vertebrate Bauplan and the evolution of morphologies through heterochrony. *Development* 1994 (suppl): 135–142.

Elinson, R.P. 1987. Change in developmental patterns: embryos of amphibians with large eggs. In R.A. Raff and E.C. Raff (eds.), *Development as an Evolutionary Process*, 1–21. New York: Alan R. Liss.

Garstang, W. 1922. The theory of recapitulation: a critical restatement of the biogenetic law. *Journal of the Linnean Society of London (Zoology)* 35:81–101.

Gilbert, S.F., Opitz, J.M., and Raff, R.A. 1996. Resynthesizing evolutionary and developmental biology. *Developmental Biology* 173:357–372.

Goodwin, B.C., Kauffman, S.A., and Murray, J.D. 1993. Is morphogenesis an intrinsically robust process? *Journal of Theoretical Biology* 163:135–144.

Gould, S.J. 1977. *Ontogeny and Phylogeny*. Cambridge: Belknap Press of Harvard University Press.

Graveson, A.C., Smith, M.M., and Hall, B.K. 1997. Neural crest potential for tooth development in a urodele amphibian: developmental and evolutionary significance. *Developmental Biology* 188:34–42.

Gruss, P., and Kessel, M. 1991. Axial specification in higher vertebrates. *Current Opinions in Genetics and Development* 1:204–210.

Haeckel, E. 1866. *Generelle morphologie der Organismen: Allgemeine Grundzüge der organischen Formen-Wissenschaft, mechanisch begründet durch die von Charles Darwin reformirte Descendenz-Theorie*. Berlin: Riemer.

Hall, B.K. 1983. Epigenetic control in development and evolution. In B.C. Goodwin, N.J. Holder, and C.C. Wylie (eds.), *Development and Evolution*, BSBD Symposium 6, 353–379. Cambridge: Cambridge University Press.

———. 1984. Developmental processes and heterochrony as an evolutionary mechanism. *Canadian Journal of Zoology* 62:1–7.

———. 1987. Tissue interactions in the development and evolution of the vertebrate head. In P.F.A. Maderson (ed.), *Developmental and Evolutionary Aspects of the Neural Crest*, 215–259. New York: John Wiley and Sons.

———. 1990. Heterochronic change in vertebrate development. *Seminars in Developmental Biology* 1:237–243.

———. 1994. *Homology: The Hierarchical Basis of Comparative Biology*. San Diego: Academic Press.

———. 1996a. Evolutionary developmental biology. In *McGraw-Hill Yearbook of Science and Technology*, 102–112. New York: McGraw-Hill.

———. 1996b. *Baupläne*, phylotypic stages, and constraint: why there are so few types of animals. *Evolutionary Biology* 29:215–261.

————. 1997. Phylotypic stage or phantom: is there a highly conserved embryonic stage in vertebrates? *Trends in Ecology and Evolution* 12:461–463.

————. 1998a. *Evolutionary Developmental Biology,* 2nd ed. London: Chapman and Hall/Netherlands: Kluwer Academic Publishers.

————. 1998b. Epigenetics: regulation not replication. *Journal of Evolutionary Biology* 11:201–205.

————. 1998c. Germ layers and the germ-layer theory revisited: primary and secondary germ layers, neural crest as a fourth germ layer, homology, demise of the germ-layer theory. *Evolutionary Biology* 30:121–186.

————. 1999. *The Neural Crest in Development and Evolution.* New York: Springer-Verlag New York.

————. 2000. Evolution of the neural crest in vertebrates. In L. Olsson and C.-O. Jacobson (eds.), *Regulatory Processes in Development: The Legacy of Sven Hörstadius,* 101–113. London: Portland Press.

Hall, B.K., and Hörstadius, S. 1988. *The Neural Crest.* Oxford: Oxford University Press.

Hall, B.K., and Miyake, T. 1992. The membranous skeleton: the role of cell condensations in vertebrate skeletogenesis. *Anatomy and Embryology* 186:107–124.

————. 1995a. How do embryos measure time? In K.J. McNamara (ed.), *Evolutionary Change and Heterochrony,* 3–20. Chichester: John Wiley and Sons.

————. 1995b. Divide, accumulate, differentiate: Cell condensation in skeletal development revisited. *International Journal of Developmental Biology* 39:881–893.

Hall, B.K., and Wake, M.H. (eds.). 1999. *The Origin and Evolution of Larval Forms.* San Diego: Academic Press.

Hanken, J. 1993. Model systems versus outgroups: alternative approaches to the study of head development and evolution. *American Zoologist* 33:448–456.

Hardy, M.H. 1992. The secret life of the hair follicle. *Trends in Genetics* 8:55–61.

Hart, M.W. 1996. Evolutionary loss of larval feeding: development, form and function in a facultatively feeding larva *Brisaster latifrons. Evolution* 50:174–187.

Holland, P.W.H., Holland, L.Z., Williams, N.A., and Holland, N.D. 1992. An amphioxus homeobox gene: sequence conservation, spatial expression during development and insights into vertebrate evolution. *Development* 116:653–661.

Huxley, J.S. 1932. *Problems of Relative Growth.* London: MacVeagh.

Lankester, E.R. 1911. *The Kingdom of Man.* London: Watts and Co.

Lewis, E.B. 1978. A gene complex controlling segmentation in *Drosophila. Nature* 276:565–570.

Liem, K.F., and Wake, D.B. 1985. Morphology: current approaches and concepts. In M. Hilderbrand, D.M. Bramble, K.F. Liem, and D.B. Wake (eds.), *Functional Vertebrate Morphology,* 366–377. Cambridge: Harvard University Press.

Lumsden, A. 1988. Spatial organization of the epithelium and the role of neural crest cells in the initiation of the mammalian tooth germ. *Development* 103 (suppl): 15–169.

Lumsden, A., and Graham, A. 1996. Death in the neural crest: implications for pattern formation. *Seminars in Cell and Developmental Biology* 7:169–174.

Macho, G.A., and Wood, B.A. 1995. The role of time and timing in hominid dental evolution. *Evolutionary Anthropology* 4:17–31.

Martin, J.F., Bradley, A., and Olson, E.N. 1995. The paired-like homeobox gene *MHox* is required for early events of skeletogenesis in multiple lineages. *Genes and Development* 9:1237–1249.

Mayr, E. 1982. *The Growth of Biological Thought: Diversity, Evolution, and Inheritance.* Cambridge: Belknap Press of Harvard University Press.

Minugh-Purvis, N. 1996. The modern human origins controversy, 1984–1994. *Evolutionary Anthropology* 4:140–147.

Miyake, T., Cameron, A.M., and Hall, B.K. 1996a. Detailed staging of inbred C57BL/6 mice between Theiler's [1972] stages 18 and 21 (11–13 days of gestation based on craniofacial development). *Journal of Craniofacial Genetics and Developmental Biology* 16:1–31.

———. 1996b. Stage-specific onset of condensation and matrix deposition for Meckel's and other first arch cartilages in inbred C57BL/6 mice. *Journal of Craniofacial Genetics and Developmental Biology* 16:32–47.

———. 1997. Variability of embryonic development among three inbred strains of mice. *Growth, Development and Aging* 61:141–155.

Moore, J.A. 1986. Science as a way of knowing—developmental biology. *American Zoologist* 27:1–159.

Nüsslein-Volhard, C., and Wieschaus, E. 1980. Mutations affecting segment number and polarity in *Drosophila. Nature* 287:795–801.

Osborn, H.F. 1894. *From the Greeks to Darwin: An Outline of the Development of the Evolution Idea.* New York: Macmillan.

Purugganan, M.D. 1996. Evolution of development: molecules, mechanisms and phylogenetics. *Trends in Ecology and Evolution* 11:5–7.

Raff, R.A. 1996. *The Shape of Life: Genes, Development, and the Evolution of Animal Form.* Chicago: University of Chicago Press.

Richardson, M.K. 1995. Heterochrony and the phylotypic period. *Developmental Biology* 172:412–421.

Richman, J., and Mitchell, P.J. 1996. Craniofacial development: knockout mice take one on the chin. *Current Biology* 6:364–367.

Rivera-Pérez, J.A., Mallo, M., Gendron-Maguire, M., Gridley, T., and Behringer, R.B. 1995. *goosecoid* is not an essential component of the mouse gastrula organizer but is required for craniofacial and rib development. *Development* 121:3005–3012.

Rowe, T. 1996. Coevolution of the mammalian middle ear and neocortex. *Science* 273:651–654.

Satokata, I., and Maas, R. 1994. *Msx*1 deficient mice exhibit cleft palate and abnormalities of craniofacial and tooth development. *Nature Genetics* 6:348–356.

Seilacher, A. 1988. Vendozoa: organismic construction in the Proterozoic biosphere. *Lethaia* 22:229–239.

Shea, B.T. 1992. Developmental perspective on size change and allometry in evolution. *Evolutionary Anthropology* 1:125–133.

Simpson, G.G. 1944. *Tempo and Mode in Evolution.* New York: Columbia University Press.

Slack, J.M.W., Holland, P.W.H., and Graham, C.F. 1993. The zootype and the phylotypic stage. *Nature* 361:490–492.

Smith, B.H. 1992. Life history and the evolution of human maturation. *Evolutionary Anthropology* 1:134–142.

Smith, K.K. 1996. Integration of craniofacial structures during development in mammals. *American Zoologist* 36:70–79.

Smith, M.M., and Hall, B.K. 1993. A developmental model for evolution of the vertebrate exoskeleton and teeth: the role of cranial and trunk neural crest. *Evolutionary Biology* 27:387–448.

Templeton, A.R. 1982. Why read Goldschmidt? *Paleobiology* 8:474–481.

Thesleff, I., and Sahlberg, C. 1996. Growth factors as inductive signals regulating tooth morphogenesis. *Seminars in Cell and Developmental Biology* 1:185–193.

Thompson, D'A.W. 1942. *Growth and Form,* 2nd ed. New York: Macmillan.

Valentine, J.W. 1990. Molecules and the early fossil record. *Paleobiology* 16:94–95.

Valentine, J.W., and May, C.L. 1996. Hierarchies in biology and paleontology. *Paleobiology* 22:23–33.

Vogel, K. 1991. Concepts of constructional morphology. In N. Schmidt-Kittler and K. Vogel (eds.), *Constructional Morphology and Evolution,* 55–68. Berlin: Springer Verlag.

Wagner, G.P., and Altenberg, L. 1996. Complex adaptations and the evolution of evolvability. *Evolution* 50:967–976.

Weber, K.E. 1992. How small are the smallest selectable domains of form? *Genetics* 130:345–353.

Whitman, C.O. 1919. Orthogenetic evolution in pigeons. In H.A. Carr (ed.), *Posthumous Works of Charles Otis Whitman,* 1–194. Washington, D.C.: Carnegie Institute.

Wilson, E.B. 1896. The embryological criterion of homology. *Wood's Hole Biological Lectures for 1894,* 101–124. Boston, Ginn and Co.

Zelditch, M.L., Bookstein, F.L., and Lundrigan, B.L. 1992. Ontogeny of integrated skull growth in the cotton rat *Sigmodon fulviventer. Evolution* 46:1164–1180.

Zelditch, M.L., and Fink, W.L. 1996. Heterochrony and heterotopy: stability and innovation in the evolution of form. *Paleobiology* 22:241–254.

Chapter 2

Shape and Stage in Heterochronic Models

David M. Alba

> Embryos undergo development; ancestors have undergone evolution,
> but in their day they also were the products of development. Our first
> task must therefore be to define these two sets of events to which liv-
> ing things are subject.
>
> Gavin de Beer (1958, 1)

Heterochrony (from the Greek *heteros* and *chronos*, meaning "different time")
is one of the essential concepts in evolutionary biology relating ontogeny
and phylogeny (see McNamara 1997 and Klingenberg 1998 for the most
recent reviews). As currently used, the term is not consistent with the
Haeckelian original meaning (an exception to his "fundamental biogenetic
law"), but it is consistent with the meaning given by de Beer (1958) and the
widely accepted definition since Gould (1977) (see also Gould 1988 and
1992a for details on the history of the word *heterochrony*). According to this
modern definition, heterochrony refers to evolution by means of "changes
in the relative time of appearance and rate of development for characters
already present in ancestors" (Gould 1977, 2).

If we follow this definition, in the case when shape is not changed but
size is, we cannot speak of heterochrony. Furthermore, it could also be argued that we cannot speak of heterochrony when no parallelisms between
ontogeny and phylogeny (paedomorphosis or peramorphosis) have been
produced (Gould 1977). Ontogenies evolve, so every evolutionary change
implies some change in ontogeny. In this restricted view, heterochrony

refers only to the "relative temporal shift of features already present in ancestors, as opposed to introductions of novelties" (Gould 1992b, 275). Thus, not all evolutionary changes would be considered heterochronic, and the reorganization of the whole ontogeny (the "ontogenetic repatterning" of Wake 1989) would not be heterochrony because it involves not merely an alteration, but a disruption of the ancestral developmental timing (no parallelism between ontogeny and phylogeny is produced). In more recent years, however, several authors, notably McKinney and McNamara (1991), Reilly et al. (1997), and McNamara (1997), have taken a more "panheterochronic" view, in which they view heterochrony as embracing much evolutionary change, including the evolution of morphological novelties.

Although these different views of heterochrony should be distinguished when assessing its actual evolutionary significance, even the more restrictive one by no means precludes the possibility that heterochrony is really pervasive in evolution. In fact, heterochrony has received great attention from many authors, probably because of the suspicion that ontogeny is one of the more readily available sources of variability for evolution. It has been pointed out that heterochronic studies should investigate how ontogenetic variability is generated and later "used" by natural selection (compare McNamara 1997), trying to infer the original selective pressures and the significance of developmental constraints during the evolution of particular lineages. Although this means that heterochronic studies should go beyond the mere labeling and description of evolutionary patterns, the use of some kind of model is essential to describe and evaluate heterochronic change.

According to Gould's operational definition of heterochrony as "the evolutionary displacement of a specific feature (a 'shape') relative to a common standard of size, age or developmental stage" (Gould 1977, 252), a complete heterochronic model should ideally measure simultaneously morphogenesis (broadly understood as shape generation and change), growth (size increase), time (absolute age), and maturation (ontogenetic age), both in the ancestor and in the descendant. Shape can be measured by means of some kind of dimensionless variable or proportion (although other valid possibilities exist); growth is measured by means of a metrical variable, such as length, area, volume, mass, and so on; and time is measured by means of absolute age at different homologizable developmental stages through ontogeny. Of course, this does not mean that each of these developmental processes (morphogenesis, growth, and maturation) is independent of the other, but simply that they are better measured separately because they can be potentially dissociated when heterochrony occurs.

Some criterion is necessary to standardize comparisons between ancestor and descendant when assessing heterochronic change. As stressed by Gould (1977, 261), "developmental stage is clearly the most satisfactory criterion for comparing ancestors and descendants" because "other standardizations are 'restricted' and 'can lead to pitfalls or causally ambiguous results.'" The soundest models thus far proposed are those of Gould (1977) and Alberch et al. (1979) (Fig. 2.1). Gould's (1977) clock model compares ancestor and descendant at a given developmental stage, showing simultaneously changes in size, shape, and age at this stage. Unfortunately, this is a static model, so that a depiction of several consecutive clock models (each at a different developmental stage) would be necessary to show a continuous portion of an organism's ontogenetic trajectory. On the other hand, the model of Alberch et al. (1979) is dynamic, allowing the simultaneous comparison of the ontogenetic trajectories of ancestor and descendant at any moment, but it does not incorporate the developmental stage criterion advocated by Gould (1977) to standardize ontogenetic trajectories.

Because of these differences, these models are not entirely compatible (Klingenberg 1998), although neither seems better than the other. Although the model of Alberch et al. (1979) improved some aspects not properly considered by Gould (1977) (dynamism and changes at the beginning of the ontogenetic trajectory), it was partially flawed because of neglect of the relationship between developmental time and absolute age (standardization by developmental stage). In my opinion, both models should be combined to permit standardization using developmental stage, both at onset and at offset, as well as the simultaneous comparison of the entire ontogenetic trajectory between ancestor and descendant. In this way, the need to standardize the offset of the ontogenetic trajectory (in my opinion, the greatest contribution of Gould's clock models) could be improved by the need to consider also the onset (as pointed out by Alberch et al. 1979).

$\longrightarrow$

Fig. 2.1. Different types of heterochrony according to the clock model of Gould (1977) (A) and the model of Alberch et al. (1979) (B). Rate hypomorphosis/hypermorphosis (A) is not considered here as a pure type of heterochrony but as a particular case of acceleration/neoteny, respectively. Symbols: σ = shape; a = age; k = "growth rate" parameter; α = onset of a developmental event; β = offset of a developmental event; X = ancestral ontogenetic trajectory; Y = descendant ontogenetic trajectory. *Source:* Redrawn from Gould 1977, Shea 1983b, and Alberch et al. 1979.

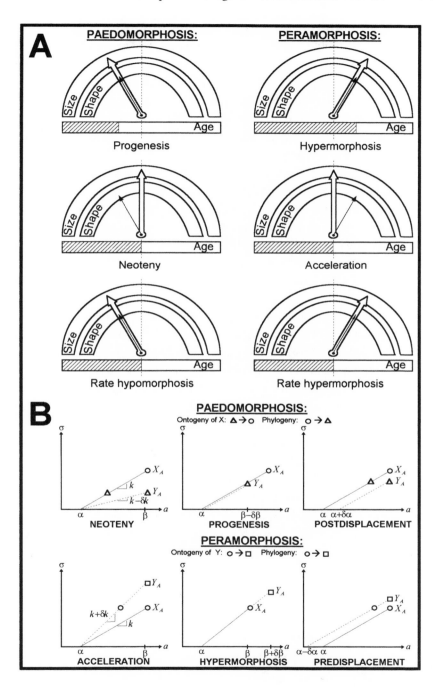

Meaningful Shape Variables, Developmental Stage Standardizations, and "Pure" Types of Heterochrony

None of the basic heterochronic terms (Fig. 2.1) should be employed without clearly stating the developmental stages and the variables used for comparison between ancestor and descendant. Two additional new terms are introduced here: *postformation* (paedomorphosis due to initial shape underdevelopment) and *preformation* (peramorphosis due to initial shape overdevelopment) (Fig. 2.2). A dichotomic key for diagnosing pure types of heterochrony, substituting that of McNamara (1986), is proposed (Fig. 2.2).

Although some partial models (those lacking shape, developmental stage, and/or absolute age information) can be used to distinguish some "pure" types of heterochrony, there can be no question that the theoretical definition of the latter should be based on the complete model, which includes shape, size, time, and maturation. Pure heterochrony, as defined here, refers to heterochronic processes that can be identified on the basis of shape and age at standard developmental stages. Since developmental stages are essential in carrying out meaningful comparisons of morphology between ancestor and descendant, partial models entirely lacking this information should probably be avoided. Ideally, heterochronic models should also include information about maturation (developmental stages) as well as morphology (size and shape at developmental stages) and (optionally) behavior, whereas the complete model should also include information about chronological time (age at developmental stages). When defining pure types of heterochrony, Gould (1977) relied heavily on the dissociation between size and shape. However, since too many possibilities exist and heterochrony occurs only when shape at some developmental stage is changed between ancestor and descendant, it is better to rely exclusively on shape, leaving size aside, when defining pure types of heterochrony. Thus, proportioned dwarfism and giantism, as defined by Gould (1977) and Alberch et al. (1979), are *not* heterochronies.

It is not yet clear how behavior should be included in a testable framework of heterochrony: can it be measured by means of behavioral shape variables, or should it be considered an alternative to maturation when defining developmental stages? In any case, there is no doubt that heterochrony plays a role in behavioral evolution, so that behavior must not be discarded when studying parallelisms between ontogeny and phylogeny (as discussed by Crockford in Chap. 6, McKinney in Chap. 8, and Parker in Chap. 14).

Some authors have argued that shape is "difficult to define and quantify" (Atchley 1987, 317). McKinney and McNamara (1991, 228) have even asserted that "the distinction between 'size' and 'shape' . . . is fundamentally artificial" (see also McKinney 1988, 25), thus recommending "the use of direct size metrics . . . rather than 'shape,' which is highly subjective and misleading" (McKinney and McNamara 1991, 365). I disagree with the latter point of view because, despite size-shape covariation, the assessment of shape is independent from that of size (e.g., a cube is always a cube and a sphere is always a sphere, no matter their size). Thus, size-age plots do not reflect morphogenesis but only growth, and therefore they are not relevant to assess heterochrony (although changes in growth can obviously be caused by heterochrony). It is possible to discuss ad infinitum how shape ought to be measured and analyzed, but it is unquestionably an inherently multivariate concept (Klingenberg 1998; Eble, Chap. 3, this volume), so that at least two metrical variables are needed to quantify it.

Although there are different valid approaches to biological shape, a first approximation can be easily accomplished by mean ratios and/or slopes of allometric plots. The consequences of treating shape as a metrical variable instead of a dimensionless one have been explored by Godfrey and Sutherland (1995b). However, when trait size is used instead of trait shape, heterochronic diagnosis is no longer compatible with the classical standards of Gould (1977) and Alberch et al. (1979), and even the meaning of paedomorphosis and peramorphosis becomes obscure.

Gould (1977) stressed the need to compare ancestor and descendant at homologizable standard developmental stages. However, this has not been recognized by many subsequent authors, possibly because of the lack of agreement about the metric to be used for standardizing developmental (intrinsic, ontogenetic) time between species (see Hall 1992 and Hall and Miyake 1995). For example, Reilly et al. (1997) used chronological instead of developmental time and defined terminal shape by the attaining of shape asymptotes. This is not surprising at all, since even Alberch et al. (1979) failed to make an explicit distinction between intrinsic and extrinsic time. Recently, Hall and Miyake (1995, 5) reviewed the four different criteria of standardization (maturity, growth, age, and morphology), concluding that "there is no objective way of selecting which criterion or combination of criteria to use," although recommending the use of morphological stages to measure the ontogenetic passage of time. This seems appropriate because stages more clearly homologizable than shape asymptotes can obviously be found.

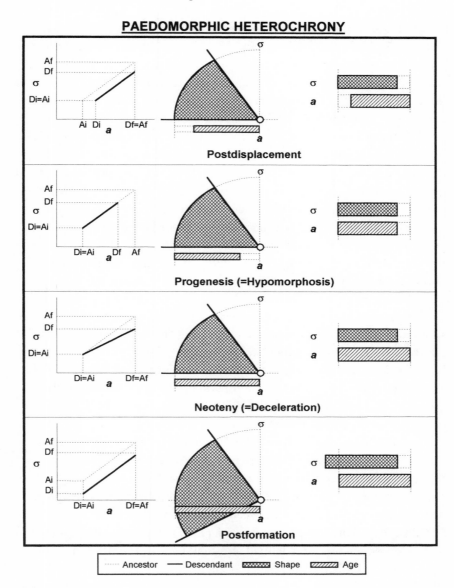

Fig. 2.2. The eight pure types of heterochrony recognized when the double standardization by developmental stage is taken into account. The types are represented by means of shape-age plots, Gould's clock models (transformed according to the double standardization), and simplified clock models. Symbols: σ = shape; *a* = age; *A* = ancestor; *D* = descendant; *i* = initial standard developmental stage; *f* = final standard developmental stage.

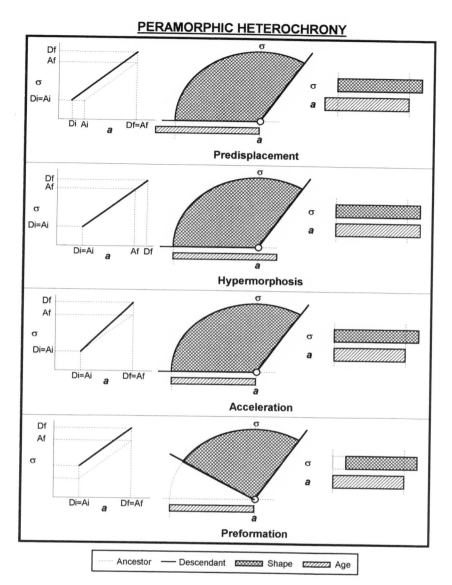

Fig. 2.2. (*Continued*)

Nevertheless, this poses the problem that shape comparisons are then carried out in stages partially defined on the basis of shape itself. Instead, I will advocate in this chapter the use of standardization by developmental stages defined on the basis of homologous shape-independent developmental events. These can be typical life-history events, such as birth or hatching, eruption of the first molar, metamorphosis, attainment of sexual maturity, and so on, depending on the taxa under investigation. In other words, maturation (the passage of intrinsic time) can be simply measured as age at several homologizable developmental stages, at which shape (and also size) comparisons can be made. Practically, in most cases, it is impossible to compare the whole ontogenetic trajectory (from the zygote's fertilization to the death of the individual) between ancestor and descendant, so that it is necessary to define precisely the span to be compared. Gould's (1977) model permitted only one standardization per clock (at the offset but not at the onset); therefore, I propose to adopt, whenever possible, a double standardization of the ontogenetic time span considered, similar to that of Alberch et al. (1979) but accomplished by means of standard homologizable developmental stages instead of absolute age, as proposed by Gould (1977). Although the significance of this kind of standardization has been disregarded by many authors since Gould (e.g., Reilly et al. 1997), it has been essential for subsequent improvements of the clock model, such as the distinction between time and rate processes made by Shea (1983b) (see Fig. 2.1).

Consequently, any useful heterochronic model, no matter whether complete or partial, should delimit the ontogenetic time span considered by means of at least two (initial and final) homologizable standard developmental stages, at which all comparisons could be made. When this double standardization by means of developmental stage is taken into account, the eight different pure types of heterochrony can be distinguished.

When shape is changed at the final standard developmental stage, then paedomorphosis or peramorphosis can be diagnosed. Many mixed types of heterochrony are possible when various heterochronic disturbances (even opposite ones) act together within the same growth field. Because different kinds of heterochrony can be produced in different parts of the body, or even simultaneously in the same part, heterochronic studies require an individualized treatment of growth fields (if not characters) rather than a global treatment of the organism as a perfectly and inextricably integrated whole. However, this is not to say that the holistic view does not remain useful when inferring original selective pressures and adaptations. Theoretically, in some cases heterochrony could not result in any parallelism between ontogeny and phylogeny (paedomorphosis or peramorphosis) because of the canceling effect of opposite heterochronic disturbances; to these special (and probably highly unusual) cases, the term *isomorphosis* could be applied, as proposed by Reilly et al. (1997).

The importance of standardization by final developmental stage is not clearly appreciated even though it was already present in Gould's (1977) clock model. However, it has not been previously noted that shape at the final standard developmental stage can change not only because of a different duration of the ontogenetic time span considered and/or a different rate of shape change but also because of a different initial shape. With Gould's classical clock model, all changes in duration were attributed to shifts in time of the final standard developmental stage (progenesis and hypermorphosis) or to different rates of shape change (neoteny and acceleration). With the subsequent model of Alberch et al. (1979), it was realized that true rate changes (neoteny and acceleration) had to be distinguished from shifts in time of the initial stage (postdisplacement and predisplacement).

Nevertheless, these latter authors failed to realize that two different possibilities existed, simply because they did not use a developmental stage-based standardization. When this is accomplished, two alternatives emerge: ancestral initial shape is attained at a different absolute age in the descendant because the initial developmental stage has been also shifted in age (true predisplacement and postdisplacement), or ancestral initial shape is attained at a different absolute age in the descendant in spite of the fact that the initial developmental stage has not been shifted in age (preformation and postformation). Since standardization is by developmental stage, when the initial shape of the ancestor is reached by the descendant at a different age, but age at the initial developmental stage is not changed, this change will be perceived as a shift in shape at the initial developmental stage (i.e., as a shift in initial shape). For example, postformation leads to paedomorphosis because the initial shape is underdeveloped in the descendant in relation to the ancestor, which in turn could be produced, among others, by neoteny before the selected initial developmental stage. As nothing can be said about this previous heterochronic change, which is not investigated, only the shift detected in the initial shape value can be documented.

Different kinds of heterochrony can be detected and diagnosed depending on the developmental stages selected for standardization, as recognized by Klingenberg (1998). This is due to the fact that shape trajectories change during ontogeny and can even reverse their polarity; therefore, there is no reason why the descendant could not be peramorphic or paedomorphic relative to the ancestor at a given developmental stage and just the reverse at a latter one (contra Reilly et al. 1997). This relativity implicit in all heterochronic studies, however, is not a fault of the models used, but a consequence of the complexity of ontogenetic trajectories and of the fact that changes can be produced at any portion of the ontogeny (not just at the end).

Preformation and Human Brain Evolution

The significance of preformation could be exemplified with human brain evolution. It is universally recognized that, when allometric considerations are taken into account, humans have brains relatively larger (in relation to body size) than those of most other primate species, thus often resembling juvenile and even fetal primates in brain size/body size proportions. This example serves to illustrate the significance of developmental stages when diagnosing types of heterochrony and their bearing on the inherent relativity of heterochronic diagnosis. Therefore, the problem of whether brain mass/body mass is a good shape variable or, instead, a bad mixture of different growth fields will not be addressed here.

It is widely accepted that our large brains result from the retention of typically high fetal rates of brain growth into the postnatal period (Gould 1977; Martin 1990; McKinney, Chap. 8, this volume). Much disagreement, however, has been generated around two opposite ways to describe this heterochronically. Some argue that the resemblance in cranial form between adult humans and juvenile apes is the result of true paedomorphosis reflecting the action of neoteny in human evolution (Gould 1977), even when *Homo sapiens* is compared with *Homo erectus* (Antón and Leigh 1998). Others have argued that this paedomorphosis is merely a superficial resemblance resulting from some kind of hypermorphosis in time (Shea 1988, 1989, 1992; McKinney and McNamara 1991; McNamara 1997). And finally, it has even been suggested that the high degree of encephalization found in humans results not from heterochrony, but from the insertion of a new developmental stage (childhood) between infancy and juvenility (Bogin 1997). The question is: Should the retention of fetal growth rates be considered a phenomenon of heterochrony and, if so, a case of neoteny; or, as McNamara (Chap. 5) argues, is it better termed *sequential hypermorphosis?*

Although Bogin's (1997) view is based on the tacit assumption that not all evolution is heterochronic, the use of behavioral criteria to define developmental stages (he defines infancy as terminating at weaning) is very doubtful. In fact, when a different definition of infancy (from birth until the emergence of the first permanent teeth) is used, it becomes clear that human brain evolution has been driven by changes in developmental timing. These changes have resulted in an extended period of growth and development (especially during early ontogeny) otherwise very similar to that of African apes (Leigh and Park 1998). Thus, it seems that new behaviors (instead of a new developmental stage) have been inserted in the ancestral ontogenetic trajectory because of a heterochronically extended interval available for brain development (see Parker, Chap. 14). Which, therefore, has been the main type of heterochrony involved in human brain evolution?

The solution to this problem becomes clear when it is realized that shape does not change monotonically during ontogeny, so that brain size/body size allometries can change dramatically (a phenomenon known as *multiphasic growth*) (Vrba 1998; Klingenberg 1998). As a result, the polarity of paedomorphosis/peramorphosis is reversed, depending on the developmental stages selected for comparison. In most primates, the brain grows with slightly positive allometry in relation to body size during prenatal development, whereas shortly after birth the relationship becomes progressively more negatively allometric. In humans, the retention of fetal brain growth rates during early postnatal development is attained by means of delaying offset time of the rapid supposedly prenatal growth phase, as already recognized by Gould (1977), Shea (1989), McKinney and McNamara (1991), McNamara (1997), and many others (see McKinney, Chap. 8). Therefore, if the focus is placed on embryonic, fetal, and early postnatal growth (from fertilization until emergence of lower first deciduous molars, for example), this alteration has to be qualified as time hypermorphosis leading to peramorphic proportions. This may reflect the more pervasive phenomenon of "sequential hypermorphosis" (McNamara, Chap. 5), since developmental stages in human ontogeny are generally progressively delayed over time (thus allowing more time for development). However, when postnatal ontogeny is taken into account, the final brain size/body size proportion in humans is truly juvenilized and therefore potentially paedomorphic, as interpreted by Gould (1977) and Shea (1988, 1989). As stated by Shea (1988, 252–253): "Our relatively large brains therefore result from time hypermorphosis, and they yield a high brain/body ratio that is paedomorphic, given the general postnatal negative allometry of brain/body growth."

This apparent paradox is due to the above-mentioned reversal of ontogenetic polarity and the consequent relativity of heterochrony, so that the high degree of encephalization in humans is a case of hypermorphosis leading to peramorphosis relative to the first ontogenetic time span discussed. However, it can also be viewed as a case of postformation (initial shape underdeveloped) leading to paedomorphosis relative to the latter. Here postformation means only that initial shape in the descendant is more juvenilized, which is known to be due to hypermorphosis in a previous ontogenetic time span (note that a previous hypermorphosis leads to paedomorphosis in a latter ontogenetic span only because the polarity has been reversed). If information was lacking about this earlier portion of ontogeny, however, only the shift in initial shape could be identified. Nevertheless, this is very important because it highlights the fact that some previous heterochronic perturbations are missing, thus indicating which portion of ontogeny should be more carefully studied. This illustrates the importance of recognizing the possibility that initial shape can be shifted because several kinds of changes in earlier portions of ontogeny can be confounded with rate changes when no changes in initial shape are allowed. Moreover, the diagnoses of paedomorphosis and peramorphosis are always relative to the standard developmental stage (compare Klingenberg 1998), as well as to the shape variable selected.

In summary, paedomorphosis and peramorphosis are not descriptive but interpretative terms, and their meaning depends on whether they are based on meaningful variables. Without developmental evidence arguing in favor of brain size/body size as a meaningful shape variable (to take this particular example), juvenilized brain proportions could be alternatively interpreted as the mixed result of different growth fields, each depending on partially different developmental basis and therefore susceptible to different heterochronic perturbations, as argued for the human chin by Gould (1977, 381–382).

Partial Models and Heterochronic Inferences

I have advocated the use of a double standardization of ontogenetic trajectories in a complete model including, at the very least, information about shape, developmental stages, and age. What happens, however, when some of these requirements cannot be met because some information is lacking? Heterochronic studies should not be abandoned, but what can and what cannot be inferred by means of the partial model should be clearly stated, in addition to those assumptions necessary to avoid the problems posed by missing data, because of the potential pitfalls already discussed by Gould

(1977). For example, when no developmental stage can be distinguished, age or size, or both, must be relied upon. However, it is then necessary to assume some hypotheses that, if false, could invalidate the inferences. The problems associated with the use of size as a proxy for age have long been recognized (McKinney and McNamara 1991) and will not be addressed here, although the same could be argued for the use of age without developmental stages and vice versa. Since I agree with Gould (1977) that comparisons must be carried out at homologous developmental stages (e.g., adult descendants should be compared with ancestral adults, regardless of whether it took them the same time to attain adulthood), partial models disregarding developmental stages will not be discussed further.

In my opinion, when the complete model is not applicable because age data are lacking (unfortunately, a very common situation, especially in the fossil record), both shape and developmental stage(s) should be relied upon simultaneously. This can be accomplished in two different ways (Fig. 2.3). First, shape variables can be plotted on a developmental stage axis; this can be used to distinguish between paedomorphosis and peramorphosis and has the advantages that shape interpretation is quite simple and that maturational information is explicitly stated. Second, one can use allometric plots of a trait y versus another trait x (being y and x the metrical variables used to calculate the shape variable y/x under consideration); this procedure is also useful in distinguishing between paedomorphosis and peramorphosis (e.g., using the line of isometry criterion proposed by Gould 1977). However, it has some disadvantages associated with shape interpretation, which becomes more difficult because of log-transformation and because maturational information is usually only implicitly incorporated; on the other hand, allometric plots allow us to see whether the descendant is an ontogenetically scaled version of the ancestor. This cannot be done so easily with shape-developmental stage plots; moreover, intrinsic age data could also be incorporated to allometric plots, as in successive data points representing means for developmental stages (B. Shea, pers. comm.).

Godfrey and Sutherland (1995a, 1995b, 1996) recently pointed out some far-reaching problems with the allometric approach to heterochrony, arguing that "growth allometries are a poor vehicle for inferring heterochronic processes" (however, see Shea, Chap. 4, for an alternative view). In fact, however, neither of the two partial models discussed here is unquestionably better than the other because they depict exactly the same information (although allometric plots remain more useful to evaluate ontogenetic scaling). The allometric diagnosis of heterochrony as proposed by McKinney

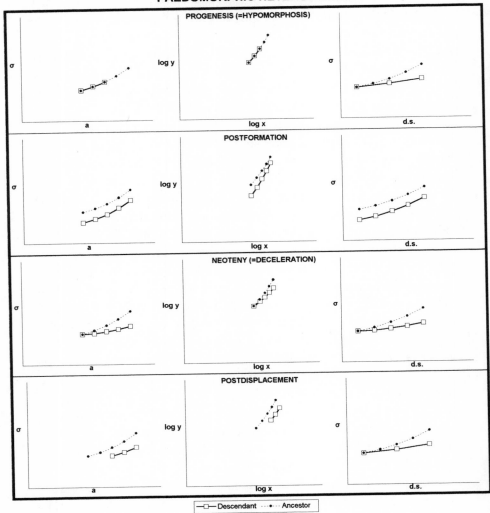

Fig. 2.3. The eight pure types of heterochrony depicted using the complete model (*left*), the model of allometric heterochrony (*center*, assuming that size is a good proxy for age), and the partial model lacking absolute age information (*right*). Only with the complete model can the eight different types of heterochrony be distinguished. The partial model confounds progenesis, neoteny, and postdisplacement, on the one hand, and hypermorphosis, acceleration, and predisplacement, on the other, in a way similar to the model of allometric heterochrony when size is not a good surrogate for age (not shown). This is only one possible example, in which a trait y grows with positive allometry in relation to trait x; if the ancestral allometric relationship were different, shifts in allometric plots would be of different type. Symbols: σ = shape (y/x); a = age; y, x = trait sizes; *d.s.* = developmental stage.

PERAMORPHIC HETEROCHRONY

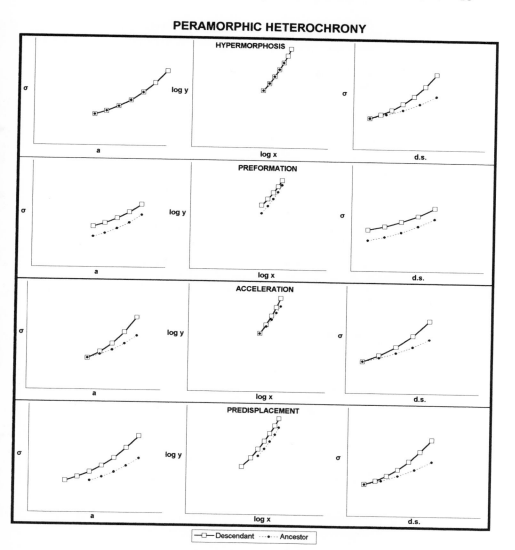

Fig. 2.3. (*Continued*)

(1986, 1988; see also McKinney and McNamara 1991), however, does have some problems because it was intended to be compatible with traditional diagnoses, provided only that size is a good proxy for time. This is not always true. When size is assumed to be a good surrogate for age, allometric plots can distinguish between all pure types of heterochrony, but the assumption remains unverified without absolute age data. Moreover, this assumption notwithstanding, the compatibility between both diagnoses holds only when the ancestral ontogenetic trajectory is positively allometric. This is because of the erroneous conceptualization of shape as a metrical instead of a proportional variable (see discussion in Godfrey and Sutherland 1995b) and also because the changes reflected by allometric plots do not depend exclusively on the disturbance produced, but also on whether the ancestral ontogenetic trajectory was positively or negatively allometric. This, however, is not an inherent flaw of allometric techniques but a mistake by omission of McKinney (1986, 1988) and McKinney and McNamara (1991). In fact, allometric techniques can be useful as long as the implicit assumptions are clearly stated and limitations of the approach are not forgotten (particularly that there is no unequivocal correspondence between types of heterochrony and changes in allometric trajectories).

When size is not assumed to be a good proxy for age, allometric plots of pure types of heterochrony are as equivocal as the other partial model considered here. For example, in Shea's assessment of the African apes discussed below, Shea could not have distinguished between rate and time processes if he had not used information of age at the final standard developmental stage (adulthood) and assumed that no changes at the beginning of ontogeny had been produced. When no additional assumptions are made or additional information is provided, neither one partial model is better than the other. Even ontogenetic scaling is not useful here; for example, although neoteny and acceleration usually imply dissociation between size and shape (as when size is a good surrogate for age), they can also be produced maintaining the ancestral allometric relationship (as in the cases of rate hypo- and hypermorphosis). In fact, ontogenetic scaling can be produced not only by progenesis and hypermorphosis, but also by rate changes (neoteny and acceleration) as well as on onset time (predisplacement and postdisplacement), so that age is essential to identify the type of heterochrony responsible for the allometric pattern. Consequently, these partial models should be relied on only when it is absolutely impossible to have absolute age data because, unlike the full model, none of them allows us to distinguish all the different pure types of heterochrony.

Allometry and Heterochrony in Great Apes

The allometric approach has been followed by Shea when assessing heterochrony in the African apes (Shea 1983a, 1983b, 1983c, 1984a, 1984b, 1988; Chap. 4, this volume). He has shown that many traits in the three living species of African apes are ontogenetically scaled, thus demonstrating the usefulness of allometric techniques for heterochronic inference. Shea's work has effectively shown that many shape differences between African apes simply result from the extension/truncation in size of common growth allometries (and hence possibly homologous developmental pathways) without disturbing the ancestral duration of the ontogenetic time span.

Particularly interesting is the evolution of a highly paedomorphic cranial anatomy (shown by the more gracile and rounded skull with less prognathism and reduced sexual dimorphism) in the pygmy chimpanzee (*Pan paniscus*). This paedomorphic cranial form (taking the common chimp as the ancestor) is attained through ontogenetic scaling of many (though not all) cranial dimensions and therefore could be attributed to rate hypomorphosis. Progenesis can be discarded because, as pointed out by Shea (1984b), no differences in the duration of postnatal ontogeny are known between chimpanzee species. Shea (1983c, 1984a, 1989) attributed this paedomorphosis to neoteny because cranial measurements (e.g., skull length) plotted against overall body size (e.g., trunk length) show a reduced slope in allometric relationship. However, this attribution is not correct; when the ancestral relationship between two traits is negatively allometric (as in skull length versus trunk length), neoteny should result in *increased* slopes (Shea 1989; Godfrey and Sutherland 1996).

This apparent incongruence does not mean that the pygmy chimp is in fact peramorphic through acceleration instead of neotenic. Rather, our classical tools for heterochronic diagnosis should not be applied to dissociation between growth fields, but only to dissociation of developmental processes (morphogenesis, growth, and maturation) within the same growth field. When comparisons are restricted to the cranium in the pygmy chimpanzee (Fig. 2.4), rates of shape change have clearly decreased. Whether this should be termed *rate hypomorphosis* or *neoteny* is another, more subjective question, although I would argue in favor of neoteny, since the duration of ontogeny is not changed and ontogenetic scaling could simply reflect that heterochrony has equally affected the whole cranial growth field.

A similar example has been documented in the Late Miocene great ape *Oreopithecus bambolii* by means of nonallometric models (Moyà-Solà and Köhler 1997; Alba et al., in press). The evolution of *Oreopithecus* cranial form has been attributed by these authors to paedomorphosis, as shown by several juvenilized characters. In particular, neoteny has been advocated as the most likely explanation, as exemplified by the zygomatic root positioned far forward in the face, which fits the inferred juvenile ancestral condition still retained in the living great apes *Pan* and *Pongo* (Fig. 2.4). In this case, however, the reduced rate of shape change has been evaluated only in relation to dental developmental stages without absolute age information, so it cannot be concluded unequivocally that the total amount of chronological time available for development has not been reduced in *Oreopithecus*. Although postformation can be rejected because the descendant shape trajectory departs from the ancestral one, progenesis and postdisplacement are still potential explanations that could only be undoubtedly rejected on the basis of absolute age at developmental stages. Since the dimensionless variable measuring the position of the zygomatic root is not calculated as a ratio of two metrical measures (but simply as position of the zygomatic root in relation to the upper tooth row), the allometric approach cannot be applied in this case, unless another (metrical) variable is introduced in the analysis. The similarity between the cases of *Oreopithecus* and *P. paniscus* is schematically shown in Figure 2.4 and deserves further study.

Summary and Conclusions

Unraveling the heterochronic processes involved in evolutionary change is essential if we are to transcend the mere descriptive patterns of paedomorphosis and peramorphosis because the same heterochronic results can be produced by means of different ontogenetic changes. Identifying these changes is essential to infer original selective pressures and adaptations. This is not straightforward, however, because selection can operate not only on morphology (shape and/or size) but also on behavior and/or life history. Since a different name cannot be given to every combination of features experiencing heterochrony, pure types of heterochrony must be defined and used simultaneously to describe more complex mixed changes. Although different partial models are available for those situations when some data are lacking, I prefer the use of a complete model including shape and absolute age at homologous developmental stages and ideally also size and behavior. It is my contention that ontogenetic trajectories must be standardized by means of two (initial and final) standard developmental stages.

The real significance of heterochrony in evolution is still being assessed in the natural world. During the last twenty years, however, enough has been learned to strongly support the suspicion that heterochrony has

HYPOTHESES OF CRANIAL NEOTENY IN HOMINOIDS

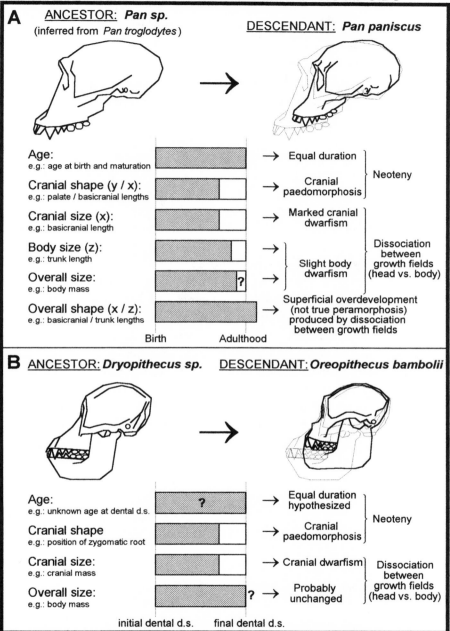

Fig. 2.4. Schematic depiction of hypotheses of neoteny. *A: Pan paniscus* (ancestral values inferred from *Pan troglodytes,* assuming no changes in initial values). *Source:* After Shea (1983c, 1984a, 1992). *B: Oreopithecus bambolii* (ancestral values inferred from *Dryopithecus* and living great apes, assuming that absolute time was not dissociated from developmental stages [*d.s.*]). *Source:* After Moyà-Solà and Köhler (1997).

played a very pervasive role in evolution because of limits to the range of potentially adaptive variations. Thus, instead of strolling freely through a universe of almost endless possibilities, evolution more often travels along roads that, to a certain extent, have already been predetermined (channeled and constrained) by ancestral developmental pathways. Quantitative models and unambiguous definitions of pure types of heterochrony, like those discussed in this chapter, are absolutely essential for rigorous assessment of heterochrony among living organisms as well as in the fossil record.

Acknowledgments

I am very grateful to the editors of this book, Nancy Minugh-Purvis and Ken McNamara, for inviting me to contribute and to Brian Shea for initiating the contact. I also thank all the people who have read any previous version of this manuscript for their comments, criticisms, encouragement, and helpful discussion: Jaume Baguñà, Christian Klingenberg, Ken McNamara, Salvador Moyà Solà, and Brian Shea. The latter is deeply acknowledged for insightful discussion about the significance of partial models during the gestation of this chapter. Moreover, I express my boundless gratitude to Salvador Moyà Solà for giving me the opportunity to devote myself to the study of heterochrony in the hominoid fossil record. Anna Fernández is sincerely acknowledged for philological advice. This chapter has been partially supported by an FI grant from the Generalitat de Catalunya. It is dedicated to Meritxell Mas Aceves because of her unlimited patience, comprehension, and support.

References

Alba, D.M., Moyà-Solà, S., Köhler, M., and Rook, L. In press. Heterochrony and the cranial anatomy of *Oreopithecus*: some cladistic fallacies and the significance of developmental constraints in phylogenetic analysis. In G. Koutos, L. de Bonis, and P. Andrews (eds.), *Hominoid Evolution and Climatic Change in Europe*, vol. 2. Cambridge: Cambridge University Press.

Alberch, P., Gould, S.J., Oster, G.F., and Wake, D.B. 1979. Size and shape in ontogeny and phylogeny. *Paleobiology* 5:296–317.

Antón, S.C., and Leigh, S.R. 1998. Paedomorphosis and neoteny in human evolution. *Journal of Human Evolution* 34:A2.

Atchley, W.R. 1987. Developmental quantitative genetics and the evolution of ontogenies. *Evolution* 41:316–330.

Bogin, B. 1997. Evolutionary hypotheses for human childhood. *Yearbook of Physical Anthropology* 40:63–89.

de Beer, G.R. 1958. *Embryos and Ancestors*, 3d ed. Oxford: Clarendon Press.

Godfrey, L.R., and Sutherland, M.R. 1995a. Flawed inference: why size-based tests of heterochronic process do not work. *Journal of Theoretical Biology* 172:43–61.

———. 1995b. What's growth got to do with it? Process and product in the evolution of ontogeny. *Journal of Human Evolution* 29:405–431.

———. 1996. The paradox of peramorphic paedomorphosis: heterochrony and human evolution. *American Journal of Physical Anthropology* 99:17–42.

Gould, S.J. 1977. *Ontogeny and Phylogeny*. Cambridge: Belknap Press of Harvard University Press.

———. 1988. The uses of heterochrony. In M.L. McKinney (ed.), *Heterochrony in Evolution: A Multidisciplinary Approach*, 1–13. New York: Plenum Press.

———. 1992a. Heterochrony. In E.F. Keller and E.A. Lloyd (eds.), *Keywords in Evolutionary Biology*, 158–165. Cambridge: Harvard University Press.

———. 1992b. Ontogeny and phylogeny: revisited and reunited. *BioEssays* 14: 275–279.

Hall, B.K. 1992. *Evolutionary Developmental Biology*. London: Chapman and Hall.

Hall, B.K., and Miyake, T. 1995. How do embryos measure time? In K.J. McNamara (ed.), *Evolutionary Change and Heterochrony*, 3–20. Chichester: John Wiley and Sons.

Klingenberg, C.P. 1998. Heterochrony and allometry: the analysis of evolutionary change in ontogeny. *Biological Reviews* 73:79–123.

Leigh, S.R., and Park, P.B. 1998. Evolution of human growth prolongation. *American Journal of Physical Anthropology* 107:331–350.

Martin, R.D. 1990. *Primate Origins and Evolution: A Phylogenetic Reconstruction*. Princeton: Princeton University Press.

McKinney, M.L. 1986. Ecological causation of heterochrony: a test and implications for evolutionary theory. *Paleobiology* 12:282–289.

———. 1988. Classifying heterochrony: allometry, size, and time. In M.L. McKinney (ed.), *Heterochrony in Evolution: A Multidisciplinary Approach*, 17–34. New York: Plenum Press.

McKinney, M.L., and McNamara, K.J. 1991. *Heterochrony: The Evolution of Ontogeny*. New York: Plenum Press.

McNamara, K.J. 1986. A guide to the nomenclature of heterochrony. *Journal of Paleontology* 60:4–13.

———. 1997. *Shapes of Time: The Evolution of Growth and Development*. Baltimore: Johns Hopkins University Press.

Moyà-Solà, S., and Köhler, M. 1997. The phylogenetic relationships of *Oreopithecus bambolii* Gervais, 1872. *Comptes Rendus de l'Academie des Sciences, Paris, IIa* 324:141–148.

Reilly, S.M., Wiley, E.O., and Meinhardt, D.J. 1997. An integrative approach to heterochrony: the distinction between interspecific and intraspecific phenomena. *Biological Journal of the Linnean Society* 60:119–143.

Shea, B.T. 1983a. Size and diet in the evolution of African ape craniodental form. *Folia Primatologica* 40:32–68.

———. 1983b. Allometry and heterochrony in the African apes. *American Journal of Physical Anthropology* 62:275–289.

———. 1983c. Paedomorphosis and neoteny in the pygmy chimpanzee. *Science* 222:521–522.

———. 1984a. An allometric perspective on the morphological and evolutionary relationships between pygmy (*Pan paniscus*) and common (*Pan troglodytes*) chimpanzees. In R.L. Susman (ed.), *The Pygmy Chimpanzee: Evolutionary Biology and Behavior,* 89–130. New York: Plenum Press.

———. 1984b. Ontogenetic allometry and scaling: a discussion based on growth and form of the skull in African apes. In W.L. Jungers (ed.), *Size and Scaling in Primate Biology,* 175–205. New York: Plenum Press.

———. 1988. Heterochrony in primates. In M.L. McKinney (ed.), *Heterochrony in Evolution: A Multidisciplinary Approach,* 237–266. New York: Plenum Press.

———. 1989. Heterochrony in human evolution: the case for neoteny reconsidered. *Yearbook of Physical Anthropology* 32:69–101.

———. 1992. Neoteny. In S. Jones, R. Martin, and D. Pilbeam (eds.), *The Cambridge Encyclopedia of Human Evolution,* 401–404. Cambridge: Cambridge University Press.

Vrba, E.S. 1998. Multiphasic growth models and the evolution of prolonged growth exemplified by human brain evolution. *Journal of Theoretical Biology* 190:227–239.

Wake, D.B. 1989. Phylogenetic implications of ontogenetic data. *Geobios, mémoire spécial* 12:369–378.

Chapter 3

Multivariate Approaches to Development and Evolution

GUNTHER J. EBLE

Development is the usual arena of evolutionary change in morphology, and much of evolution obviously depends on it. It is very common, however, for discussions of human evolution and developmental change to focus on individual characters (brain size, body weight, bipedalism, etc.). This particularist approach has been enormously useful in understanding how individual phenotypic modules have changed in the course of human evolutionary history. But it has also led to dilemmas when, on the basis of the pattern of developmental evolution of individual features, broad statements about evolution in general are made (e.g., paedomorphosis vs. peramorphosis).

Dissociated heterochrony (McKinney and McNamara 1991) and heterotopy (Zelditch and Fink 1996) are natural consequences of the modular nature of most organisms, and they have often been documented. The relative dissociability of phenotypes (quasi-independence of Lewontin 1978) justifies a more comprehensive treatment of developmental change whenever synthetic statements are sought. Conversely, high morphological integration may well entail uniform developmental change across characters. A multivariate perspective permits the discernment of such alternatives and the production of average descriptions of development and evolution in multidimensional space. Two approaches are immediately obvious by contrast to the particularist approach: the tabulation of how developmental change occurs across individual characters (a useful line of attack that often figures in the literature but that soon becomes intractable as the number of characters increases and as redundancy confounds the potential for synthetic inferences) and the direct summarization of multivariate variation with statistical techniques. This chapter will address the latter.

McKinney and McNamara (1991) and Tissot (1988) give good overviews of the advantages of multivariate analysis in the context of heterochrony, and Shea (1985) and Klingenberg (1996) discuss multivariate allometry. While I will elaborate on traditional multivariate methods in this chapter, my intent is to eventually move beyond approaches that have now become de rigueur, such as principal components analysis, and explore alternative ways of assessing patterns and processes of human developmental change in a multivariate framework.

Evolutionary anthropology as a discipline, in both biological and paleontological terms, shares much with evolutionary biology and paleobiology and yet, for a plethora of reasons (both sociological and historical), has remained self-contained and somewhat isolated from methodological and conceptual developments in those fields. The reverse is also true, but from the standpoint of anthropology this isolation has meant a rather idiosyncratic application of such developments. This is because they are relevant only to the extent that they can be fruitfully adapted to the empirical situations with which evolutionary anthropology is usually faced and tied to the tradition of empirical research in a straightforward fashion. Landmark-based approaches, for example, are widely accepted within anthropology (e.g., Lynch et al. 1996; Chaline et al. 1998). As a result, recent advances in multivariate biology and paleobiology have not yet been explicitly applied in evolutionary anthropology.

Most discussions of evolutionary change in development focus on the fundamental notion of ontogenetic trajectory—how it is modified from ancestor to descendant. If one had access to a fully resolved phylogenetic tree and if complete ontogenetic sequences were available for all species, one would proceed to interpret transition after transition in terms of how particular trajectories were modified. This tradition, while defensible (it will be applied below), hides the possibility that other observables might be informative in understanding the relationship between development and evolution. For example, developmental constraints are usually taken as given and their potential breaking deemphasized, since heterochrony is usually thought to occur along an established trajectory. A focus on individual transitions also impedes a comprehensive appraisal of sweeping cross-phylogenetic predictions, such as von Baer's laws.

I argue below that an ensemble approach, explicitly directed at identifying statistical regularities across a clade, may give insights that would not be immediately obvious to a student of individual transitions. From the perspective of human evolution studies, this implies incorporation of principles and methods most commonly found in invertebrate evolutionary paleobiology. In particular, the concepts of morphospace and disparity (Raup 1966; Foote 1997) are implicit in much of the data and aims of human evolution studies—their explicit recognition could add new dimensions of interpretability. By adapting the notion of morphospace and disparity to the arena of development (see Eble 1998a, 2001), I intend to suggest how such concepts can be valuable in the study of human developmental change and, by extension, human evolution in general.

Heterochrony is a powerful mechanism of evolutionary change in development. However, as much as in a univariate context (Rice 1997), in a multivariate context the question of whether additional ways of studying developmental evolution are possible or useful deserves more attention. Ultimately, all changes in development must involve changes in rate or timing (whether uniform or nonuniform, along a single trajectory or leading to novelty, having a spatial component or not)—the question is whether any given formalism should be used by default or whether a more pluralistic attitude is desirable. Current discontent with the standard nomenclature of heterochrony has been for the most part cast in terms of univariate situations, in line with the received typology (Alberch et al. 1979). Exceptions include Zelditch and Fink (1996) and Godfrey and Sutherland (1996). While multivariate heterochrony will inevitably surface in the discussions, this chapter will mostly be concerned with the general utility of multivariate approaches to developmental change, on the one hand, and with the possibility of, through them, studying evolution and development in novel ways, on the other.

Morphospace and Disparity

In paleobiology and macroevolution, the use of multivariate approaches has, over the years, become encapsulated in two related concepts: morphospace and disparity. Morphospace studies have a rather long history. Any representation of variation in state space (uni- or multivariate) is tantamount to a morphospace representation. In many fields, including biological anthropology, this is already implicit in the very application of statistical methods to describe and summarize variation, but in paleobiology and macroevolution the idea of morphospace came to symbolize a concern for the relationship between actual and possible and the mechanisms underlying overall distributions in ordination space. Hence the distinction between theoretical (Raup 1966; McGhee 1998) and empirical (McGhee 1991; Foote 1997) morphospaces. Generally conceived, the idea of morphospace often demands a multivariate perspective because the usual goal is the representation of whole organisms, or representative samples of overall morphology, in state space. As in multivariate statistics, each original variable is treated equivalently to allow many variables to contribute to the characterization of objects in morphospace (Foote 1991).

The concept of disparity, in turn, grew out of phenetics and the notion of distance in state space (Sneath and Sokal 1973; Van Valen 1974; Foote 1991). Rather than a summarization of morphological distributions through ordination, the goal is to produce state variables that represent the average spread and spacing of forms in morphospace. Many morphospace studies nowadays thus incorporate the quantification of disparity as a major goal (e.g., Foote 1997; Roy 1996; Wagner 1997; Eble 1998b, 2000). A body of theory regarding expectations for the dynamics of disparity vis-à-vis diversity is also beginning to grow (Foote 1996).

In the end morphospace and disparity studies, beyond the reliance on multivariate statistics, seek to understand the evolution of whole clades through time, with taxa treated as equivalent entities subject to quantitative representation. While this has usually implied sampling fossils over several geological intervals, the concepts of morphospace and disparity are equally applicable to neontological data and to situations where time is measured differently (say, with phylogenies). Because morphospace and disparity studies can generate much insight into the mechanisms of macroevolution (including selection, developmental constraints, and chance), the challenge of extending them to other disciplines seems a worthy one.

A Multivariate Description of Variation

As implied, the concepts of morphospace and disparity do not in themselves demand a multivariate perspective. A multivariate approach does not supersede univariate ones. It is best seen as a complement that allows reduction of redundancy, power of summarization, or both. In addition, certain outliers are never revealed in an univariate context because they are by their very nature multivariate (see Barnett and Lewis 1994). While there are disadvantages to multivariate analysis (see Pimentel 1979; McKinney and McNamara 1991), they are far outweighed by the synthetic insight that usually accrues. And because form is multidimensional, morphospaces and disparity are best studied with multivariate methods.

Ordination

Multivariate ordination techniques can be used to produce space diagrams of one's data, thus effectively approximating a morphospace representation. While some distortion of actual similarities can occur, points that are close together can be taken as similar entities and vice versa. Many ordination techniques can isolate differences on different axes (e.g., size on one axis and shape on others). What is usually referred to as *traditional morphometrics* (the multivariate analysis of linear distance measurements) relies heavily on ordination to gain insight into data structures. Methods that rely on landmark data, such as thin-plate splines, also have ordination as one main goal of the analysis (see Bookstein 1991). Here I focus on principal components analysis because it is a straightforward technique that is widely used and immediately applicable to both linear distance measurements (on which most studies in evolutionary anthropology are based) and landmark data. Bookstein (1991) and Marcus et al. (1996) contain discussions of other ordination techniques, including those designed to landmark data. More traditional methods can be found in modern textbooks on multivariate statistics (e.g., Pimentel 1979; Morrison 1990; Reyment and Jøreskog 1993).

Principal components analysis (PCA) is probably the most widely used ordination technique in multivariate biology and paleobiology. Its appeal derives from its effectiveness in reducing the dimensionality of the data while at the same time summarizing much of the variance in the few resultant dimensions. The elegance of the underlying logic, which is based on linear algebra, allows a variety of extensions, elaborations, and applications. In this sense PCA is often the method of choice when one needs to organize a multivariate array of data and then choose one of several pathways of further inquiry.

PCA is an eigenvector technique, and as such it shares many of the features of a whole class of methods, including factor analysis, canonical analysis, and correspondence analysis, which attempt to unravel simple patterns in a set of multivariate observations (Davis 1986). It relies on the assumption that the original data are correlated in some way, so that its variance-covariance or correlation structure can be used as the basis for a dimensionality reduction process that retains as much as possible of the original variation (Jolliffe 1986).

Algebraically, the main objective is to create a set of new and uncorrelated variables (the principal components) that can be ranked according to the amount of variation explained. The iterative process of finding new variables involves maximizing the original variance explained by each of them. Geometrically, what PCA does is to define the principal axes of an ellipsoid encompassing the whole cloud of points in multidimensional space. In conformity with the algebraic formulation, one creates a set of new and orthogonal axes (the principal component axes), which are successively oriented according to the largest variance among individuals or variables (Neff and Marcus 1980).

Operationally, the finding of new variables is achieved by calculation of eigenvalues and eigenvectors from a square symmetric matrix representing the data in some form. This is done by solution of the characteristic equation. Each eigenvalue represents a proportion of the total variance. The elements of each eigenvector (loadings) are the coefficients of the linear equation that lead to the creation of a new variable from the linear weighted compound of the original variables. The eigenvectors give the orientations, and the eigenvalues give the magnitudes of the principal axes (Neff and Marcus 1980; Davis 1986). The original objects or variables can be projected onto each principal component (PC) axis as scores, which represent the importance of each axis in the objects or variables. The new variables and axes are latent (Bookstein et al. 1985; Morrison 1990), in the sense that they constitute abstract constructs that themselves might represent, in potentia, possible underlying causes.

The success of a PCA is more than anything else proportional to the degree of reduction of dimensionality. In other words, when the first few axes account for most of the total variance (exactly how much is a matter of technical debate), we are justified in presuming that a parsimonious representation of the original data structure was achieved (Morrison 1990). Although PCA may lead to the formulation of hypotheses, it is best conceived as a mathematical, not a statistical method (Davis 1986). Its primordial role is to represent data in simple form; whatever possible causal relationships are indicated, they should not arise as a result of the logic of the method. Biological interpretability should be the ultimate guide in inferences from PCA (Tissot 1988).

Cluster Analysis

Cluster analysis is one of the simplest multivariate techniques. This stems in part from an intuitive underlying logic and an easily interpretable graphic output. Apart from its weaknesses (sensitivity to choice of similarity measure and method of clustering) and relative primitiveness, its popularity implies that more researchers (sometimes even in different areas) can readily understand results from a cluster analysis. This means that, whenever there is a suspicion that pattern is present in a strong way, one can benefit from at least preliminarily undertaking such an analysis.

Given a multivariate space, one may be interested in the possibility of its segregation into groups and the way those groups can be compared and associated. Cluster analysis allows such an organization of hyperspace by segregating entities or variables into groups, or clusters, that imply greater phenetic resemblance within than among groups. In other words, the elements of each particular group produced are more closely related (in a phenetic sense) than elements of different groups (Sneath and Sokal 1973; Pimentel 1979; Neff and Marcus 1980).

Usually, cluster techniques are sequential, agglomerative, hierarchical, and nonoverlapping (hence the acronym of SAHN techniques). They are sequential because there is an iteration of certain algorithmic calculations; agglomerative because elements are added in such a way that progressively fewer and larger clusters are formed; hierarchical because nested groups are formed, so that for N elements $N - 1$ clusters are formed; and nonoverlapping, since elements at any particular level are mutually exclusive—clusters at the same level never share elements (Sneath and Sokal 1973; Pimentel 1979).

Simply put, a cluster analysis involves an assessment of distance of objects or variables on the basis of their values in a data matrix, the creation of clusters by using a particular cluster algorithm, and the generation of a scaled treelike graphic output (i.e., a phenogram). Once this is done, the two-dimensional output can be interpreted in light of external data. Details of different similarity measures and clustering algorithms can be found in Sneath and Sokal (1973).

Disparity Measures

Several disparity measures are possible, by reference to an extensive literature on taxonomic distances (e.g., Sneath and Sokal 1973). Measures of multivariate distance are most useful and intuitive when the variables are equally weighted (Van Valen 1974), such that transformations of the data may be necessary. Because different disparity metrics are meant to reflect the same underlying morphospace, one can often find equivalence relations among different metrics (Van Valen 1974; Foote 1995; M. Foote, pers. comm.), with general patterns emerging regardless of metric (e.g., Foote 1995).

Still, at its most basic level, two fundamentally different aspects of morphospace occupation, and thus different disparity measures, should be recognized when estimates of dispersion are sought: the variance and the range. The variance captures the average dissimilarity among forms in morphospace; the range reflects the amount of morphospace occupied (Foote 1991). In multivariate terms, these measures can be generalized as the total variance (the sum of the univariate variances) and the total range (the sum of the univariate ranges) (see Van Valen 1974). Although empirically the total variance and the total range tend to correlate (Foote 1992), the relationship is not always proportional or monotonic, since the total range is very sensitive to sample size (Foote 1991, 1993). When sample size differs among samples under comparison, procedures like morphological rarefaction (Foote 1992) should be used. Other multivariate statistics of variation exist (see Van Valen 1974; Foote 1990), but in general the total variance and the total range can effectively summarize dispersion in morphospace.

Developmental Morphospaces and Developmental Disparity

Following Eble (1998a, 1998c, 2001), I here expand the notions of morphospace and disparity to the realm of development. A developmental morphospace is any morphospace that reflects development by reason of either the variables or the entities depicted in such morphospace carrying meaningful developmental information (see also McGhee 1998). As such, the concept of developmental morphospace is a natural morphological extension of the notions of epigenetic landscape (Waddington 1957), epigenetic space (Goodwin 1963), and parameter space (Alberch 1989). It also appears explicitly in the work of Ellers (1993), although there the focus was on theoretical parameters, not measurements. Implicitly, developmental morphospace is also discussed by Rice (1998a, 1998b) and Zelditch and

Fink (1996). The explicit construction of developmental morphospaces is a necessary step in assessing the importance of development in structuring adult morphological distributions and in identifying phenomena like heterochrony, heterotopy, and developmental constraints. Use of the term brings into focus the very aims of evolutionary developmental biology and encourages the incorporation of ontogenetic data in multivariate studies of evolution.

Accordingly, developmental disparity can be formalized as the disparity among objects in any subregion of a developmental morphospace. Developmental disparity can be quantified either across taxa or within taxa in developmental time, depending on the purpose. Instead of changes in amount of variation in evolutionary time, the focus is shifted to changes in amount of variation in ontogenetic time.

At the limit, evolutionary and developmental renditions of morphospaces and disparity can be brought together, for one might be interested in tracking how changes in disparity and morphospace occupation through ontogeny themselves change in evolutionary time. For sufficiently well-sampled paleontological time series, this can be determined directly against a time scale, but in many cases the indirect route of mapping on a phylogeny can be used. An emphasis on developmental morphospaces and developmental disparity, especially from a multivariate perspective, generalizes the playing field of evolutionary developmental biology by allowing, in addition to standard analysis of changes in rate and timing, the description of clade shape in ontogenetic time, the assessment of global developmental trends, the testing of broad predictions such as von Baer's laws, and the description of potential links between ontogeny and phylogeny (Eble 2001).

An Empirical Illustration: Reanalyzing Heinz's (1966) Hominoid Dataset

Below I explore methods and approaches previously discussed with an empirical illustration involving ontogenetic data for extant hominoids extracted from Nicole Heinz's (1966) study. The dataset, while incorporating observations on fossil species and a discussion of postcranial variation, is mostly concerned with ontogenetic variation in recent skulls of the superfamily Hominoidea (*Homo, Pan, Gorilla, Pongo,* and *Hylobates*). Ontogenetic variation is partitioned into dental age classes (no teeth erupted; deciduous or milk teeth; first permanent molar erupted, or M1; second permanent molar erupted, or M2; third permanent molar erupted, or M3). In the dataset, only the human sample yields measurements for the no-teeth-erupted class. The dataset is explicitly restricted to the genus level, with intraspecific variation disregarded in the case of modern humans and intrageneric variation disregarded for gibbons and the great apes. No malformations or pathological individuals are considered. Sexual dimorphism is also disregarded and reputed to be uninfluential in an analysis at the genus level. (Heinz further argues that sexual differences play a role only rather late in development, after the onset of puberty, and that intergeneric differences are mostly laid out earlier; however, see German and Stewart, Chap. 10.) In any case, the present data analyses and interpretations are limited by the resolution of the data.

While Heinz's study was mostly concerned with levels of variability, the dataset suffers from a rather pronounced unevenness in sample sizes across species. Thus, only average values will be used in the present analyses, even though variance and coefficient of variation estimates are reported in Heinz's study as well.

Thirty-five distance measurements are presented in the dataset, but only twenty-nine of these are consistently reported across species. Six of these measurements were not reported for the M3 dental stage in *Gorilla*, *Pongo*, and *Hylobates* and were accordingly not considered. Thus, twenty-three variables were consistently represented. However, the number of objects is restricted to twenty-one (four dental stages per species plus one stage with no teeth erupted in humans). Whenever the number of objects is smaller than the number of variables, procedures such as PCA give only approximate scores because of computational complexities. For the present purposes, four additional variables were thus removed; such variables correspond to the set of "nonclassical measurements" in Heinz's work. The final data matrix, then, consists of twenty-one objects and nineteen variables.

In exploratory fashion, several "developmental morphospace" and "developmental disparity" analyses were performed: a principal components analysis on log variables, with size clearly expressed on PC I; a principal components analysis on row-normalized log variables, emphasizing shape and substantially removing size effects; cluster analyses, to identify phenetic propinquity among adults and juveniles of different taxa; juvenile-adult comparisons of disparity; mapping of developmental disparity against a phylogeny; and comparisons of ontogenetic trajectories of disparity relative to a *Hylobates* juvenile standard.

Principal Components Analysis on Log Variables

A principal components analysis was run on the correlation matrix of log variables. The choice between variance-covariance and correlation matrices is not trivial. Even though the difference between the two is only a matter of standardization (making the variance-covariance matrix a correlation matrix), they imply different manifestations of variation and therefore lead to PCA results that are not interchangeable (Jolliffe 1986). While the use of variance-covariance matrices has become standard in many multivariate growth studies, it also gives greater importance to variables with high variance. The scale of measurement can thus significantly affect the results, either because of incommensurate units of measurement or because of widely different variances, as was the case here. In such situations, principal components (PCs) are less informative—the first few will summarize little more than the relative sizes of variances. In addition, informal comparison of PCA results from different studies is more straightforward when correlation matrices are used (Jolliffe 1986).

The first three principal components were retained for further analysis, based on (1) inspection of natural breaks in the decay of the variance of ranked eigenvalues, (2) amount of variance explained (98.6%), and (3) identification of PCs with more than one variable displaying relatively large values. (While the retention of only the first two PCs would achieve the goal of variance summarization—95 percent—intepretation of the third PC is biologically meaningful, as will be seen below.)

PC I (84.2% of the variance) has all variables loading highly and positively on it. PC I is thus a size-allometry axis, which, while also summarizing shape, is the best overall summary of size variation. PC II (10.8%) has palate length, basion-prosthion length, maximal biparietal width, porion-bregma height, and minimum frontal width loading highly. PC III (3.6%) is influenced mostly by variation in interorbital width and to a lesser extent by foramen magnum length.

The resulting ordinations (Fig. 3.1), using scores on the rotated principal components, can be seen as representations of size-shape developmental morphospaces. PC I is here plotted against PC II and PC III. The PC I-PC II ordination clearly separates the ontogenetic trajectories of *Hylobates* and *Homo* from those of *Pongo, Gorilla,* and *Pan,* which overlap to a

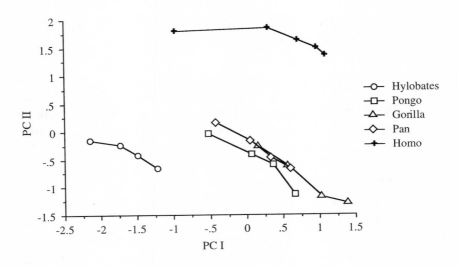

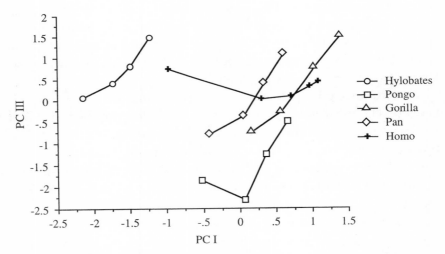

Fig. 3.1. Principal component ordinations based on the correlation matrix of log variables. Age runs from *left* to *right*. Successive points correspond to the milk teeth, M1, M2, and M3 stages. In *Homo*, the no-teeth stage is also represented.

substantial degree. It is only at larger sizes that the overlap on PC II is reduced (note positions of *Pongo* M3 and *Gorilla* M2 and M3). Note, on the PC I-PC II ordination, the isolation of *Homo* at the no-teeth stage. In contrast, the PC I-PC III ordination approximates *Homo* at the no-teeth stage to *Hylobates*, on the one hand, and *Homo* at the milk teeth through M3 stages to the more derived apes. There is also less overlap among the ontogenetic trajectories of *Pongo, Gorilla,* and *Pan. Pongo* and *Pan* at the milk stage, for example, are quite separate in PC III, in contrast to their proximity in PC II.

On the assumption that the timing of molar eruption in apes is equivalent and by taking into account the timing of molar eruption in humans and our current understanding of phylogenetic relationships among extant hominoids (for overviews, see Futuyma 1998 and Lewin 1998), one can advance some heterochronic inferences (Fig. 3.2). On PC I, great apes and humans differ from *Hylobates* mostly by predisplacement, *Gorilla* from *Pongo* by predisplacement, *Pan* from *Gorilla* by postdisplacement, and *Homo* from *Pan* by a combination of predisplacement and hypermorphosis (and, to a limited extent, neoteny). On PC II, *Gorilla* differs from *Pongo* mostly by acceleration and *Pan* differs from *Gorilla* by neoteny and postdisplacement. The difference between *Homo* and *Pan* is describable in terms of a combination of neoteny and postdisplacement. Finally, on PC III, great apes differ from gibbons mostly by postdisplacement. *Pongo,* in addition, differs through nonlinearity of the underlying ontogenetic trajectory; in other words, a different growth function implying nonuniform change from ancestor to descendant applies (see Rice 1997). A change into a qualitatively different growth function thus would characterize the difference between *Gorilla* and *Pongo,* although much of the difference is also attributable to predisplacement. *Pan* differs from *Gorilla* by neoteny, and *Homo* differs from *Pan* by neoteny and predisplacement.

Principal Components Analysis on Row-Normalized Log Variables

Row normalization was used to partially remove the effects of size. Row normalization renders the sum of squares of variates for each object equal to one; it retains the proportionality of variables within objects and destroys differences in magnitude between objects (Reyment and Jøreskog 1993).

As with the analysis on nonnormalized log variables, the first three principal components were retained for further analysis, once again based on (1) inspection of natural breaks in the decay of the variance of ranked eigenvalues, (2) amount of variance explained (89%), and (3) identification of PCs with more than one variable displaying relatively large values (the first three PCs collectively contained the highest loadings for all variables). In contrast to the analysis of log variables, PC I here summarizes a considerably smaller proportion of the total variance (48%). Several of the loadings

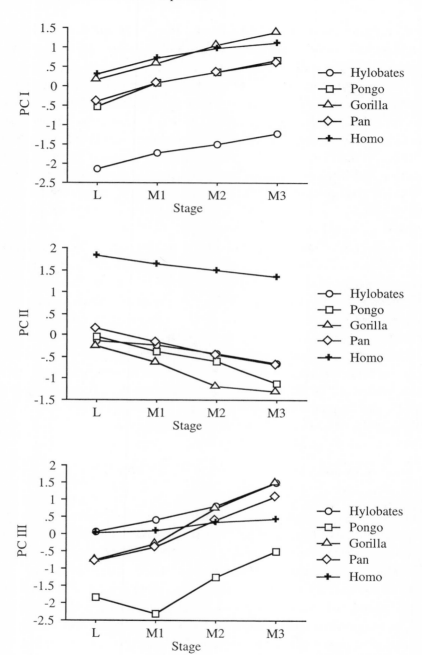

Fig. 3.2. Principal components based on the correlation matrix of log variables, plotted against age. The no-teeth stage in *Homo* is not shown. *L,* the milk teeth stage.

have low values, and several are negative. Thus, PC I in this case is hardly a size-allometry axis in the conventional sense or a good proxy for overall size. The contribution of shape variation to it is more conspicuous. PC I has nasion-opisthion length, maximal biparietal width, minimum frontal width, basion-bregma height, porion-bregma height, nasion prosthion height, palate length, and nose height loading highly (notice that some of these are the variables that contributed importantly to variation on PC II in the analysis of log variables). PC II (25.6%) has bi-auricular width, palate width, orbital height, and interorbital width with important contributions. PC III (15.4%) seems to reflect mostly variation in basion-nasion length and foramen magnum length and width.

Corresponding ordinations (again based on rotated axes) are thus best interpreted not as guides to multivariate allometry per se (although they do reflect allometric differences), but as developmental morphospaces relatively unconfounded by size. Figure 3.3 shows PC I plotted against PC II and PC III. The PC I-PC II projection isolates the ontogenetic trajectory of *Homo* from the rest. The trajectories of *Hylobates, Pan,* and *Gorilla* follow a similar gradient, with that of *Pongo* somewhat offset. In the PC I-PC III ordination, much as with the previous analysis, *Homo* at the milk teeth stage is close to *Hylobates.* In the PC I-PC II ordination, juveniles at the milk teeth stage are rather disjunct, but PC III approximates juveniles of *Pongo* and *Homo* to *Gorilla* and *Pan.*

In terms of the timing of molar eruption in apes and humans and in extant hominoid phylogeny, the following qualitative heterochronic inferences can be made (see Fig. 3.4). On PC I, the great apes (but not humans) differ from *Hylobates* by predisplacement, *Gorilla* from *Pongo* also by predisplacement (as well as neoteny by the M3 stage), *Pan* from *Gorilla* mostly by postdisplacement, and *Homo* from *Pan* by both postdisplacement and neoteny. On PC II, *Pongo* differs from *Hylobates* only at the M1 stage, perhaps suggesting nonuniform change; *Gorilla* differs from *Pongo* by predisplacement, and despite the substantial overlap in the trajectories, *Pan* is slightly neotenic relative to *Gorilla. Homo* is distinct from *Pan* through a combination of neoteny, postdisplacement, and hypermorphosis. On PC III, in turn, postdisplacement seems mostly responsible for the difference between great apes and gibbons, with *Pongo* in addition differing through nonuniform change in ontogeny. Accordingly, a change in growth function accounts for the difference between *Gorilla* and *Pongo,* with acceleration also being involved. The trajectories for *Pan* and *Gorilla* are almost indistinguishable. Finally, *Homo* differs from *Pan* through neoteny.

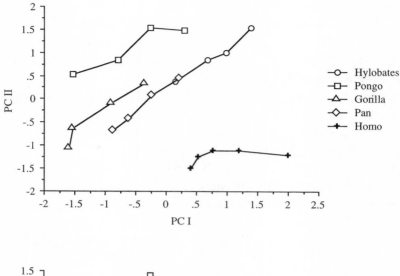

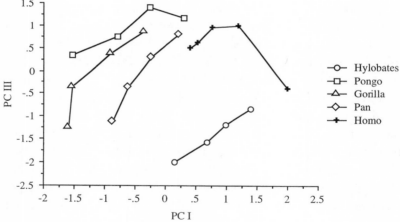

Fig. 3.3. Principal components ordinations based on the correlation matrix of row-normalized log variables. Age runs from *left* to *right*. Successive points correspond to the milk teeth, M1, M2, and M3 stages. In *Homo*, the no-teeth stage is also represented.

While these ordination results are constrained by the limitations of the dataset, they do suggest a considerable amount of dissociated heterochrony, not only in terms of different composite variables being subject to different heterochronic processes, but also because different heterochronic processes may contribute to the pattern of evolutionary change in the same composite variable. Also, no consistent heterochronic trend is apparent as one

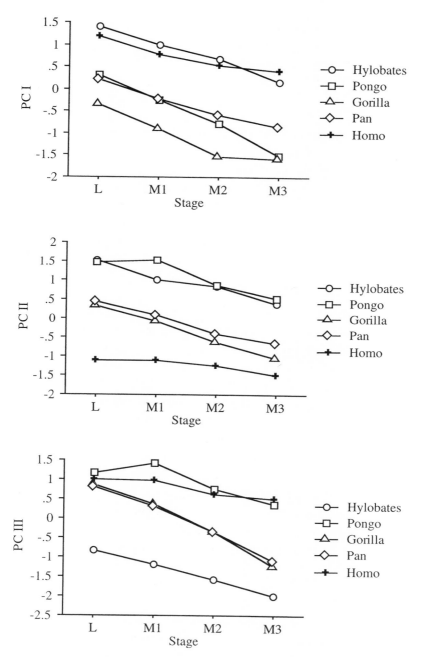

Fig. 3.4. Principal components based on the correlation matrix of row-normalized log variables, plotted against age. The no-teeth stage in *Homo* is not shown. *L,* the milk teeth stage.

moves from more primitive to more derived taxa. Could large-scale regularities be discernible by evaluating patterns of variation globally, in average terms? The analyses below address this possibility.

Cluster Analyses

Cluster analysis can be used to compress information on phenetic relationships into two-dimensional diagrams. This has its disadvantages, for some distortion of relationships implied by ordinations can occur. Its main advantage is the provision of an average characterization of groupings in morphospace, effectively accomplishing multivariate sorting of the data. Historically, cluster analysis has usually served the purpose of taxonomic discrimination. It is rather uncommon, thus, for cluster analyses to be performed on samples of both juveniles and adults. Examples include Boyce (1964) for hominoids (including fossils) and Eble (1998c, 2001) for echinoids.

Caution must be exercised in performing and interpreting cluster analyses because of sensitivity to choice of clustering algorithm and distance metric. Though such choices can usually be retrospectively justified, a sensible approach is to base inferences on patterns that emerge from several different analyses. For convenience, I will use the Euclidean distance as the index of association and explore the results of six different algorithms (see Sneath and Sokal 1973): unweighted pair-group method using arithmetic averages (UPGMA), weighted pair-group method using arithmetic averages (WPGMA), single-linkage, complete linkage, unweighted centroid, and weighted centroid clustering. z-scores on row-normalized log variables were used (thus the same data as in the second PCA analysis). Two sets of analyses were performed: (1) on all ontogenetic stages across taxa and (2) on a subset comprising the milk teeth and the M3 representative for each taxon (thus maximizing the potential distinction between juveniles and adults).

When the total sample is considered, all methods of clustering consistently isolate the whole ontogenetic trajectory of *Homo* in a separate grouping, as implied by the PCA on the correlation matrix of row-normalized log variables. *Hylobates* is also consistently separated into a cluster, as partially suggested by the ordinations. *Pongo* is generally (but not uniformly across analyses) isolated from *Pan* and *Gorilla*, with the milk teeth and M1 stages and M2 and M3 stages always conjunct, respectively. Younger stages of *Pan* tend to cluster with younger stages of *Gorilla*, with the same being true for later stages. Thus, cluster analyses of the total sample seem to accomplish a mixture of taxonomic and ontogenetic discrimination. For illustration purposes, the output of the UPGMA analysis is shown in Figure 3.5.

When the subsample of milk teeth (L) and M3 representatives is considered, a rather different picture is suggested (almost uniformly across

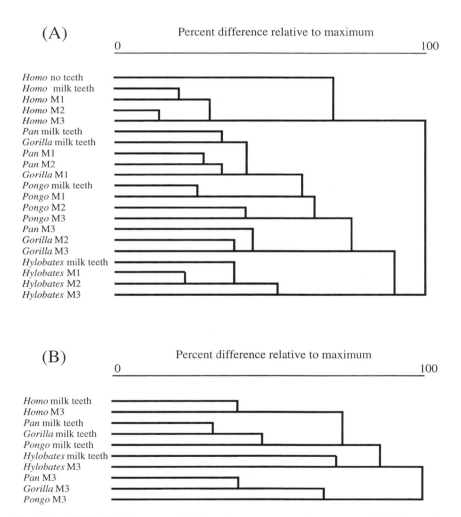

Fig. 3.5. UPGMA cluster analysis based on z-scores of row-normalized log variables. *A:* All ontogenetic stages across taxa are included. *B:* Only the milk teeth and the M3 stages are included. Notice the separation of *Homo* in *A* and its clustering with great ape juveniles in *B*. See text for discussion.

cluster analyses). Results of the UPGMA analysis are illustrated in Figure 3.5. *Homo* is still consistently separated, but it always clusters with juveniles of *Pongo*, *Gorilla* and *Pan*. The adults of these taxa form a separate cluster, away from the larger cluster that also usually incorporates a separate grouping of *Hylobates* L and M3. The clustering of the great ape juveniles is consistent with von Baer's laws, and the joining of *Homo* suggests overall paedomorphosis in human cranial evolution.

This difference between the analysis of the total sample and the analysis of extremes ultimately reflects the contrast between the structuring implied by the endpoints of ontogenetic trajectories and the fluidity (or not) of such trajectories in morphospace. It is clear from the above that both kinds of analyses can be informative. However, prudence is needed because the exclusion of ontogenetic stages can conceal the taxonomic signal of ontogenetic variation.

Developmental Disparity

In terms of developmental disparity, several relevant questions arise: how the disparity among taxa changes across ontogenetic stages (in other words, how disparity changes as one proceeds from juveniles to adults); how the disparity of such pooled samples changes with the level of phylogenetic inclusiveness (or how the addition of particular taxa affects disparity); how disparate are ontogenetic endpoints and how such endpoint disparity changes in evolution; and how disparity accumulates through ontogeny relative to a standard of comparison. I will approach these questions below. Because of the small sample sizes involved (which, upon resampling, yield comparatively large error bars), the results shown can, for the most part, be interpreted only qualitatively. They are meant to indicate possible representations (and interpretations) of data on developmental disparity.

Disparity across Stages

Changes in hominoid disparity across ontogenetic stages are presented in Figure 3.6. z-scores on row-normalized log variables were again used. Disparity was measured as the total variance of original variables, as the total variance across the first three PCs, and as the variance in each PC considered individually. Changes in disparity (both as the total variance of the original variables and of the PCs) across stages are rather subdued. One major expectation born out of evolutionary developmental biology is that

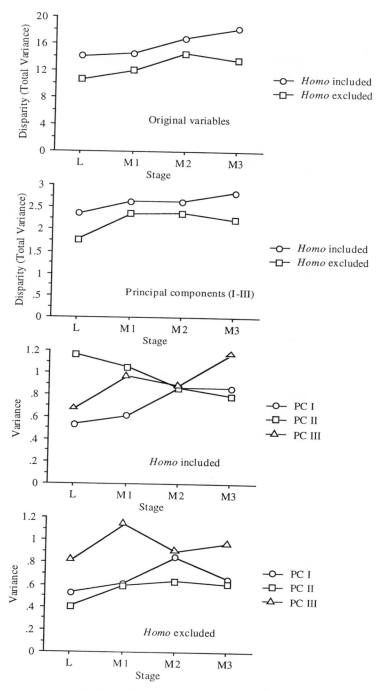

Fig. 3.6. Developmental disparity among taxa across ontogenetic stages, in terms of z-scores of row-normalized log variables, of the first three principal components based on the correlation matrix of row-normalized log variables and of individual principal components with *Homo* included and excluded, respectively.

von Baer's second law should hold—in other words, that disparity should be smaller in earlier stages than in later stages. This is generally the case, but it is interesting that the exclusion of *Homo* entails a slight reduction in disparity between M2 and M3.

In terms of the variance along individual PCs, however, disparity changes are much more irregular. When *Homo* is included, the variance along PC I changes in general conformance to von Baer's second law, but the same does not apply to PC II, where the variance decays steadily, and to PC III, where it goes up and down. When *Homo* is excluded, a different pattern accrues but no consistency emerges. The variance of PC I increases up to M2 but then decreases, that of PC II initially increases but then stabilizes, and that of PC III goes up and down much in the same way as when *Homo* is included. Thus, while in general (and on average) von Baer's second law would seem to hold in hominoid evolution, for sizable portions of the phenotype (as summarized by PC II and PC III) it seems to be broken. Similar results are reported for echinoids by Eble (2000a).

Developmental Disparity and Phylogeny

The use of a phylogenetic framework was implicit in the interpretation of several results thus far presented. In the absence of direct temporal information from the fossil record, the branching sequence implied by a phylogeny can be used as a framework for mapping of developmental disparity and for the interpretation of its evolution.

Against the hominoid phylogeny presented in Figure 3.7 (which was also used above), the disparity (once again in terms of z-scores of row-normalized log variables) at the milk teeth (L) and the M3 stages at different levels of phylogenetic inclusiveness is presented along nodes. At all nodes, the disparity of juveniles is smaller than that of near adults, in conformance with von Baer's second law. At the level of *Hylobates*, however, there is a change in the proportionality of disparity differentials: while M3 disparity is essentially the same as that of the next node, that of the milk teeth stage is higher, suggesting a somewhat larger evolutionary jump early in development between *Hylobates* and the rest.

The disparity of ontogenetic endpoints (L and M3 stages) is also presented along each branch in the phylogeny. Here a rather strong pattern emerges: the disparity of ontogenetic endpoints in *Homo* is much smaller than that in the other hominoids. In fact, bootstrapped error bars for *Homo* never overlap with those of the other taxa. The increase in ontogenetic end-

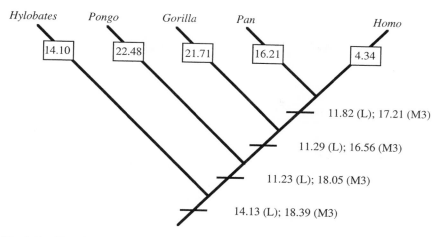

Fig. 3.7. Developmental disparity and phylogeny. Mapped along the nodes is the developmental disparity among taxa at the milk teeth (L) and M3 stages, reflecting different levels of phylogenetic inclusiveness. Along the terminal branches, the developmental disparity within taxa (disparity of ontogenetic endpoints L and M3) is displayed. Disparity is quantified in terms of z-scores of row-normalized log variables.

point disparity from *Hylobates* to *Pongo* suggests global peramorphosis. Global paedomorphosis is also apparent from *Gorilla* to *Pan* and is very clear from *Pan* to *Homo*.

Ontogenetic Trajectories of Disparity

The quantification of developmental disparity within taxa can be accomplished with a focus on the disparity of endpoints, but one may be interested instead in how developmental disparity changes along the whole of the ontogenetic trajectory of a taxon and how such trajectories of developmental disparity themselves compare from taxon to taxon. Using the milk teeth stage of *Hylobates* as a standard of comparison, the trajectories of disparity (based on z-scores of row-normalized log variables) in each taxon can be represented relative to size or age. The age-disparity diagram (Fig. 3.8) allows inferences regarding "overall" or global trends in ontogeny, without the complications of a representation of morphospatial relationships.

As ontogeny proceeds, there is a progressive increase in disparity relative to the primitive state in all taxa. It is noticeable that the rate of change in humans is very subdued, however; there is more change within the tra-

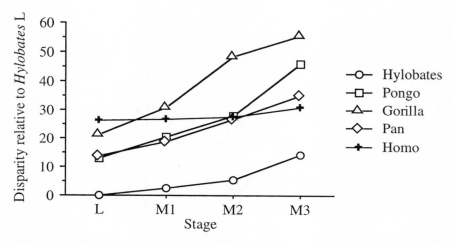

Fig. 3.8. Ontogenetic trajectory of disparity. Developmental disparity here reflects the within-taxon disparity of successive stages relative to a primitive standard of comparison, the milk teeth stage (*L*) of *Hylobates*.

jectory of *Hylobates* itself than within *Homo*. *Pan* and *Pongo* overlap substantially in their trajectories of disparity. In terms of conventional heterochronic nomenclature, and treating disparity as a variable, Figure 3.8 suggests a global peramorphic trend (by predisplacement and acceleration) from *Hylobates* to *Pongo* to *Gorilla* and a global paedomorphic trend from *Gorilla* to *Pan* (by postdisplacement and neoteny) and to *Homo* (mostly by neoteny). In terms of disparity, at least, humans are on average neotenic. Of course, this representation of ontogenetic change is silent about individual characters and not incompatible with, say, peramorphosis also playing a role in human evolution.

Conclusions

This chapter addressed various approaches to the study of multivariate developmental change in human evolution. While several of them are already current in biological anthropology, it is hoped that some of the less conventional approaches and language used (which are nonetheless more and more familiar to quantitative paleobiologists) can add a new dimension to the interpretation of the relationship between development and evolution in the context of human morphological evolution.

Perhaps noticeable in this chapter was the lack of emphasis on heterochrony per se at each and every step of the discussion. This is less an attempt to avoid current ambiguity concerning the definition and scope of heterochrony (Godfrey and Sutherland 1995a, 1995b, 1996; Zelditch and Fink 1996; Rice 1997; Alba, Chap. 2; Shea, Chap. 4)—so long as a given set of definitions and methods is provisionally accepted, there is always room for important insights to be gained, however nondefinitive—than a desire to build on past work on human evolution and developmental change and advance (however exploratorily) concepts and methodological approaches that might eventually prove useful in biological anthropology.

It is hoped that the results here presented as illustration can be reevaluated, with the application of some of the approaches outlined to other taxa, larger sample sizes, more variables, and better age resolution. Different protocols of data collection (e.g., landmark-based) are likely to yield more precision and more insight as well. Beyond the various limitations of the dataset used, the results and their interpretation stand as suggestions for future research. Multivariate approaches in general and the language of morphospaces and disparity in particular deserve to be more fully integrated into the analytical arsenal available to studies of human evolution and development.

Acknowledgments

I thank J. Clark, R. Potts, L. Profant, and B. Senut for useful advice, comments, and suggestions on topics relating to human evolution and B. David, D. Erwin, M. Foote, D. Jablonski, D. Rasskin-Gutman, and M. Zelditch for encouragement and feedback on topics relating to developmental disparity and developmental morphospaces. Research was supported by postdoctoral fellowships from the Smithsonian Institution and the Santa Fe Institute.

References

Alberch, P. 1989. The logic of monsters: evidence for internal constraint in development and evolution. *Geobios* (mémoire spécial) 12:21–57.
Alberch, P., Gould, S.J., Oster, G.F., and Wake, D.B. 1979. Size and shape in ontogeny and phylogeny. *Paleobiology* 5:296–317.

Barnett, V., and Lewis, T. 1994. *Outliers in Statistical Data,* 3d ed. Chichester: Wiley.

Bookstein, F.L. 1991. *Morphometric Tools for Landmark Data.* Cambridge: Cambridge University Press.

Bookstein, F., Chernoff, B., Elder, R., Humphries, J., Smith, G., and Strauss, R. 1985. *Morphometrics in Evolutionary Biology.* Special Publication 15. Philadelphia: Academy of Natural Sciences of Philadelphia.

Boyce, A.J. 1964. The value of some methods of numerical taxonomy with reference to hominoid classification. In V. H. Heywood and J. McNeill (eds.), *Phenetic and Phylogenetic Classification,* 47–65. London: Systematics Association.

Chaline, J., David, B., Magniez-Jannin, F., Malass, A.D., Marchand, D., Courant, F., and Millet, J-J. 1998. Quantification de l'évolution morphologique du crâne des Hominidés et hétérochronies. *Compte Rendu Academie Science Paris, Sciences de la terre et des plantes* 326:291–298.

Davis, J.C. 1986. *Statistics and Data Analysis in Geology,* 2nd ed. New York: Wiley and Sons.

Eble, G.J. 1998a. The role of development in evolutionary radiations. In M.L. McKinney and J.A. Drake (eds.), *Biodiversity Dynamics: Turnover of Populations, Taxa, and Communities,* 132–161. New York: Columbia University Press.

———. 1998b. Diversification of disasteroids, holasteroids and spatangoids in the Mesozoic. In R. Mooi and M. Telford (eds.), *Echinoderms through Time,* 629–638. Rotterdam: A. A. Balkema Press.

———. 1998c. Developmental disparity and developmental morphospaces. *Geological Society of America Meeting Abstracts* 30:A-327.

———. 2000. Contrasting evolutionary flexibility in sister groups: disparity and diversity in Mesozoic atelostomate echinoids. *Paleobiology* 26:56–79.

———. 2001. Developmental morphospaces and evolution. In J.P. Crutchfield and P. Schuster (eds.), *Evolutionary Dynamics: Exploring the Interplay of Selection, Neutrality, Accident, and Function.* Oxford: Oxford University Press.

Ellers, O. 1993. A mechanical model of growth in regular sea urchins: predictions of shape and a developmental morphospace. *Proceedings of the Royal Society of London* B254:123–129.

Foote, M. 1990. Nearest-neighbor analysis of trilobite morphospace. *Systematic Zoology* 39:371–382.

———. 1991. Analysis of morphological data. In N.L. Gilinsky and P.W. Signor (eds.), *Analytical Paleobiology,* 59–86. Knoxville: University of Tennessee, Paleontological Society.

———. 1992. Rarefaction analysis of morphological and taxonomic diversity. *Paleobiology* 18:1–16.

———. 1993. Human cranial variability: a methodological comment. *American Journal of Physical Anthropology* 90:377–379.

———. 1995. Morphological diversification of Paleozoic crinoids. *Paleobiology* 21:273–299.

————. 1996. Models of morphological diversification. In D. Jablonski, D. Erwin, and J. Lipps (eds.), *Evolutionary Paleobiology*, 62–86. Chicago: University of Chicago Press.

————. 1997. The evolution of morphological diversity. *Annual Review of Ecology and Systematics* 28:129–152.

Futuyma, D.J. 1998. *Evolutionary Biology*, 3d ed. Sunderland, Mass.: Sinauer.

Godfrey, L.R., and Sutherland, M.R. 1995a. What's growth got to do with it? Process and product in the evolution of ontogeny. *Journal of Human Evolution* 29:405–431.

————. 1995b. Flawed inference: why size-based tests of heterochronic processes do not work. *Journal of Theoretical Biology* 172:43–61.

————. 1996. Paradox of peramorphic paedomorphosis: heterochrony and human evolution. *American Journal of Physical Anthropology* 99:17–42.

Goodwin, B.C. 1963. *Temporal Organization in Cells*. London: Academic Press.

Heinz, N. 1966. Le crâne des anthropomorphes: croissance relative, variabilité, évolution. *Musée Royal de l'Afrique Centrale-Tervuren, Belgique. Annales, Nouvelle Série, Sciences Zoologiques*, 6.

Jolliffe, I.T. 1986. *Principal Component Analysis*. New York: Springer.

Klingenberg, C.P. 1996. Multivariate allometry. In L.F. Marcus, M. Corti, A. Loy, G.J.P. Naylor, and D.E. Slice (eds.), *Advances in Morphometrics*, 23–49. NATO ASI Series A: Life Sciences vol. 284. New York: Plenum Press.

Lewin, R. 1998. *Principles of Human Evolution*. Oxford: Blackwell.

Lewontin, R.C. 1978. Adaptation. *Scientific American* 239:156–169.

Lynch, J.M., Wood, C.G., and Luboga, S.A. 1996. Geometric morphometrics in primatology: craniofacial variation in *Homo sapiens* and *Pan troglodytes*. *Folia Primatologica* 67:15–39.

Marcus, L.F., Corti, M., Loy, A., Naylor, G.J.P., and Slice, D.E. (eds.). 1996. *Advances in Morphometrics*. NATO ASI Series A: Life Sciences vol. 284. New York: Plenum Press.

McGhee, G.R. 1991. Theoretical morphology: the concept and its applications. In N.L. Gilinsky and P. W. Signor (eds.), *Analytical Paleobiology*, 87–102. Knoxville: Paleontological Society.

————. 1998. *Theoretical Morphology*. New York: Columbia University Press.

McKinney, M.L., and McNamara, K.J. 1991. *Heterochrony: The Evolution of Ontogeny*. New York: Plenum Press.

Morrison, D.F. 1990. *Multivariate Statistical Methods*, 3d ed. New York: McGraw-Hill.

Neff, N.A., and Marcus, L.F. 1980. *A Survey of Multivariate Methods for Systematics*. New York: privately published.

Pimentel, R.A. 1979. *Morphometrics*. Dubuque, Iowa: Kendall/Hunt.

Raup, D.M. 1966. Geometric analysis of shell coiling: general problems. *Journal of Paleontology* 40:1178–1190.

Reyment, R., and Jøreskog, K.G. 1993. *Applied Factor Analysis in the Natural Sciences*. Cambridge: Cambridge University Press.

Rice, S.H. 1997. The analysis of ontogenetic trajectories: when a change in size or shape is not heterochrony. *Proceedings of the National Academy of Sciences, USA* 94:907–912.

———. 1998a. The bio-geometry of mollusc shells. *Paleobiology* 24:133–149.

———. 1998b. The evolution of canalization and the breaking of von Baer's laws: modeling the evolution of development with epistasis. *Evolution* 52:647–656.

Roy, K. 1996. The roles of mass extinction and biotic interaction in large-scale replacements: a reexamination using the fossil record of stromboidean gastropods. *Paleobiology* 22:436–452.

Shea, B.T. 1985. Bivariate and multivariate growth allometry: statistical and biological considerations. *Journal of Zoology* 206:367–390.

Sneath, P.H.A., and Sokal, R.R. 1973. *Numerical Taxonomy*. San Francisco: Freeman.

Tissot, B.N. 1988. Multivariate analysis. In M.L. McKinney (ed.), *Heterochrony in Evolution: A Multidisciplinary Approach*, 35–51. New York: Plenum Press.

Van Valen, L. 1974. Multivariate structural statistics in natural history. *Journal of Theoretical Biology* 45:235–247.

Waddington, C.H. 1957. *The Strategy of the Genes*. London: Allen and Unwin.

Wagner, P.J. 1997. Patterns of morphologic diversification among the Rostroconchia. *Paleobiology* 23:115–145.

Zelditch, M.L., and Fink, W.L. 1996. Heterochrony and heterotopy: stability and innovation in the evolution of form. *Paleobiology* 22:241–254.

Chapter 4

Are Some Heterochronic Transformations Likelier than Others?

Brian T. Shea

Studies of evolution via heterochrony have proliferated in evolutionary biology since the seminal works of Gould (1977) and Alberch et al. (1979). Many aspects of heterochrony have been investigated and debated in depth, including morphological characterizations, genetic and developmental underpinnings, and ecological contexts. Certainly, one area where much attention has been focused involves the recognition and definition of the myriad types and categories of morphological heterochrony, which is evidenced by the detailed classificatory tables and figures that were a central concern in Gould (1977), Alberch et al. (1979), and many subsequent papers by various authors (Fig. 4.1). For example, one contribution of my own was predominantly centered on recognizing and identifying new subcategories of heterochronic transformation, specifically rate and time hypo- and hypermorphoses (Shea 1983a).

These debates over appropriate definitions and classifications are certainly a necessary and productive component of heterochronic research, even if they often seem, to insiders and outsiders alike, as exercises that dwell excessively on ever-growing mounds of tongue-twisting jargon of uncertain biological relevance. In this chapter, I focus less on these traditional realms of recognition, definition, naming, and classification, which are covered in Chapter 2, and instead ask a more general question not often dealt with in the literature on heterochrony. Quite simply, I consider whether what we have learned about heterochrony to date indicates that certain types of transformations are *likelier* than other types. And, if so, I further

79

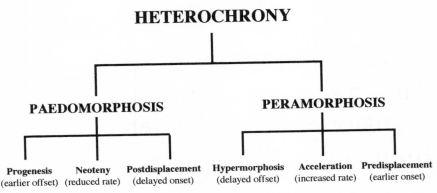

Fig. 4.1. One depiction of the two major patterns of heterochronic transformation classified according to underlying shifts in duration of growth, rates of shape change, and onset proportions. *Source:* Based on McNamara 1993, Figure 1.

consider why this might be. Such a question, even framed as a general hypothesis, of course begs issues of context and elaboration (i.e., "likelier" in precisely what sense). Possibilities for greater likelihood include higher frequencies of observed morphological patterns (e.g., hypermorphosis vs. neoteny), or more likely in relation to known genetic and developmental underpinnings, or more likely in regard to selective scenarios and ecological contexts.

Previous studies dealing with issues of relative frequency and likelihood can be cited. One of the best examples, explicitly rooted in underlying growth dynamics, is McNamara's (1988a) investigation of heterochrony in echinoids. He concluded that progenesis (time hypomorphosis) was more common in the evolution of regular as opposed to irregular echinoids, the latter being characterized by more frequent dissociative neoteny and acceleration. In regular echinoids, growth changes predominantly involve rates of plate production; in irregulars, growth is characterized by variation in individual plate allometries, and it is this divergence that leads to the biases in types of heterochronic transformations observed, according to McNamara. In other work, McNamara (1986) documented that peramorphosis (largely through acceleration) was considerably more common than paedomorphosis in early Ordovician trilobites and also documented relative frequencies in several, mainly invertebrate, groups (McNamara 1988b). Other studies noted various biases in the patterning of allometric relations or heterochronic categories (e.g., Voss et al. 1990; Bjørklund 1994; Klingenberg and Ekau 1996), and McKinney and McNamara (1991) discussed possible ecological contexts for biases in heterochronic transformations. I hope that additional theoretical and empirical efforts will further clarify the issues of relative frequency and likelihood of change in various organisms and environments.

The Frequency of Observed Patterns

Growth Allometry in the Analysis of Heterochrony

All analyses of quantitative morphological heterochrony must assess comparable or homologous shapes and proportions at a given developmental stage in a presumed or hypothetical ancestor-descendant pair (Shea 2000). There are any number of acceptable approaches to depicting such morphological proportions, just as there are innumerable methods for analyzing size and shape in the biometrics literature (e.g., see Klingenberg 1996, 1998, for discussion). The use of allometry in studies of heterochrony dates from the early work of Huxley (1932), Rensch (1959), and others through the classic studies of Gould (e.g., 1966, 1968, 1971, 1977) and many subsequent workers. Comparative studies of growth allometry (Huxley's "relative growth") are Gould's (1968, 83; 1977, 238–239) favored metric for determining whether ancestral patterns of ontogenetic size/shape covariation have been dissociated or changed versus whether they have been maintained in descendant forms. In this framework, the classic heterochronic categories of paedomorphosis via neoteny and peramorphosis via acceleration are characterized by dissociation of ancestral allometric covariance, while paedomorphosis via progenesis and peramorphosis via hypermorphosis are produced by the truncation and extension, respectively, of such ontogenetic allometries (see Gould 1977).

In the minor elaboration I proposed (Shea 1983a), allometric dissociation (including position or slope differences) is associated with paedomorphosis via neoteny and peramorphosis via acceleration, while allometric concordance (ontogenetic scaling—Gould 1975) is characteristic of (rate and time) hypomorphosis and hypermorphosis. The allometric patterning associated with the standard types of heterochronic categories (dissociated vs. ontogenetically scaled) is illustrated in Figure 4.2. This follows the basic framework proposed by Gould (1977) and Alberch et al. (1979), though they also offer additional ways to depict size and shape covariance and dissociation. Devoid of any additional input, such presentation makes no im-

A. Paedomorphosis via Hypomorphosis

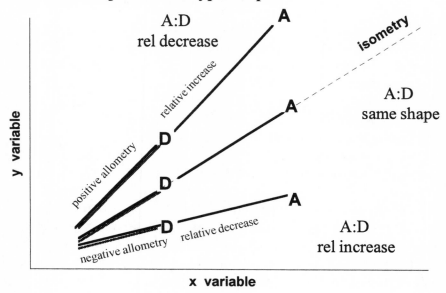

B. Paedomorphosis via Neoteny (Allometries dissociated)

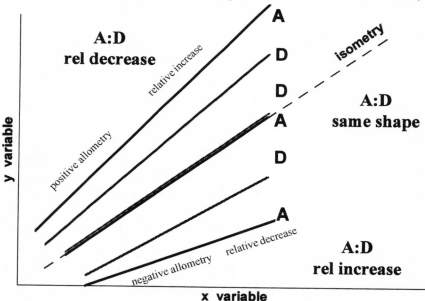

Fig. 4.2. *A:* Schematic representation of juvenilization (paedomorphosis) via simple truncation (ontogenetic scaling) for three bivariate growth trajectory comparisons, of which one is positively allometric, one is isometric, and a third is negatively allometric. The directionality of the shape changes along the ancestral trajectory plus the direct adult-to-adult shape transformation (A:D) are noted. *B:* Schematic representation of juvenilization (paedomorphosis) via allometric dissociation for positively and negatively allometric ancestral trajectories. A represents the ancestral trajectory of relative growth, and D represents the descendant trajectory. The case for ancestral isometric trajectories is shown as unchanged in descendants. *Source:* Based on Vrba 1994.

plicit suggestions as to which type of transformation is more likely (in genetic, developmental, morphometric, and/or ecological terms) than the next. Is it in fact the case that all might be considered equally likely in evolutionary transformations? At this point, I consider only the morphometric implications.

I would argue that, for cases of the isolated bivariate comparison, paedomorphosis and peramorphosis via ontogenetic scaling are indeed more likely outcomes than comparable shape transformations via allometric dissociation. This argument is simply based on the degree of change involved (i.e., a dissociation requires a fundamental reorganization of the ancestral pattern of feature covariance, while ontogenetic scaling involves no such shifts). This is the basic premise of the analysis of covariance in statistics. In a plot of $y{:}x$ dimensions during relative growth in related species, this is reflected in coincident versus divergent regression lines in these respective cases. However, if this consideration argues for some bias in the frequency of observed categories of transformation for the individualized trajectories, we would predict an even greater disparity in favor of a higher frequency of ontogenetic scaling when multivariate suites of comparisons are analyzed. This fundamental asymmetry was clearly and explicitly made previously in my discussions of the potential role of neoteny in human evolution (Shea 1989), but I elaborate further on it here.

Let us assume we have ten features or proportions, represented quantitatively as ontogenetic allometries of trait lengths relative to body length, in an ancestral form. All exhibit "simple allometry," or linear relations in log space. Further assume that four of these traits scale with positive allometry, or relative enlargement, ranging from slight (with a bivariate slope of 1.25) to moderate (1.33) to marked (1.50) to extreme (2.0); an additional four demonstrate negative allometry, or relative diminution (with slopes of 0.85, 0.66, 0.33, and 0.15). The precise slope values are, of course, irrelevant for our example. The final two traits exhibit isometric slope values (or 1.0 in this linear-linear proportion comparison). Taken together, these ten proportions reveal a complex trajectory of allometric shape change (and no change for the two isometric traits) during ontogeny; individual proportions are changing, and to varying degrees, but the pattern is underlain by regular and coordinated transformation, as Huxley (1932) explicated so well in his classic work.

The asymmetry between the alternative pathways to paedomorphosis (and peramorphosis) emerges on the morphometric level when we understand that, *while the same type of change is required of all ten trajectories in the cases of global paedomorphosis (and peramorphosis) via rate and time hypomorphosis (and hypermorphosis), divergent and opposite changes are required to yield a concordant paedomorphosis (and peramorphosis) in the case of neoteny (and acceleration),* associated as they are with allometric dissociations or uncoupling of ancestral patterns of size/shape covariance (Gould 1977). Shorn of some of the labels, complexities, and parallels, the asymmetry reduces to this: to produce a global juvenilization across the ten proportions, all are simply truncated through ontogenetic scaling, but quite different and opposing changes are needed to effect such a global juvenilization in the presence of allometric dissociations, *depending on whether the ancestral trajectory is positively or negatively allometric.* This is depicted schematically in a contrast between Parts *A* and *B* of Figure 4.2. The same characterization holds for the "overgrowth" associated with peramorphosis, and this is illustrated in Figure 4.3. The asymmetry, then, contrasts a situation where the same changes are involved with one where divergent and opposing (in terms of morphometric patterns) shifts are required.

Consider further the specific proportion changes depicted by the allometric slopes themselves. All of these trajectories, whether positively allometric, negatively allometric, or isometric, change in the same fashion in heterochronic transformations associated with ontogenetic scaling or maintenance of ancestral patterns of size/shape covariance during ontogeny. That is, they change globally by extrapolation (in the case of peramorphosis via hypermorphosis) or truncation (in the case of paedomorphosis via hypomorphosis). In contradistinction, in heterochronic transformations associated with ancestral-to-descendant size/shape dissociation, the trajectory shift required varies, depending on whether the ancestral trajectory is positively allometric, negatively allometric, or isometric (see Figs. 4.2*B* and 4.3*B*). For features that are *positively allometric* in ancestral forms, the change required is intuitive. In such cases, the proportion (ratio of) y/x is increasing throughout ancestral ontogeny, so a derived descendant that is paedomorphic in shape at a comparable developmental stage must exhibit a *decreased slope or downward transposition of allometric trajectory.* This yields a relatively smaller y for a given x at the homologous developmental stage in the descendant form, resembling the juvenile shape of the ancestor. This is straightforward and directly corresponds to intuitive notions of the shape retardation explicit in neotenic (dissociated) paedomorphosis.

Negatively allometric ancestral trajectories require shifts in the opposite direction, however, which may be counterintuitive to some. In the case of negative allometry, the proportion (ratio of) y/x is decreasing throughout

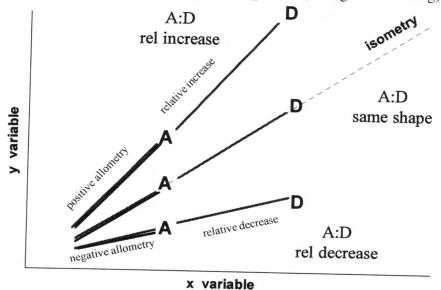

A. Peramorphosis via Hypermorphosis (Ontogenetic scaling)

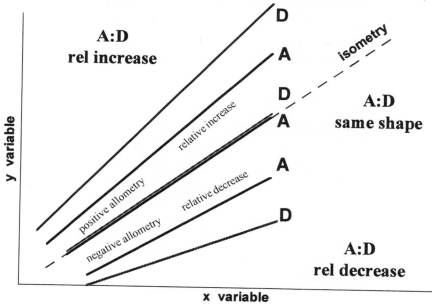

B. Peramorphosis via Acceleration (Allometries dissociated)

Fig. 4.3. *A:* Schematic representation of peramorphosis via extension and ontogenetic scaling. *B:* Schematic representation of peramorphosis via allometric dissociation for positively and negatively allometric ancestral trajectories. Ancestral-descendant trajectories and transformations are as in Figure 4.2. The case for ancestral isometric trajectories is shown as unchanged in descendants.

ancestral ontogeny, so that a derived descendant paedomorphic in shape at a comparable developmental stage must exhibit *an increased slope or upward transposition of the allometric trajectory* (see Fig. 4.2A). This results in a relatively larger y for given x at the homologous developmental stage in the descendant, once again characteristic of the juvenile shape of the ancestor. The key here is that, while the absolute value of the slope or position shifts is divergent and opposing in these cases, the resultant "shape" change effected via the allometric transformations is not. In other words, retardation of a morphogenetic trajectory of relative diminution involves increasing the relative size of y-to-x via a reduction in the degree or strength of such negative allometry (via slope increase in terms of absolute values, such as 0.50 to 0.75, for example, or via an upward transposition). On the other hand, decreasing the absolute value of the slope of an already negatively allometric trajectory only further exaggerates (accelerates) the ancestral shape transformation.

The parallel argument, of course, applies to cases of peramorphosis via acceleration, where allometric dissociation again occurs. In this case, for positively allometric slopes, acceleration is produced through increased slope or upward transposition of the allometric trajectory; with negatively allometric slopes, acceleration is produced through decreased absolute slope values or downward transpositions, thus accelerating the degree of negative allometry (Fig. 4.3B).

The above discussion assumes for the sake of simplicity and clarity a situation where the descendant's size (or x value in the bivariate comparisons) is not profoundly different from the ancestral value at comparable terminal developmental stages. If large-scale extrapolation into new size ranges does in fact occur, then the above relationships and predictions can be modified just by virtue of this distant extrapolation. For example, extrapolation of an upwardly transposed trajectory of negative allometry in descendants can yield peramorphosis (through increased negative allometry in the adult-to-adult comparison) at extremely disparate body sizes, while it will result in paedomorphosis in size ranges at or near the ancestral values (which is why Gould 1977 depicts neoteny and acceleration as dissociated allometries with descendants unchanged in terminal body size). Arguments of developmental processes versus static morphological results aside for the moment, this is an additional way in which the heterochronic categories generated via dissociation of allometric covariance are fundamentally more complex and problematic than those underlain by concordance of allometric covariance. In the latter, no amount of extrapolation, no matter how distant, alters the morphological result: the homologous developmental stages of ancestor-descendant share the same slope as the ancestral trajectory.

Left unaddressed in the foregoing hypothetical scenario are those trajectories that exhibit isometry, or no change in proportions. This complicates even further the transformations of paedomorphosis via neoteny and peramorphosis via acceleration. Technically speaking, such isometric transitions do not even lie within the legitimate realm of heterochrony, since shape, proportion, and morphology have not changed (Gould 1977; Shea 2000). But we must also bear in mind that, in a multivariate suite of characters such as I offered in the preceding scenario, selected pairwise comparisons will probably follow geometric similarity if only by chance alone, even while other pairwise comparisons using one or both of these dimensions will change allometrically.

Any significant slope or position shifts from ancestral isometries will inevitably result in proportions that cannot be described as paedomorphic or peramorphic in the descendants at comparable developmental stages. Rather, the ancestral-to-descendant transformation in this case will inevitably involve a novel shape change that departs from the ancestral trajectory of no shape change across ontogeny. This, in essence, is why Gould (1968, 83) noted that there can be no morphological heterochrony when geometric similarity is preserved throughout ontogeny; heterochrony has allometry as a necessary condition of its occurrence. Therefore, a global paedomorphosis via neoteny (or peramorphosis via acceleration—Figures 4.2*B* and 4.3*B*) would, in fact, require ontogenetic scaling between descendants and ancestors for such isometric trajectories, in opposition to the dissociation typically associated with retardation and acceleration. Additional discussion of the likelihood of "evolution via geometric similarity," in the sense of Gould (1971, 1977) and others, is given below.

The preceding discussion allows us to summarize the morphometric and allometric pattern asymmetry for suites of characters as follows. Paedomorphosis via neoteny and peramorphosis via acceleration involve multiple and divergent shifts in ancestral trajectories (depending on their direction of allometric shape change and degree of overall size change), whereas paedomorphosis via hypomorphosis and peramorphosis via hypermorphosis exhibit uniform shifts of general allometric truncation and extension, respectively, across characters. The multiple and divergent changes required for the dissociated categories of heterochrony make these a more complex scenario and therefore a less likely observation on a regional or global level than the simple cases of ontogenetic scaling (i.e., the hypo- and hypermorphoses). On the level of pattern comparison, this is no different than acknowledging as the appropriate null hypothesis a finding of no slope or y-intercept (position) differences between groups (see Sokal and Rohlf 1981, 522–526 and Fig. 14.17). Genetic, developmental, and ecological arguments will be layered on this morphometric observation in the ensuing discussions.

Genetic and Developmental Controls

Our understanding of the genetic and developmental controls of allometric and associated heterochonic transformations is in its infancy, though recent discoveries in developmental genetics are continually providing new insight. I offer the following oversimplified characterization as a working hypothesis for the control of quantitative allometric tranformations during growth and evolution in mammals and other vertebrates. The growth of any given feature might be viewed as the potential composite result of general systemic controls, plus regional growth factors, plus specific localized controls, as modeled by Wright (1932) many years ago. Palate length, for example, might be increased as a result of general overall body growth, the enlargement of the facial region as a whole, or the localized increase in solely the skeletal palate itself. I have elsewhere (Shea 1988, 1992a) built on Katz's (1980) discussion of allometry and growth controls to further explicate such a framework. Katz (1980), in fact, proposed that allometric extrapolations might be developmentally produced by general systemic factors that merely expand the patterns (rates) of growth via cell division, in keeping with relative proportions initially set by the ancestral system. In contradistinction, he argued that specific, localized changes (as, e.g., in stem cell populations or particular rates of cell differentiation or multiplication) would underlie allometric dissociations such as transpositions and slope differences.

Huxley (1932) originally described allometric growth in the bivariate case as being comparable with two sums of money placed in the bank at divergent interest rates, with the ratio of these sums progressively and regularly changing according to the ratio of the respective interest rates, this ratio being equivalent to the coefficient of relative or allometric growth. We can easily generalize from Huxley's bivariate example and envision a multivariate system, changing progressively according to such a preset (i.e., ancestral) configuration of myriad allometric coefficients. In this model, simultaneously pouring more money into all accounts initially will merely extrapolate the pattern of changing ratios upward. This would be analogous to a systemic change affecting all features in concert but in proportion to their baseline growth allometric coefficients. By contrast, selectively increasing the initial amount or interest rate of one or several monetary sums would result in particular, specific shifts. The same holds for regional or specific growth changes affecting the allometry of an individual feature or series of features.

Such a system of hierarchical growth controls would seem to provide maximum evolutionary flexibility for both integrated transformations (e.g., Frazzetta 1975; Shea 1985a) and the localized dissociations required for specific adaptive responses through non-size-related shape change. It would also bias the patterning of morphological differentiation toward global and coordinated allometric transformations (ontogenetic scaling) any time that shifts in growth rates or terminal sizes were genetically and developmentally mediated by such systemic growth controls. Elsewhere (Shea 1992a), I suggested a key role for growth hormone (GH), insulin-like growth factor 1 (IGF-1), and growth-hormone binding proteins (GHBPs) in such size differentiation among closely related mammalian species.

We do not have the requisite information on physiological growth control in this example, but the detailed work by the Grants on Darwin's finches provides a model case study and example here (Grant 1986). They show that several species (*Geospiza fuliginosa* and *Geospiza fortis*) seem to represent primarily size variants, characterized by common patterning of allometric covariance during growth, whereas a closely related species, *Geospiza scandens,* exhibits a beak morphology clearly indicative of complex selection for a specific shape configuration related to a dietary shift. Grant further hypothesizes that the speciation associated with the axis of size differentiation and allometric patterning took markedly less time than that involving *G. scandens,* with its more complex genetic, selective, and dissociative shape transformations. This is a wonderful example of how ontogenetic allometric covariance can be used in ancestor-descendant comparisons as a "criterion of subtraction" to differentiate between size-correlated and size-independent shape in evolutionary transformations.

How are we to characterize the allometric patterning described above? Such biases might be attributed to historical factors (Shea 1985a) or even constraints (Gould 1989), though the latter term must be used judiciously (e.g., Antonovics and van Tienderen 1991), as such allometric patterning seems fairly easily restructured, precisely as one would predict in a model of multiple systemic and local growth controls and precisely as one observes in the case of beak morphology in *Geospiza* species. It is only in those cases where virtually all of the covariance between features (e.g., feature y and body size x) results from shared pleiotropic controls that it would be unlikely for a restructuring of ancestral allometries to evolve (Price and Langen 1992). Therefore, it is probably most accurate and heuristic to simply view these examples of allometric covariance as merely a special case of general morphological covariance. Price and Langen (1992) discuss some of these general issues, building on previous work by Lande and Arnold (1983) and many others.

The previous hierarchical growth control system with systemic, regional, and local inputs is not the only one that can be envisioned, however. Significantly, alternative growth control systems among ancestors would predict different patterns of morphometric differentiation in descendants. For example, imagine that the changing proportions of the skeletal system during morphogenesis were controlled by two primary inputs, one affecting global isometric transformations and a second controlling regional shape differentiation (as, in fact, suggested by Gould 1971, 1977). Or we might envision a system with no global systemic controls at all, large-scale transformations being merely the epiphenomenal sum of many localized and regional shifts. The first example would predict the frequent occurrence of geometric giants and dwarfs (Gould 1971) accompanying size diversification, along with the independent evolution of specific local shape changes in many lineages. In the second example, overall size differences would be merely a reflection of the sum total of whatever localized changes occurred, and there would probably be no regular patterning of size-related shape transformations. These are but two possible examples where phyletic patterning would not follow trajectories of ancestral allometries and where heterochrony would have a reduced role in morphological differentiation.

Which type of growth control system does our current knowledge support? Here I believe a growing body of research from selection experiments, mutant strains, transgenes and knock-outs, and various other evidence clearly support the hierarchical model of systemic, regional, and local growth controls, producing generalized allometric patterning during size diversification (Shea 1992a). As noted above, it is unlikely in the extreme that any single systemic growth control (genes, gene products) would induce a generalized set of neotenic changes in quantitative features, in particular because negative allometrically growing structures would need to be shifted in the opposite direction from positive allometrically growing structures (i.e., the former would need to start growth relatively larger and the latter relatively smaller). This stands in sharp contrast to the ontogenetic scaling transformation, where all structures are similarly affected (extrapolated or truncated) in the presence of a common growth control substance. I am unaware of any evidence of gene mutations or products that have been shown to result in regional or global heterochronic transformations via neoteny or acceleration. Such would be most likely in regional cases where all features grow in concert with positive allometry. Here we can envision genetic and developmental shifts that would retard rates of shape change (or decrease initial sizes, or both) and yield a neotenic paedomorphosis (as in the case of facial neoteny in the pygmy chimpanzee [Shea 1983b, 1984, 2000]). It is precisely because such uniformity in allometric growth (e.g., all features positively allometric or all features negatively allometric) is so unlikely that the generalized, global dissociative changes required for neotenic paedomorphosis and accelerative peramorphosis are not expected or observed.

Qualitative cases of neoteny, such as the retention of a suite of juvenile features in non- or partially metamorphosing salamanders, represent alternative cases that perhaps must be assessed separately as their underlying developmental genetics and physiology become better known. Last, none of the above mitigates against the possibility of a given multivariate transformation (whether globally dissociative and neotenic or not) if selection is intense enough for the particular morphological configuration. But such would probably require precise and intense selective monitoring of individual features and proportions and almost certainly could not be underlain by simple genetic and physiological shifts, as is the case with systemic hormonal controls.

Precisely the same types of argument can be forwarded in response to the global geometric transformations hypothesized by Gould (1971) in his paper on "evolution via geometric similarity" to be commonplace in evolution and underlain by simple genetic and developmental (hormonal) controls. Gould (1971) suggested that generalized isometric patterning was common in evolutionary series and, indeed, that it should be expected precisely because genetic and hormonal data supported a model of growth control with independent inputs for global geometric size and localized shape factors (he further elaborated on this model in his 1977 book when he hypothesized distinct underpinnings for isometric enlargement [his "growth"] and shape differentiation [his "proportion" or morphogenesis]). But here again, positively allometric trajectories would need to be altered in opposite directions from negatively allometric ones to yield an isometric ancestor-descendant transformation. Moreover, each trajectory would probably require a different degree of alteration to approximate the common trajectory of geometric similarity.

Such global isometric transformations are unlikely in the extreme, a point recently also made strongly by Klingenberg (1998, 84). The myriad mutants and other known genetic syndromes in mice and other model systems also do not support such hypothesized shifts. For example, the giant transgenic mouse, originally described as having the same proportions as its smaller littermate controls (Palmiter et al. 1983), was in fact subsequently shown to be an allometrically enlarged morph differing in various proportions (Shea et al. 1987, 1990), as would be predicted in the case of overgrowth via GH and IGF-1. African pygmies, offered by Gould (1971) as an example of hormonally mediated geometric dwarfism, are in fact predominantly allometrically scaled along ontogenetic trajectories of varying degrees of allometry as well as isometry (Shea and Bailey 1996). No known systemic growth controls (hormones, etc.) have been shown to induce geometric transformations in skeletal or other bodily proportions in mammals. Contrary to Gould's (1977) attempts both to conceptually distinguish growth (isometric enlargement) from development (proportional changes) and to seek independent underlying genetic and developmental bases for such a distinction, no evidence supporting such a view has emerged in the past two decades.

Those who follow Gould (1977) in this endeavor (e.g., Godfrey and Sutherland 1995a, 1995b, 1996; Godfrey et al. 1998; see also Shea 2000) do so in isolation from developmental biologists and growth investigators who have long rejected any such simplistic notions as the dissociability of global isometric size from global or regional shape. There is simply no accumulating genetic or developmental evidence for such a distinction and dissociation in mammals and other vertebrates. There is also little biological basis for continued attempts by some morphometricians to separate some general composite and isometric size component from residual generalized or localized shape components in multivariate analyses, as has been advocated by Jungers and colleagues (e.g., Falsetti et al. 1993; Jungers et al. 1995). I could not agree more with Blackstone (1987, 77) in his statement that there is no evidence that size and shape factors (*sensu* Bookstein et al. 1985) underlie animal development. On the other hand, the generalized allometric scaling induced by shifts in systemic growth controls, such as GH and IGF-1, do provide biological credence for models that distinguish between global and more localized multivariate allometric patterning (Shea 1985b; Klingenberg 1998). This suggests that Wright's (1932) classic partitioning of quantitative feature covariance into general, group, and special factors can be productively translated into systemic, regional, and local allometric growth, which ultimately may find explication in particular genes, hormones, and growth factors.

Selective Scenarios and Ecological Contexts

Size change is one of the commonest themes in evolutionary transformations, as has been documented and discussed extensively (e.g., Gould 1966; Jungers 1985; Jablonski 1996). It is frequently observed in island mammals, where larger forms tend to dwarf and smaller ones to increase in size (e.g., Maurer et al. 1992). This is thought to be related to changes in resource utilization, predator pressure, and other interspecific interactions. In other contexts, selection for rapid growth and larger size is often associated with niche partitioning and increased use of high-fiber, lower-quality herbiage (e.g., Janson and von Schaik 1993; Leigh 1994). Similarly, size decrease may be associated with an increased concentration on high-quality fruits or other such food items.

The ecological conditions that commonly select for size diversification will tend to result in the selection of global peramorphosis via hypermorphosis and paedomorphosis via hypomorphosis (for discussion of "select for" vs. "selection of," see Sober 1984), and therefore we would expect to find a high frequency of such morphological patterning in the fossil record and comparative series. These morphological transformations and trends can be best viewed as consequences of selection for altered growth rates and terminal sizes, though the issue of potentially maladaptive consequences of hypermorphic extrapolation is always present (Gould 1966), and balancing selection may often need to winnow out individual problematic transitions. There is no reason whatsoever to expect a high frequency of neoteny or acceleration on a global (or local) level as a consequence of such size-based shifts. The nondissociative peramorphosis and paedomorphosis that are predicted will be underlain by global hyper- and hypomorphic transformations, or ontogenetic scaling. This pattern of size and allometric differentiation is very commonly observed in vertebrate clades, and we might generally predict a multivariate axis of coordinated allometric transformation associated with such peramorphic and paedomorphic patterning (e.g.,

Bjørklund 1993). Departures from this general axis are common and profoundly important in adaptation and differentiation, but these changes are all the more clearly identified and explicated against the backdrop of such transformations.

In what contexts might paedomorphosis via neoteny or peramorphosis via acceleration be predicted to occur? Simply put, these transformations are predicted in cases of selection for specific, localized morphological configurations. Examples include the retention of the swimming tail and larval gill structures in aquatic environments in salamanders (Gould 1977) and the allometric dissociations accompanying dietary shifts in the terrestrial transition in tiger salamanders investigated by Reilly and Lauder (1990). The latter example is particularly informative, since these biomechanically required skull proportions stand out as allometric dissociations against a general background of ontogenetic scaling of other body proportions in the transition to terrestriality (Ashley et al. 1991). Within primates, the neotenic paedomorphosis of the facial region associated with reduced sexual dimorphism and corresponding sociosexual relations in the pygmy chimpanzee, *Pan paniscus*, may represent such a localized adaptive shift (Shea 1983b, 1984).

Selection for Size versus Developmental Duration in Hypo- and Hypermorphoses

The preceding discussion established the frequency of size differentiation and global allometric patterning. But Gould (1977) clearly emphasized that such size and allometric transitions might result from selection for truncated or prolonged growth periods, as opposed to terminal size itself. In further drawing the distinctions between rate and time hypo- and hypermorphoses, I emphasized that closely related and size-differentiated species of mammals tended to differ more in rates of overall growth-in-size over a roughly comparable duration of chronological time than in the absolute duration of those time periods (Shea 1983a). This seems to be generally true of the anthropoid primates that have been studied, where species of African apes (Shea 1983a) and cercopithicine monkeys (Shea 1992b) appear more "rate-differentiated" than "time-differentiated" on the whole. This presumably reflects selection for different growth rates and terminal size in response to dietary specialization, predator defense, and other factors, perhaps combined with selection against major changes in life-history timing.

Significant extensions or truncations in the time duration of growth and life-history periods may prove to be more difficult to evolve than size differences via rate differentiation. Interestingly, it is in the examination of sexual dimorphism in these anthropoids (e.g. Shea 1986; Leigh 1992; Leigh and Shea 1995) that time differentiation is more commonly observed (see German and Stewart, Chap. 10), as one might predict from previous observations of sexual bimaturism (e.g., Jarman 1983). This may result from selection for prolonged male growth, in association with age- and size-graded hierarchies, or selection for early maturation and higher reproductive turnover in females (Shea 1986; Leigh 1992). Traditional heterochronic labels are merely heuristic and somewhat problematic in such cases, where we compare males and females within a single species, since they obviously do not represent hypothetical evolutionary transformations (Reilly et al. 1997).

Our preliminary considerations here suggest a working hypothesis of size diversification via predominant rate changes (rate hypo- and hypermorphoses) rather than time changes (time hypo- and hypermorphoses). Approaches distinguishing between selection for changes in body size and rates of growth, in contradistinction to duration of growth periods, might contribute to resolving these issues (e.g., Lande and Arnold 1983). Alternatively, compilation of which pattern predominates within closely related and size-differentiated clades would provide a rough test of this working hypothesis.

Application to Human Evolution

In reconsidering the heterochronic transformations that predominate in human evolution, I am led to much the same conclusion I reached in my 1989 review. That is, no single pattern emerges. While a few selected traits may fit with the expectations of neotenic paedomorphosis, on the one hand, and hypermorphic peramorphosis, on the other, there is no central component of heterochronic transformation that predominantly accounts for the bulk of the morphogenetic and evolutionary transitions in human morphology. Certainly, at present no emergent data support any genetic or developmental basis for a global or generalized neoteny (and the carp-gut cocktail of the vain youth-seeker in Aldous Huxley's *After Many a Summer Dies the Swan* remains as fanciful as ever—see Gould 1977, 352–353).

The arguments presented here suggest that global transformations based on neoteny or acceleration are highly unlikely in any lineage or clade; dissociative allometric shape retardation (neoteny) and acceleration are instead predicted to occur differentially in regional and localized changes. Nor, of course, does the broad pattern of human evolution accord with a generalized hypermorphosis mediated by growth factors and hormones (though such may indeed characterize the global paedomorphosis revealed in the microevolutionary dwarfism of the human pygmies; see Shea and Bailey 1996). Recent attempts to revitalize the view that hominid evolution has predominantly involved a generalized neotenic transformation (e.g., Godfrey and Sutherland 1996) are unconvincing, to say the least.

Conclusions

The literature on heterochrony has tended to avoid consideration of the key issue of which types of heterochronic transformations are more and less likely and in what contexts. I show here that, on a widespread regional or global level, the dissociative allometric patterns characteristic of neotenic paedomorphosis and accelerative peramorphosis are less likely to be observed in most organisms than are the simpler, coordinated allometric extensions and truncations characteristic of hypomorphic paedomorphoses and hypermorphic peramorphoses. This is a direct result of the fact that divergent and opposing changes are required in (neotenic and accelerative) dissociative allometric shifts, and there is no known genetic and developmental basis for such changes in mammals. In contrast, a growing body of genetic and developmental data suggests a hierarchical model of quantitative growth control in mammals; this model combines levels of systemic, regional, and local inputs and accords with the observed pattern of frequent ontogenetic scaling and hypo/hypermorphoses in the presence of size diversification via systemic hormones (Shea 1992a). This model argues against widespread global neotenic or accelerative changes in morphology, as well as global evolution via geometric similarity, contra Gould (1971). However, the localized allometric dissociations characteristic of neoteny and acceleration (and geometric similarity) may often be produced in response to selection for particular morphological configurations and functional requirements.

The frequent size diversification observed in the adaptive radiation of clades and generated developmentally in large part via systemic growth controls may largely account for the longstanding phenomena of allometric patterning and trends noted by virtually all major observers of evolutionary processes and patterns (e.g., Huxley 1932; Simpson 1953; Rensch 1959; Gould 1977; Levinton 1988; Raff 1996). Evolution proceeding along an axis of size differentiation and allometric and heterochronic patterning is a common theme, though one associated primarily not with major evolutionary novelties but with minor variations on adaptive themes and the parallels between ontogeny and phylogeny that have given so much impetus to the study of heterochrony in evolution (Thomson 1988).

Acknowledgments

I thank Michael McKinney, Nancy Minugh-Purvis, and Ken McNamara for the opportunity to participate in the American Association for the Advancement of Science symposium and this subsequent volume. Thanks to Eleanor Weston for pointing out several key examples of differential heterochronic transformations and to Christian Klingenberg, Lorna Profant, Steve Leigh, Sandra Inouye, and Matt Ravosa for various illuminating discussions and comments.

References

Alberch, P., Gould, S.J., Oster, G.F., and Wake, D.B. 1979. Size and shape in ontogeny and phylogeny. *Paleobiology* 5:296–317.
Antonovics, J., and van Tienderen, P.H. 1991. Ontoecogenophyloconstraints? The chaos of constraint terminology. *Trends in Ecology and Evolution* 6:166–168.
Ashley, M.A., Reilly, S.M., and Lauder, G.V. 1991. Ontogenetic scaling of hindlimb muscles across metamorphosis in the tiger salamander, *Ambystoma tigrinum. Copeia* 1991:767–776.
Bjørklund, M. 1993. Phenotypic variation of growth trajectories in finches. *Evolution* 47:1506–1514.
———. 1994. Allometric relations in three species of finches (Aves: Fringillidae). *Journal of Zoology, London* 233:657–668.
Blackstone, N.W. 1987. Allometry and relative growth: pattern and process in evolutionary studies. *Systematic Zoology* 35:76–78.
Bookstein, F.L., Chernoff, B., Elder, R.L., Humphries, J.M., Smith, G.R., and Strauss, R.E. 1985. Morphometrics in evolutionary biology. *Academy of Natural Science, Philosophy Special Publication* 15.
Falsetti, A.B., W.L. Jungers, and Cole, T.M. 1993. Morphometrics of the cal-

litrichid forelimb: a case study in size and shape. *International Journal of Primatology* 14:551–572.

Frazzetta, T.H. 1975. *Complex Adaptations in Evolving Populations.* Sunderland, Mass.: Sinnauer.

Godfrey, L.R., and Sutherland, M.R. 1995a. What's growth got to do with it? Process and product in the evolution of ontogeny. *Journal of Human Evolution* 29:405–431.

———. 1995b. Flawed inference: why size-based tests of heterochronic processes do not work. *Journal of Theoretical Biology* 172:43–61.

———. 1996. Paradox of peramorphic paedomorphosis: heterochrony and human evolution. *American Journal of Physical Anthropology* 99:17–42.

Godfrey, L.R., King, S.J., and Sutherland, M.R. 1998. Heterochronic approaches to the study of locomotion. In E. Strasser, J. Fleagle, A. Rosenberger, and H. McHenry (eds.), *Primate Locomotion: Recent Advances,* 277–307. New York: Plenum Press.

Gould, S.J. 1966. Allometry and size in ontogeny and phylogeny. *Biological Reviews* 41:587–640.

———. 1968. Ontogeny and the explanation of form. In D.B. Macurda (ed.), *Paleobiological Aspects of Growth and Development: A Symposium.* Paleontological Society Memoir 2. *Journal of Paleontology* 42(5):81–98.

———. 1971. Geometric scaling in allometric growth: a contribution to the problem of scaling in the evolution of size. *American Naturalist* 105:113–136.

———. 1975. Allometry in primates, with emphasis on scaling and the evolution of the brain. In F.S. Szalay (ed.), *Approaches to Primate Paleobiology. Contributions in Primatology* 5:244–292. Basel: Karger.

———. 1977. *Ontogeny and Phylogeny.* Cambridge: Belknap Press of Harvard University Press.

———. 1989. A developmental constraint in *Cerion,* with comments on the definition and interpretation of constraint in evolution. *Evolution* 43:516–539.

Grant, P.R. 1986. *Ecology and Evolution of Darwin's Finches.* Princeton: Princeton University Press.

Huxley, J.H. 1932. *Problems of Relative Growth.* London: MacVeagh.

Jablonski, D. 1996. Body size and macroevolution. In D. Jablonski, D.H. Erwin, and J.H. Lipps (eds.), *Evolutionary Paleobiology,* 256–289. Chicago: University of Chicago Press.

Janson, C.H., and von Schaik, C.P. 1993. Ecological risk aversion in juvenile primates: slow and steady wins the race. In M.E. Pereira and L.A. Fairbanks (eds.), 57–76. Oxford: Oxford University Press.

Jarman, P. 1983. Mating system and sexual dimorphism in large, terrestrial, mammalian herbivores. *Biological Reviews* 58:485–520.

Jungers, W.L. (ed.). 1985. *Size and Scaling in Primate Biology.* New York: Plenum Press.

Jungers, W.L., Falsetti, A.B., and Wall, C.E. 1995. Shape, relative size, and size-adjustments in morphometrics. *Yearbook of Physical Anthropology* 38:137–162.

Katz, M.J. 1980. Allometry formula: a cellular model. *Growth* 44:89–96.

Klingenberg, C.P. 1996. Multivariate allometry. In L.F. Marcus, M. Corti, A. Loy, G.J.P. Naylor, and D.E. Slice (eds.), *Advances in Morphometrics*, 23–49. New York: Plenum Press.

———. 1998. Heterochrony and allometry: the analysis of evolutionary change in ontogeny. *Biological Reviews* 73:79–123.

Klingenberg, C.P., and Ekau, W. 1996. A combined morphometric and phylogenetic analysis of an ecomorphological trend: pelagization in Antarctic fishes (Perciformes: Nototheniidae). *Biological Journal of the Linnean Society* 59:143–177.

Lande, R., and Arnold, S.J. 1983. The measurement of selection on correlated characters. *Evolution* 37:1210–1226.

Leigh, S.R. 1992. Patterns of variation in the ontogeny of primate body size dimorphism. *Journal of Human Evolution* 23:27–50.

———. 1994. Ontogenetic correlates of diet in anthropoid primates. *American Journal of Physical Anthropology* 94:499–522.

Leigh, S.R., and Shea, B.T. 1995. Ontogeny and the evolution of adult body size dimorphism in apes. *American Journal of Primatology* 36:37–60.

Levinton, J. 1988. *Genetics, Paleontology, and Macroevolution.* Cambridge: Cambridge University Press.

Maurer, B.A., Brown, J.H., and Rusler, R.D. 1992. The micro and macro in body size evolution. *Evolution* 46:939–953.

McKinney, M.L., and McNamara, K.J. 1991. *Heterochrony: The Evolution of Ontogeny.* New York: Plenum Press.

McNamara, K.J. 1986. The role of heterochrony in the evolution of Cambrian trilobites. *Biological Reviews* 61:121–156.

———. 1988a. Heterochrony and the evolution of echinoids. In C.R.C. Paul and A.B. Smith (eds.), *Echinoderm Phylogeny and Evolutionary Biology,* 149–163. Oxford: Clarendon Press.

———. 1988b. The abundance of heterochrony in the fossil record. In M.L. McKinney (ed.), *Heterochrony in Evolution: A Multidisciplinary Approach,* 287–325. New York: Plenum Press.

———. 1993. Inside evolution: 1992 presidential address. *Journal of the Royal Society of Western Australia* 76:3–12.

Palmiter, R.D., Norstedt, G., Gelinas, R.E., Hammer, R.E., and Brinster, R.L. 1983. Metallothionein-human GH fusion genes stimulate growth in mice. *Science* 222:809–814.

Price, T., and Langen, T. 1992. Evolution of correlated characters. *Trends in Ecology and Evolution* 7:307–310.

Raff, R.A. 1996. *The Shape of Life.* Chicago: University of Chicago Press.

Reilly, S.M., and Lauder, G.V. 1990. Metamorphosis of cranial design in tiger salamanders (*Ambystoma tigrinum*): a morphometric analysis of ontogenetic change. *Journal of Morphology* 204:121–137.

Reilly, S.M., Wiley, E.O., and Meinhardt, D.J. 1997. An integrative approach to heterochrony: the distinction between interspecific and intraspecific phenomena. *Biological Journal of the Linnean Society* 69:191–143.

Rensch, B. 1959. *Evolution above the Species Level.* New York: Columbia University Press.

Shea, B.T. 1983a. Allometry and heterochrony in the African apes. *American Journal of Physical Anthropology* 62:275–289.

———. 1983b. Paedomorphosis and neoteny in the pygmy chimpanzee. *Science* 222:521–522.

———. 1984. An allometric perspective on the morphological and evolutionary relationships between pygmy (*Pan paniscus*) and common (*Pan troglodytes*) chimpanzees. In R.L. Susman (ed.), *The Pygmy Chimpanzee: Evolutionary Biology and Behavior,* 89–130. New York: Plenum Press.

———. 1985a. Ontogenetic allometry and scaling: a discussion based on the growth and form of the skull in African apes. In W.L. Jungers (ed.), *Size and Scaling in Primate Biology,* 175–206. New York: Plenum Press.

———. 1985b. Bivariate and multivariate growth allometry: statistical and biological considerations. *Journal of Zoology, London* 206:367–390.

———. 1986. Ontogenetic approaches to sexual dimorphism in anthropoids. In M. Pickford and B. Chiarelli (eds.), *Sexual Dimorphism in Living and Fossil Primates,* 93–106. Florence: Il Sedicesimo.

———. 1988. Heterochrony in primates. In M.L. McKinney (ed.), *Heterochrony in Evolution: A Multidisciplinary Approach,* 237–266. New York: Plenum.

———. 1989. Heterochrony in human evolution: the case for neoteny reconsidered. *Yearbook of Physical Anthropology* 32:69–101.

———. 1992a. A developmental perspective on size change and allometry in evolution. *Evolutionary Anthropology* 1:125–134.

———. 1992b. Ontogenetic scaling of skeletal proportions in the talapoin monkey. *Journal of Human Evolution* 23:283–307.

———. 2000. Current issues in the investigation of evolution by heterochrony, with emphasis on the debate over human neoteny. In S. Parker, J. Langer, and M.L. McKinney (eds.), *The Evolution of Behavioral Ontogeny,* 181–214. Santa Fe: School of American Research Press.

Shea, B.T., and Bailey, R.C. 1996. Allometry and adaptation of body proportions and stature in African pygmies. *American Journal of Physical Anthropology* 100: 311–340.

Shea, B.T., Hammer, R.E., and Brinster, R.L. 1987. Growth allometry of the organs in giant transgenic mice. *Endocrinology* 121:1–7.

Shea, B.T., Hammer, R.E., Brinster, R.L., and Ravosa, M.J. 1990. Relative growth of the skull and postcranium in giant transgenic mice. *Genetical Research, Cambridge* 56:21–34.

Simpson, G.G. 1953. *The Major Features of Evolution.* New York: Columbia University Press.

Sober, E. 1984. *The Nature of Selection: Evolutionary Theory in Philosophical Focus.* Cambridge: MIT Press.

Sokal, R.R., and Rohlf, F.J. 1981. *Biometry: The Principles and Practice of Statistics in Biological Research,* 2nd ed. New York: W.H. Freeman and Co.

Thomson, K.S. 1988. *Morphogenesis and Evolution.* Oxford: Oxford University Press.

Voss, R.S., Marcus L.F., and Escalante, P. 1990. Morphological evolution in muroid rodents: I. Conservative patterns of craniometric covariance and their ontogenetic basis in the neotropical genus *Zygodontomys. Evolution* 44:1568–1587.

Vrba, E.S. 1994. An hypothesis of heterochrony in response to climatic cooling and its relevance to early hominid evolution. In R.S. Corruccini and R.L. Ciochon (eds.), *Integrative Paths to the Past: Paleoanthropological Advances in Honor of F. Clark Howell,* 345–375. Englewood Cliffs, N.J.: Prentice Hall.

Wright, S. 1932. General, group, and special size factors. *Genetics* 17:602–619.

Chapter 5

Sequential Hypermorphosis
Stretching Ontogeny to the Limit

Kenneth J. McNamara

A perennial point of contention in debates concerning the role of developmental change in evolution is the definition and scope of heterochronic processes. While few would argue with the basic premise that heterochronic processes give rise either to paedomorphosis (retention of juvenile ancestral traits in descendant adults) or to peramorphosis (extension of ancestral ontogenetic trajectories beyond the ancestral condition), the nature of the processes, and even what you call them, have evoked much discussion. While this may seem to be little more than pedantic semantics, incorrect use of terms can result in fundamental misunderstandings of the underlying processes that control developmental changes and lead to evolutionary change. This has been particularly so in human evolution.

In the last two decades, there have been several attempts (see McKinney and McNamara 1991) to clarify the nature of the processes and to refine the heterochronic nomenclature, with the intention of encouraging more researchers to examine their material from a heterochronic perspective. However, there has been an unfortunate tendency of late to introduce a host of new terms (e.g., Reilly et al. 1997). To a large degree this has arisen, I believe, because of confusion about and misunderstanding of the nature of many heterochronic processes. The attitude seems to be: "I don't understand it, so let's introduce a new term." Herein the terminology of McKinney and McNamara (1991) is followed (see "What Is Heterochrony?" and Alba, Chap. 2, both this volume). Moreover, the processes that result in intraspecific heterochrony are not seen to be any different from those that result in interspecific heterochrony, despite recent claims to the contrary (Reilly et al. 1997).

One term in particular that has caused a fair degree of confusion is *neoteny*. Given the frequency with which human evolution has been regarded in terms of neoteny, it is paramount to clarify exactly what the term means. Neoteny, as Reilly et al. (1997) have pointed out, has changed its meaning over the years. It was originally introduced by Kollman (1885) as a descriptor synonymous with the term *paedomorphosis*, to describe the retention of larval features into adult salamanders. De Beer (1930) used *neoteny* to describe accelerated gonadal development and retardation of somatic development. Gould (1977) described *neoteny* as "paedomorphosis induced by retardation of somatic development." However, despite Gould's work and that of Alberch et al. (1979), McNamara (1986, 1988), McKinney (1988), McKinney and McNamara (1991), Godfrey and Sutherland (1996), and Klingenberg (1998) having quite clearly defined this process as one that involves a reduction in the *rate* of somatic development, the term has still been used by some as a synonym for paedomorphosis.

However, the misuse and misunderstanding of a term by some researchers is no justification for casting the term into the nomenclatural trash can and then burdening us with another term, as some have done (e.g., Reilly et al. 1997). Yes, *deceleration,* the term introduced by Reilly et al. (1997), may intuitively and with hindsight seem to be the natural corollary to *acceleration,* but the term *neoteny* has gained widespread acceptance as a descriptor of a reduced rate of somatic development. As such, its use is retained herein.

There has also been an unfortunate tendency to equate neoteny with the attainment of a larger body or trait size, as if the two must always go together. However, with "pure" neoteny, there will be no change to body or trait size (McKinney and McNamara 1991). What is actually happening in such cases, when an apparent paedomorphic shape is associated with large size, is the operation of more than one heterochronic process. Furthermore, in other cases what has been assumed to be neoteny and increased size is not neoteny at all, but something quite different—the stretching out of particular developmental stages. Thus, rather than developmental rates being phylogenetically adjusted, developmental timing is changed. Although some authors have described this fundamental difference as "semantic obfuscation" (Godfrey and Sutherland 1996), I will show that rate and timing heterochronies are brought about by fundamentally different processes.

Perhaps one of the more classic examples of confusion over the role of rate and timing heterochronies is in hominid evolution. Many authors have argued that humans are demonstrably (e.g., Gould 1977) or "essentially" (Godfrey and Sutherland 1996) neotenic. This would imply that human evolution has been driven by a reduction in growth rates. Yet this reduction in growth rates is accompanied by an increase in body size. But, as I have argued, neoteny is a rate, not a timing, heterochrony. Consequently, any associated size change will be brought about by different timing heterochronies. Moreover, the association of any apparent rate reduction with size increase is likely, in many cases, to be a misinterpretation of the underlying cause—not a rate reduction, but an extension of preadult growth stages.

Such extensions of preadult growth stages, without a concomitant change in the time of terminal offset of growth, have been termed *sequential hypermorphosis* (McKinney and McNamara 1991). Recently, the same process has been termed *proportional growth prolongation* by Vrba (1998). It can, in some instances, have the curious effect of reducing the number of growth phases, causing mitotic paedomorphosis (*sensu* McKinney and McNamara 1991) and what Godfrey and Sutherland (1996) have termed the "paradox of peramorphic paedomorphosis"—a contradiction in terms if there ever was one. Yet the actual rate of somatic development within the reduced number of phases may be the same, or less, or more. If the rate remains the same but the length of the growth stage is prolonged, extensions of ancestral allometries will result in peramorphic attributes. Reduced rate but a longer period spent in that growth stage can, theoretically, actually result in no phenotypic difference. Acceleration and extension of the growth stage can produce pronounced peramorphosis.

In humans, for example, there is no evidence for any reduction in growth rates of many traits (although there are some exceptions, along with some increases in growth rates). On the contrary, many traits are extended by sequential hypermorphosis, producing peramorphosis arising from scaling effects. Extending preadult growth stages for longer than in the ancestor is not paedomorphosis. Viewed in terms of extensions of successive growth stages, it becomes possible to explain many morphological and behavioral changes in hominid evolution as being peramorphic and having arisen from sequential hypermorphosis. However, developmental trade-offs have resulted in some traits being truly paedomorphic.

What Is Sequential Heterochrony?

Sequential heterochrony can be defined as the prolongation or compression of ontogenetic growth or life-history stages in the descendant relative to the ancestor. If the growth stage or stages are prolonged, this is termed *sequential hypermorphosis*. If the growth stages are compressed, this is known as *sequential progenesis*.

However, our basic definitions of the manifestations of heterochrony, *paedomorphosis* and *peramorphosis,* historically have been used to describe the adult characteristics in a descendant, compared with ancestral ontogenetic stages. They are not processes in themselves, but the product of various processes. Extensions or contractions of the end of growth are called *hypermorphosis* and *progenesis,* respectively. These are usually viewed in a global context, based on changes to the time of onset of sexual maturity and cessation of somatic growth, with the two frequently coinciding (although, as Reilly et al. 1997 have pointed out, this need not always be the case). However, extensions and contractions can occur during ontogeny at transitions between particular life-history stages. Furthermore, McKinney and McNamara (1991) pointed out that local growth fields can also be affected by these processes. Rate changes tend to affect primarily local growth fields, but again there is often the tendency to compare the descendant adult condition with that of the ancestral ontogeny.

This preoccupation with the time at which growth ceases in a descendant, compared with the ancestral condition (termed *terminal heterochrony* [McNamara 1983; McKinney and McNamara 1991]), has masked the fact that heterochrony can operate at any time during ontogeny, from fertilization until the cessation of growth. Indeed, it can operate in some instances very early in ontogeny and have significant evolutionary implications (Richardson 1995). As they develop, most organisms pass through distinct growth phases. In mammals, for example, the embryonic and postembryonic growth phases can be characterized; the postembryonic can further be subdivided into infantile, juvenile, adolescent, and adult phases. In many invertebrates the larval phase can be distinct from the juvenile and adult phases not only morphologically but also in terms of life history and behavior: a butterfly looks very different from a caterpillar; it also interacts with its environment and behaves in a very different way.

The same is true in other vertebrates, such as amphibians, where a distinct metamorphosis takes place during ontogeny, involving both profound morphological and behavioral changes. The time of metamorphosis can be extended or contracted, with resultant impacts on life history, as well as morphology. For examples, sequential hypermorphosis can also occur when the onset of metamorphosis is delayed in the descendant, as in lampreys (McNamara 1997). Such a change in somatic development can be, and quite often is, divorced from gonadal development. Consequently, metamorphosis that may take place before the onset of sexual maturity in an ancestor can be delayed, such that it occurs after the onset of sexual maturity, or may not even occur at all. This can produce paedomorphosis by sequential hypermorphosis, as in one of the classic cases of "neoteny," the axolotl.

Such episodic growth, where there is a major, sudden outward manifestation of morphological change—a morphological jump, if you like—is most clearly demonstrated in arthropods that undergo periodic moltings. Because each instar represents the level of somatic growth attained shortly after the molt, any changes to the intervals between molts, either by prolonging (sequential hypermorphosis) or contracting (sequential progenesis) the time of change between stages, can have obvious phenotypic effects, not only down the track on the descendant adult but also on successive preadult ontogenetic growth stages. Thus, heterochrony can occur when there are no changes in growth rates, between ancestor and descendant, nor any change to the initial time of onset of growth of a trait or the cessation of somatic growth, but when there are changes to the time of transition between successive growth stages.

Sequential heterochrony can take many forms. If we take a theoretical organism that passes through ontogenetic stages A to C, a delay in the transition from growth stage A to growth stage B is sequential hypermorphosis. The transition from B to C may be the same as in the ancestor, in which case the period spent in stage B will be reduced. Alternatively, the transition from B to C may also be delayed relative to the ancestor, in which case, because the onset of B is relatively delayed, the duration of B could be either the same as in the ancestor or longer. If the time of termination of growth, T3, is the same as in the ancestor, growth stage C will be foreshortened (Fig. 5.1*B*). In both the ancestor and descendant, the final form will be similar; therefore, the descendant cannot be described as being paedomorphic. If, however, stage B is prolonged by sequential hypermorphosis for so long that growth ceases before the transition to stage C, then the descendant adult can be described as being paedomorphic, as it never attains stage C but is in stage B at the termination of growth (Fig. 5.1*A*). Of

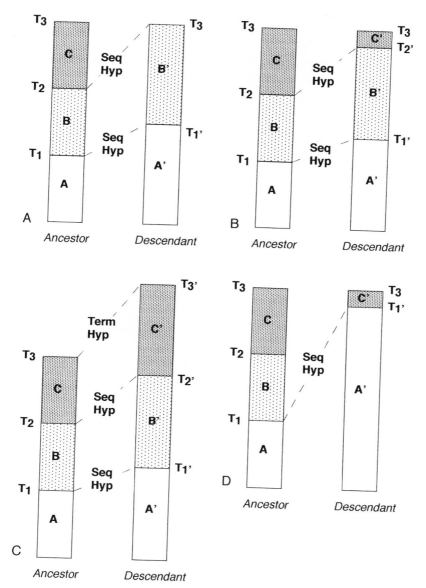

Fig. 5.1. Diagrammatic representations of the ontogeny of ancestral and descendant species that pass through distinct growth stages (A–C) at times T or T'. *A:* Greater sequential hypermorphosis results in failure of the descendant to pass into stage C before cessation of growth at time T3. *B:* Sequential hypermorphosis (*Seq Hyp*) at both A and B result in the period spent in stage C being much shorter in the descendant. *C:* Sequential hypermorphosis and terminal hypermorphosis (*Term Hyp*) result in all descendant growth stages being extended. *D:* Extreme sequential hypermorphosis of A results in stage B being omitted in the descendant.

particular significance in this scenario is that sequential hypermorphosis can result in paedomorphosis, especially if transitions between growth or life-history stages are punctuated events. However, if transitions between growth stages are more gradual and a longer period is spent in each of the ontogenetic stages, the resultant morphology may be peramorphic, because of scaling effects, as ancestral allometries are extended.

If sequential hypermorphosis affects a series of growth stages, including the offset of growth (T3 in this case)—termed *terminal hypermorphosis* by McNamara (1983)—then the likely consequences are greater development within each growth stage due to scaling effects and probable increase in size, either of the part or of the whole (Fig. 5.1C). In such cases there will be periods (e.g., after the expiration of a certain time during ontogeny in both the ancestor and the descendant but before offset of growth), that the descendant will be in a relatively more juvenile state compared with the ancestor. Some authors have described this as paedomorphosis. Yet paedomorphosis is defined in terms of adult characteristics. In this situation the adults may not show any paedomorphosis. Failure to appreciate the significance of seqeuntial heterochrony has resulted in confusion in the interpretation of many cases of heterochrony.

One of the extreme outcomes of prolongations of one or more growth stages can be a contraction in the following growth stage, resulting in the loss of entire growth stages. This can have profound effects on the embryonic life history of the organism. Thus, if stage A is prolonged by sequential hypermorphosis for a sufficiently long period, stage B might be lost entirely from the ontogeny (Fig. 5.1D). Stage B can also be lost by sequential progenesis of the transition from stage B to stage C. Thus, through a series of intermediates, the transition from B to C is progressively more precocious and stage B can be lost altogether (Fig. 5.2). Stage C might still terminate at the same time as in the ancestor; thus the adults may look alike, but preadult ontogenetic development will be quite different.

For each sequential hypermorphosis scenario, there is an equal but opposite sequential progenesis scenario. Thus, the transition from stage A to stage B occurs relatively earlier in the descendant, as does the transition from B to C, then, if offset at the end of C is the same in ancestor and descendant, either C will be prolonged or an extra stage, D, might appear. In such a case the descendant adult will be peramorphic, at least in terms of the number of stages. However, if terminal progenesis accompanies the sequential progenesis, the descendant will be smaller and may be paedomorphic in growth traits.

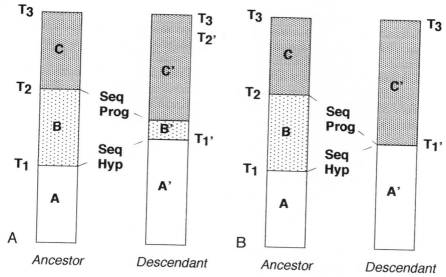

Fig. 5.2. Diagrammatic representation of the ontogeny of ancestral and descendant species that pass through distinct growth stages (A-C) at times T or T'. *A:* Growth stage B in the descendant is reduced by a combination of sequential hypermorphosis (*Seq Hyp*) from A to B and sequential progenesis (*Seq Prog*) from B to C. *B:* Carried to its extreme, stage B is omitted altogether.

In these theoretical examples, ancestral rates of growth are assumed to be the same in the descendant. Of course, the rates may change, as well as the time of transition between stages. However, quite clearly "slow development" by neoteny is very different from "slow development" by sequential and terminal hypermorphosis.

Sequential Hypermorphosis and Human Evolution

Views on the role of heterochrony in human evolution have been strongly polarized. Some (e.g., Gould 1977; Montagu 1981; Wolpert 1991; Vrba 1994; Godfrey and Sutherland 1996) have argued that humans show overwhemingly paedomorphic traits. Others (e.g., Shea 1989; Gibson 1991; McKinney and McNamara 1991; Verhulst 1993; McNamara 1997; McKinney 1998; Vrba 1998) have argued that, on the contrary, we show predominantly peramorphic features. Why such profound differences in interpretation? I would argue that the failure to appreciate the significance of sequential heterochrony—in particular, sequential hypermorphosis combined with terminal hypermorphosis—has led to many misunderstandings of the underlying intrinsic factors involved in hominid evolution.

Gould (1977) argued that humans are paedomorphic, considering "that human beings are 'essentially' neotenous, not because I can enumerate a list of important paedomorphic features, but because a general temporal retardation of development has clearly characterized human evolution." In such an interpretation even delays in the onset of changes from one state to another, such as the time of eruption of teeth and the onset of maturity, were regarded by Gould as retardation and thus paedomorphosis. Gould (1977) promoted the ideas of Bolk (1926), who considered that humans are distinguished from all other primates by many "foetal conditions that have become permanent." These include a flat face; a generally hairless state compared with other primates; loss of pigmentation in the skin, eyes, and hair; the form of the external ear; the central position of the foramen magnum in the skull; a relatively high brain weight; the form of the hand and foot; and the structure of the pelvis.

Others, such as Montagu (1981), identified other characteristics of adult humans as being paedomorphic, including the absence of brow ridges, thin skull bones, small teeth, teeth that erupt relatively late, a prolonged period of infantile dependency, a prolonged period of growth, a long life span, and large body size. However, to interpret these last four features as paedomorphic is to misrepresent, misinterpret, and misunderstand heterochrony. These features of life history are anything but paedomorphic; they are peramorphic features. Yet within them lies the key to unraveling the fundamental relationship between our ontogenetic development and our evolutionary history, for the two are inexorably linked.

Bolk argued that "foetal" growth rates were a consequence of "delayed development," producing "retardation." In his opinion, humans grow very slowly, compared with other primates, and consequently take much longer to reach their final form. If this were so, all these alleged paedomorphic features would have arisen by neoteny. Shea (1989), Gibson (1990, 1991), and McKinney (1998) have argued strongly that paedomorphosis has not been the main driving force in human evolution. For instance, comparing human growth rates with those of the chimpanzee *Pan troglodytes* shows that there is no appreciable difference in most traits (although there are some exceptions, such as rate of dental root growth—see below). But what has happened has been a delay in the transition from one growth stage to another during ontogeny. Moreover, the time of onset of sexual maturity and the accompanying reduction and then cessation of high growth rates are delayed, allowing longer periods to be spent undergoing higher growth rates. As Fiorello and German (1997) argued, "Subtle factors such as time spent growing at higher rates are the important ones."

As with other organisms, there has been a basic mistake in equating delays in transition from one growth phase to another in humans with reduction in growth rate. Yet, had humans been the product of neoteny, body size would have been smaller, as would limbs and, significantly, the brain—a reduced rate produces a smaller, morphologically less complex structure in descendants. This is hardly the case in hominid evolution. Delays in transition from one growth phase to the next and neoteny are, as I have argued, empirically different processes, each yielding fundamentally different results. The pattern of hominid evolution is characterized by long, drawn-out growth phases caused by sequential hypermorphosis, producing many peramorphic traits, such as increased body size; an enlarged cranium to house its enlarged, complex brain; highly specialized lower limbs, including a foreshortened pelvis; a unique laryngeal morphology, and so forth.

It is strange that those who still argue for a paedomorphic origin for humans view the prolongation of early juvenile growth rates as producing paedomorphosis, because the juvenile growth rates are continued for a longer time to a later stage. This is not paedomorphosis. The result of prolonging a growth stage that might have a high growth rate for a longer period than in the ancestor will be the development of larger, more complex structures. At a later stage of development, they may bear little resemblance to the same feature in the ancestral juvenile yet, curiously, this may still be perceived in some quarters as being paedomorphosis. Sequential hypermorphosis has resulted in the evolution of important anatomical, physiological, and behavioral characteristics during hominid evolution that bear no resemblance to juvenile primate characters.

It has been asserted that the relative size of the human brain is similar to that of our ancestors' juvenile brain—we look superficially more like a juvenile than an adult chimpanzee. However, as Shea (1989) argued, this is quite misleading. When viewed in terms of time (which, after all, is what heterochrony is all about) and not just allometry, the human brain grows for a longer period than the brain in other primates; it is peramorphic. It attains a larger size and is more complex, especially in terms of neural proliferation (see McKinney, Chap. 8) and the development of specialized regions, like the cortical areas involved in speech production, information processing, and mental constructional skills (Parker 1994). The human brain is not paedomorphic. As Parker and Gibson (1979) showed, we have the greatest neural information-processing capacity of any primate.

Many postcranial structures are also peramorphic. Berge (1996, 1998; Chap. 18, this volume) has shown that some pelvic traits of adult *Australopithecus* resemble those of neonate *Homo,* in particular features of the ilium, and that a unique feature of species of *Homo* has been the prolonged growth in length of the lower limb and pelvis after sexual maturity. Moreover, Berge's evidence indicates that a combination of three processes—predisplacement, acceleration, and hypermorphosis—contributed to the peramorphic pelvic features in species of *Homo* compared with *Australopithecus.*

However, it would be a mistake to argue that all human traits are peramorphic—some have been interpreted as being paedomorphic. For example, it has been argued that we produce less body hair than other primates and that this is a paedomorphic trait. Yet the body hair of the late human fetus resembles the body hair of other primates postnatally. Some, although not all, elements of the toes and jaw are paedomorphic. However, selection is unlikely to have focused particularly on these paedomorphic characters. The three fundamental aspects of human anatomy that have contributed most to our evolutionary "success" (i.e., those under strong selection pressure) are the anatomical specializations related to bipedalism—locomotion on long, powerful legs supported by large foot bones that develop more robustly through ontogeny than in other primates because of the longer infantile and juvenile periods of high growth rates. The possession of a large brain allowing complex communication arises from the extended juvenile period. This had the added advantage of producing a prolonged period of learning (see Parker, Chap. 14).

Significance of Time and Rate to Tooth Eruption

To test the hypothesis that hominid evolution involved sequential and terminal hypermorphosis, one must assess measures of the time of transitions between growth stages in all hominid species. Estimates of the time of tooth eruption are one useful tool in evaluating whether the ontogenetic stages of later hominids were longer than those of earlier species (see Anemone, Chap. 12, and Kuykendall, Chap. 13). Teeth provide distinct markers of life history and allow categorization of different developmental stages (Beynon and Dean 1988). By using the age at which teeth erupt, we can classify hominid postnatal growth into distinct phases: infantile, juvenile, adolescent, and adult. These phases equate to periods before, during, and after the eruption of permanent teeth, respectively.

Smith (1991) has stressed that, in animals, the time when teeth erupt is crucial for food-processing abilities, with direct consequences on growth of the body as a whole. Timing of tooth eruption also has a close correspondence with other life-history factors, notably gestation period, timing of onset of sexual maturity, and length of life. Consequently, the later that permanent teeth erupt, the longer the gestation period and the later the weaning, the longer the periods of infantile and juvenile dependency, the later the onset of sexual maturity, the larger the brain and body size, and the longer the life span (Fig. 5.3). Thus, analysis of relative ages of tooth eruption in fossil hominids has the potential to provide important information

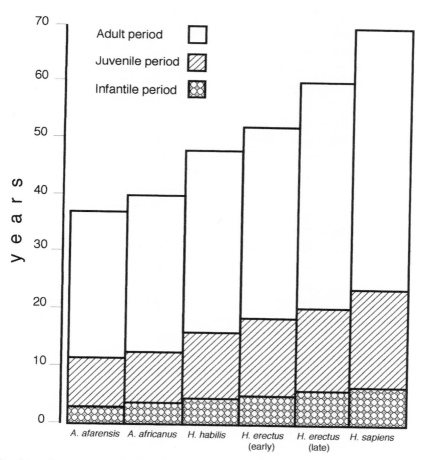

Fig. 5.3. Sequential hypermorphosis in hominid evolution, as shown by predicted temporal delays in the transitions between infantile, juvenile, and adult growth stages. *Source:* Redrawn from Smith 1991.

on all life-history parameters. Dental development is relatively insensitive to fluctuating environmental factors (Smith 1991) and so can act as a good proxy for assessment of the timing of sexual maturity in extinct species. Consequently, the extent of sequential heterochrony between hominid species can be assessed.

One method indicating that species of *Australopithecus* had shorter infantile growth periods than species of *Homo* is to count enamel perikymata. Using this method Bromage and Dean (1985) proposed that the three young australopithecines they analyzed had died just as their first molars were erupting. The analysis suggested death at 3.2–3.3 years, matching closely predictions based on evaluations made from brain capacity. Support for an earlier time of dental eruption in australopithecines, compared with species of *Homo*, comes from analyses by Conroy and Vannier (1991). Beynon and Dean (1988) also concluded that, on the basis of tooth development data, early hominids had shorter growth periods than later hominids, and thus a shorter infancy.

Interestingly, the stretching out of growth periods can be accompanied by paedomorphosis. Such is the case with dental root formation. Dean and Benyon (1991) showed that the rate of root extension in modern humans is about four times slower than that in great apes. Root extension occurs at about 2.5–3.8 μm per day, compared with 13.3 μm per day in great apes (Beynon et al. 1991). The shorter periods of greater growth in earlier hominids, resulting in earlier tooth eruption, indicate that, during hominid evolution, dental root growth is neotenic.

There is a very high correlation between the time of eruption of the first molar tooth and brain weight; consequently, the time of tooth eruption can be reasonably accurately predicted from brain weight (Smith 1991). Thus, one of the smallest living primates, *Cheirogaleus medius*, with a brain weight of 180 g, has one of the earliest emergence times of the first molar of all primates and one of the earliest onsets of maturity, smallest body size, and shortest life span. *Homo sapiens* is at the other end of the scale, having the largest brain of any primate, the latest eruption of the first molar, the latest onset of maturity, one of the largest body sizes, and longest life span. With such correlations it is possible to interpret life-history patterns of extinct hominid species and predict the heterochronic processes that operated. Although such correlations can be quantified on a broad scale, they are sometimes less evident at a narrow scale, as in Upper Pleistocene hominids.

There has been much debate on how closely brain size and life-history parameters can be correlated (see Anemone, Chap. 12; Kuykendall, Chap. 13; and Parker, Chap. 14). Smith (1991), for example, predicted that the first molars erupted in early australopithecines when they were between three and three and a half years old, while these early hominids lived until they were thirty-five to forty years old. In later hominids, such as *Homo habilis* and early forms of *Homo erectus*, the first molar erupted between four and four and a half years of age. Potential life expectancy was about fifty years. In more recent *H. erectus*, living a few hundred thousand years ago, the first molars probably erupted at about five and a half years of age. In "modern" *Homo sapiens* the first molars erupt at about six years of age. Life span is in excess of seventy years. Such estimates have been calculated on the basis of increasing brain size found in hominids (Fig. 5.4), ranging from about 400–500 cm^3 in *Australopithecus afarensis* to 430–480 cm^3 in *Australopithecus africanus*. Brain capacity jumped to 580–750 cm^3 in *H. habilis*, increasing to 900–1,100 cm^3 in later *H. erectus*, and ranging, in 90 percent of cases, from 1,040 to 1,595 cm^3 in modern *H. sapiens* (Gould 1981).

Dissociated Heterochrony and Developmental Trade-offs

Despite the major part played by sequential hypermorphosis, hominid evolution is characterized by a mixture of peramorphic and paedomorphic traits. Extension of faster embryonic, infantile, and juvenile growth rates in structures has produced peramorphic features in the skull and brain, pelvis, lower limbs, descended larynx, and flattened rib cage, as well as larger body size (McHenry 1992). Other traits, however, such as growth of the jaws, dental roots, upper limbs, and lesser digits in the lower limbs, are paedomorphic. Such mosaics of peramorphic and paedomorphic characters arise, in many instances, from developmental trade-offs (McNamara 1997). In the case of hominids, these have overwhelmingly favored peramorphic traits. Perhaps the most significant of these are the trade-offs that compensate for the phylogenetic increase in brain size and complexity.

The apparent change in diet during hominid evolution, from predominantly vegetarian in australopithecines (Blumenschine 1987, 1991; Blumenschine and Cavallo 1992; Stanley 1996) to omnivorous in species of *Homo*, may also have its roots in sequential hypermorphosis. A more than threefold increase in brain volume occurred during hominid evolution in a little over 4 million years (Fig. 5.4). However, growing brain tissue is meta-

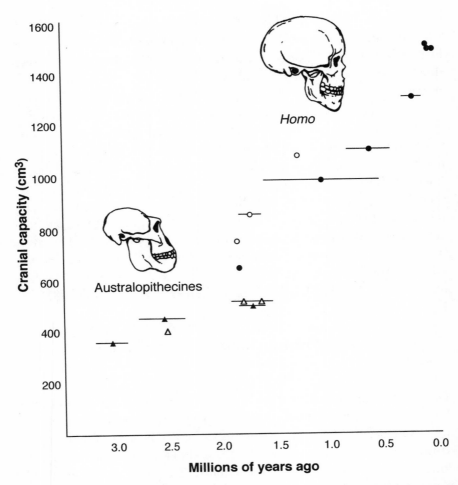

Fig. 5.4. The evolution of hominid cranial capacity. *Triangles* represent species of *Australopithecus;* circles, species of *Homo. Solid symbols* indicate more than one individual per sample; *open symbols* represent only one individual in the sample. *Source:* Data from Falk 1998.

bolically very expensive. Despite using about 17 percent of the adult body's energy, the brain weighs only about 2 percent of the total body weight. The energy to produce more brain tissue may have come from a developmental trade-off.

Aiello and Wheeler (1995) argued that such a developmental trade-off in hominid evolution existed between increased brain size and decreased gut size. In addition to the brain, the splanchnic organs (the liver and gastrointestinal tract) are also metabolically expensive. As a proportion of body size, the gut of *H. sapiens* is very small, being 60 percent of what it should be in terms of body size. As Aiello and Wheeler (1995) pointed out, gut size is highly correlated with diet. Generally, small guts function only with high-quality, easy-to-digest food, such as highly carnivorous diets. Significantly, such diets are very high in the fats needed to myelinate large brains. Large guts, on the other hand, are required to process large amounts of less nutritional vegetative matter. This "gut for brain" trade-off was termed by Aiello and Wheeler the "expensive-tissue hypothesis." The developmental trade-off producing a paedomorphic gut occurred in response to the strong selection pressure for an extension of growth phases that produced a large, more complex brain.

Phylogenetic changes in shape of the hominid rib cage support the view that there was gut reduction during hominid evolution. The rib cage of *A. afarensis* is thought to have been funnel-shaped, wider at the base than at the top, to incorporate a larger gut. *Homo ergaster* was the first hominid to have a more barrel-shaped rib cage, implying a smaller gut and a change in diet. Accompanying this change in diet was the development of the use of tools, which, to a large extent, were probably initially used in butchering. This manifestation of a more complex behavior by species of *Homo*, incorporating tool use and, in all likelihood, complex methods of catching food (Blumenschine 1987, 1991; Blumenschine and Cavallo 1992), was interconnected with the peramorphic evolution of a larger brain.

Gibson (1990) and Parker and McKinney (1999) pointed out that the corollary to the evolution of a large, complex brain was the evolution of a correspondingly wider array of behaviors. As they emphasized, the peramorphic development of the brain has produced greater growth of the cerebral cortex—the seat of conscious thought, memory, intelligence, and speech. The longer this part of the brain has to develop, the more complex its cognitive capabilities. Delay in cessation of embryonic growth is particularly crucial here because it is during this period that many of the cortical brain cells are formed. Thus, at birth, the brain has grown to about 25 percent of its adult weight. By five years of age, it has grown to 90 percent, then to 95 percent at ten years (Tanner 1992). Extending infantile and juvenile phases of growth results in an even more complex brain, for it is at this time that dendritic growth of the neurons and synaptogenesis are occurring. Of all primate species *H. sapiens* has easily the greatest number of interconnecting neurons. Moreover, this longer period of preadult growth also allows more time for the generation of synapses, myelination, and growth of blood supply. At the end of each growth phase, the human brain is larger and more complex than that of any other primate, living or extinct (Gibson 1990).

Of critical importance in the maturation of memory, intelligence, and language skills is myelination, greater development of synapses, and the time for learned behavior to selectively utilize certain pathways, resulting in synaptic death of unused or unreinforced neural pathways (Gibson 1990). The duration of myelination varies between species. For instance, in rhesus monkeys it persists until about 3.5 years. However, in humans it extends well past adolescence (Gibson 1991). Similarly, the cessation of dendritic growth occurs at a much later age in humans than in any other primate. The sequential delay in transition between growth phases in later hominids has resulted in a longer childhood, the period when learning occurs most rapidly. The physical effect of this has been the evolution in humans of a larger, more complexly interconnected neocortex with the ability to store greater amounts of information, articulate complex language, construct intricate tools, and create complex social systems.

Parker (1994), Parker and McKinney (1999), and Parker (Chap. 14, this volume) point out that analysis of patterns of cognitive development has shown that, compared with other primates, humans pass through more developmental stages and spend longer in each stage. Thus, adult great apes have a range of cognitive abilities similar to those in two- to four-year-old humans. Adult cebus monkeys attain cognitive abilities like those of a two-year-old human child. On the basis of the complexity of stone tool development, Parker and Gibson (1979) suggested that, through the 4 million years or so of hominid evolution, cognitive ability became more complex as life-history stages became progessively longer. Australopithecine adults, they argued, attained a level of cognitive development equivalent to that of a two- to three-year-old *H. sapiens;* adults of early species of *Homo,* to a five- to six-year-old; adult *H. erectus,* to a six- to eight-year-old; and adult early *H. sapiens* (or *Homo heidelbergensis* and *Homo neanderthalensis*), to a ten- to twelve-year-old. Body size, brain weight, timing of tooth eruption, and maturation times all indicate that the evolution of cognitive abilities in hominids was, like physical development, one of a steady stretching out of mental and intellectual capacities.

Acknowledgments

I thank Nancy Minugh-Purvis for her constructive comments and Danielle West for assistance with some of the illustrations.

References

Aiello, L.C., and Wheeler, P. 1995. The expensive-tissue hypothesis. *Current Anthropology* 36:199–221.

Alberch, P., Gould, S.J., Oster, G.F., and Wake, D.B. 1979. Size and shape in ontogeny and phylogeny. *Paleobiology* 5:296–317.

Berge, C. 1996. The evolution and growth of the hominid pelvis: a preliminary thin-late spline study of ilium shape. In L.F. Marcus, M. Corti, A. Loy, G.J.P. Naylor, and D.E. Slice (eds.), *Advances in Morphometrics,* 441–448. New York: Plenum Press.

———. 1998. Heterochronic processes in human evolution: an ontogenetic analysis of the hominid pelvis. *American Journal of Physical Anthropology* 105:441–459.

Beynon, A.D., and Dean, M.C. 1988. Distinct dental development patterns in early fossil hominids. *Nature* 335:509–514.

Beynon, A.D., Dean, M.C., and Reid, D.J. 1991. Histological study on the chronology of the developing dentition in gorilla and orangutan. *American Journal of Physical Anthropology* 86:189–203.

Blumenschine, R.J. 1987. Characteristics of an early hominid scavenging niche. *Current Anthropology* 28:383–407.

———. 1991. Breakfast at Olorgesailie: the natural history approach to early stone age archaeology. *Journal of Evolution* 21:307–327.

Blumenschine, R.J., and Cavallo, J.A. 1992. Scavenging and human evolution. *Scientific American* 265(10):90–96.

Bolk, L. 1926. *Das Problem der Menschwerdung.* Jena, Germany: Gustav Fischer.

Bromage, T.G., and Dean, M.C. 1985. Re-evaluation of the age at death of immature fossil hominids. *Nature* 317:525–527.

Conroy, G.C., and Vannier, M.W. 1991. Dental development in South African australopithecines: Part II. Dental stage assessment. *American Journal of Physical Anthrolopogy* 86:137–156.

De Beer, G.R. 1930. *Embryology and Evolution.* Oxford: Clarendon Press.

Dean, M.C., and Beynon, A.D. 1991. Histological reconstruction of crown formation times and initial root formation times in a modern human child. *American Journal of Physical Anthrolopogy* 86:215–228.

Falk, D. 1998. Brain evolution: looks can be deceiving. *Science* 280:1714.

Fiorello, C.V., and German, R.Z. 1997. Heterochrony within species: craniofacial growth in giant, standard, and dwarf rabbits. *Evolution* 51:250–261.

Gibson, K.R. 1990. New perspectives on insights and intelligence: brain size and the emergence of hierarchical mental constructional skills. In K.R. Gibson and A. Petersen (eds.), *"Language" and Intelligence in Monkeys and Apes,* 97–128. Cambridge: Cambridge University Press.

———. 1991. Myelination and behavioral development: a comparative perspective on questions of neoteny, altriciality and intelligence. In K.R. Gibson and A.C. Petersen (eds.), *Brain Maturation and Cognitive Development,* 29–64. New York: De Gruyter.

Godfrey, L.R., and Sutherland, M.R. 1996. Paradox of peramorphic paedomorphosis: heterochrony and human evolution. *American Journal of Physical Anthropology* 99:17–42.

Gould, S.J. 1977. *Ontogeny and Phylogeny.* Cambridge: Belknap Press of Harvard University Press.

―――. 1981. *The Mismeasure of Man.* New York: Norton.

Klingenberg, C.P. 1998. Heterochrony and allometry: the analysis of evolutionary change in ontogeny. *Biological Reviews* 73:79–123.

Kollman, J. 1885. Das Ueberwintern von europäischen Frosch- und Tritonlarven und die Umwandlung des mexikanischen Axolotl. *Verhandlungen der Naturforschenden Gesellschaft in Basel* 7:387–398.

McHenry, H.M. 1992. Body size and proportions in early hominids. *American Journal of Physical Anthropology* 87:407–431.

McKinney, M.L. 1988. Heterochrony in evolution: an overview. In M.L. McKinney (ed.), *Heterochrony in Evolution: A Multidisciplinary Approach*, 327–340. New York: Plenum Press.

―――. 1998. The juvenilized ape myth—our overdeveloped brain. *Bioscience* 48:109–116.

McKinney, M.L., and McNamara, K.J. 1991. *Heterochrony: The Evolution of Ontogeny.* New York: Plenum Press.

McNamara, K.J. 1983. Progenesis in trilobites. *Special Papers in Palaeontology* 30:59–68.

―――. 1986. A guide to the nomenclature of heterochrony. *Journal of Paleontology* 60:4–13.

―――. 1988. The abundance of heterochrony in the fossil record. In M.L. McKinney (ed.), *Heterochrony in Evolution: A Multidisciplinary Approach*, 287–325. New York: Plenum Press.

―――. 1997. *Shapes of Time: The Evolution of Growth and Development.* Baltimore: Johns Hopkins University Press.

Montagu, A. 1981. *Growing Young.* New York: McGraw Hill.

Parker, S.T. 1994. Using cladisitic analysis of comparative data to reconstruct the evolution of cognitive development in hominids. Paper presented at the Animal Behavior Society Meetings Symposium on Phylogenetic Comparative Methods, Seattle, July 1994.

Parker, S.T., and Gibson, K.R. 1979. A developmental model for the evolution of language and intelligence in early hominids. *Behavioral and Brain Science* 2:367–408.

Parker, S.T., and McKinney, M.L. 1999. *Origins of Intelligence: The Evolution of Cognitive Development in Monkeys, Apes, and Humans.* Baltimore: Johns Hopkins University Press.

Reilly, S.M., Wiley, E.O., and Meinhardt, D.J. 1997. An integrative approach to heterochrony: the distinction between interspecific and intraspecific phenomena. *Biological Journal of the Linnean Society* 60:119–143.

Richardson, M.K. 1995. Heterochrony and the phylotypic period. *Developmental Biology* 172:412–421.

Shea, B.T. 1989. Heterochrony in human evolution: the case for human neoteny. *Yearbook of Physical Anthropology* 32:69–101.

Smith, B.H. 1991. Dental development and the evolution of life history in Hominidae. *American Journal of Physical Anthropology* 86:157–174.

Stanley, S.M. 1996. *Children of the Ice Age: How a Global Catastrophe Allowed Humans to Evolve.* New York: Harmony Books.

Tanner, J.M. 1992. Human growth and development. In S. Jones, R. Martin, and D. Pilbeam (eds.), *The Cambridge Encyclopedia of Human Evolution.* Cambridge: Cambridge University Press.

Verhulst, J. 1993. Lois Bolk revisited II—retardation, hypermorphosis and body proportions of humans. *Medical Hypotheses* 41:100–114.

Vrba, E.S. 1994. An hypothesis of early hominid heterochrony in response to climatic cooling. In R.S. Corruccini and R.L. Ciochon (eds.), *Integrative Paths to the Past: Paleoanthropological Advances in Honor of F. Clark Howell,* 345–376. Englewood Cliffs, N.J.: Prentice Hall.

———. 1998. Multiphasic growth models and the evolution of prolonged growth exemplified by human brain evolution. *Journal of Theoretical Biology* 190:227–239.

Wolpert, L. 1991. *The Triumph of the Embryo.* Oxford: Oxford University Press.

Chapter 6

Animal Domestication and Heterochronic Speciation

The Role of Thyroid Hormone

Susan J. Crockford

Interpreting speciation events from the paleontological record is as difficult in hominid lineages as in other vertebrate and invertebrate families (Jones et al. 1992). New species, the essential units of evolution, are the landmarks paleontologists use to assess changes over evolutionary time (Otte and Endler 1989). Eldridge and Gould's (1972) concept of punctuated equilibrium, for example, was presented more than twenty-five years ago to explain distinct patterns of species in the paleontological record (*punctuated equilibrium* refers to recurring bouts of very rapid speciation interspersed with periods of relative stasis). Despite indications that punctuated equilibrium is a real and significant phenomenon (Gould and Eldridge 1993; Richardson 1995), we have not yet been offered a comprehensive explanation of how such an evolutionary pattern could be attained.

It has been successfully argued that heterochrony is the most probable process by which rapid speciation changes could occur and that, as a consequence, heterochrony may be the most significant process in evolution (Alberch 1991). Heterochrony is much more common as a speciation process than is generally appreciated, with differences resulting from changes in developmental rates or timing recognized in many lineages, including hominids (McKinney and McNamara 1991). Heterochrony may affect the initiation and cessation of growth stages and implement changes in fetal and postnatal growth rates of the ancestral species to various degrees, so that heterochronic changes make possible a wide variety of shape and size differences in descendant populations. However, the precise biological mechanisms responsible for the initiation and implementation of heterochronic changes have not been determined (Voss and Shaffer 1997). Unfortunately, the inability to describe these aspects of speciation leaves serious gaps in our understanding of evolution (Mayr 1994).

We have been unable to determine the process that controls rapid speciation changes because we lack an appropriate paradigm with a suitable, testable hypothesis. The population genetic models that have served well for explaining other aspects of evolutionary change (e.g., Rice and Hostert 1993) are not adequate for addressing heterochrony because the interactions between genotypes and phenotypes that affect developmental processes are not linear, one-to-one relationships but complex webs of interdependence (Alberch 1991).

In this chapter, I present a novel approach to investigating the role in evolution that is played by heterochrony and rapid speciation by taking a critical and in-depth look at the process we call *domestication*. Prompted by evidence that domestic mammals include some of our best-known examples of heterochronic change, I argue that domesticates do not always result from one continuous process initiated by humans. Instead, the process very often comprises two distinct parts, the first of which (protodomestication) is initiated by animals themselves. Furthermore, animals that have undergone protodomestication share certain morphological, physiological, and behavioral similarities indicating that thyroid hormone played a pivotal role in mediating the developmental changes that occurred. Last, I compare the heterochronic changes that characterize domestic taxa with heterochronic speciation in wild taxa.

The model presented here to explain both domestication and speciation is based on the testable hypothesis that particular thyroid hormone phenotypes within species are often the actual targets of natural selection. Because of the pleiotropic influences that thyroid hormone exerts on embryonic and postnatal growth, many distinctive morphological and behavioral traits characteristic of new species may simply be inevitable consequences of selection for certain physiological phenotypes of thyroid hormone metabolism. The end result of reassessing domestication in this evolutionary context is an elegant paradigm for heterochronic speciation that explains the truly dynamic relationship among individual variation, adaptation, and speciation in all vertebrate taxa, including humans.

Reassessing Domestication

Protodomestication versus Classic Domestication

Any discussion of domestication requires definition of both domesticates and species. Considering domesticates first, there seems little consensus among authors in defining domestic animals as a group (O'Connor 1997). For example, Reed (1984) defines domesticates simply, as those animals whose breeding is, or can be, controlled by humans. Clutton-Brock (1992a) defines a domestic animal as one that has been bred in captivity for purposes of economic profit to a community that maintains a mastery over its breeding, organization of territory, and food and insists that animals cannot be domesticated unless they are owned (1992b). Isaac (1970, 20) lists five criteria that define fully domestic taxa:

1. the animal is valued and there are clear purposes for which it is kept;
2. the animal's breeding is subject to human control;
3. the animal's survival depends, whether voluntarily or not, upon man;
4. the animal's behavior has changed as a result of domestication;
5. morphological characteristics in individuals of the domestic species occur rarely, if at all, in the wild.

However, all traditional definitions of domestication either state explicitly or imply that domesticates are derived from animals deliberately removed from the wild.

The morphological, physiological, and behavioral differences between many domestic mammals and their wild ancestors (alluded to by Isaac's definition above) are well documented and amazingly parallel, even between unrelated lineages and ecologically diverse taxa (Hemmer 1990). However, as domesticates do not conform to any single, all-encompassing definition, the biological changes generally associated with domestication (Fig. 6.1) do not apply unambiguously to every taxon. Nevertheless, the biological differences between wild taxa and their natural domestic descendants that are discussed in this chapter can be clearly established for dogs, pigs, cattle, sheep, goats, and the lesser-known Asian bovids—water buffalo and mithan. These examples are adequate for illustrating the range of ecological and physical types of animals that have undergone similar biological change as a result of the protodomestication process.

The major distinctive differences between wild ancestors and many of

ANCESTRAL WILD SPECIES

↓

1. Overall body size reduction
2. Shortening of the facial bones of the skull (accompanied by changes in dentition, and horn size reduction in those species with horns)
3. Lowered age of sexual maturity
4. Increased docility
5. Increased fecundity (principally through larger litters)
6. Changes in dominant color alleles (such as piebald & non-agouti)
7. Changes in reproductive timing (frequency &/or seasonality)

↓

DESCENDENT DOMESTIC SPECIES (NATURAL DOMESTICATES)

Fig. 6.1.   A summary of the morphological, physiological, and behavioral changes associated with protodomestication in mammals.

their domestic descendants have been shown by several studies to be the result of paedomorphosis (Morey 1992; Wayne 1986). Changes in developmental timing and growth rates are implicated in many of the morphological traits common to all domesticates because of the very nature of the differences: smaller overall size, shortened snout, juvenile behavior. Other traits (such as increased fecundity, docile behavior, piebaldness, and polyestrousness), as discussed in detail below, seem to be associated consequences of paedomorphosis rather than paedomorphic traits themselves.

In an attempt to explain how these biological changes could have occurred, I suggest that the process traditionally called domestication is actually composed of two distinct parts. I propose that the term *protodomestication* be used for the natural speciation process whereby certain wild ancestors generate descendants with modified biological features (primitive natural domesticates), a process initiated by the animals themselves. *Classic domestication* is the term I propose be used to describe the processes of conscious and subconscious human selection (working in concert with natural selection) that modify any captive population, whether those animals are products of prior protodomestication or individuals deliberately removed from the wild. Although these definitions contrast sharply with the traditional view of domestication (which collapses the two stages together for all taxa), I argue that some animals (natural domesticates) have undergone both protodomestication and classic domestication, while others are products of classic domestication only (classic domesticates).

I reiterate a statement made by Morey (1994): the fact that humans have been able to control and manipulate domesticates so thoroughly and with such dramatic success over the last few thousand years does not prove that the process began with the deliberate intent to do so. Not all animals currently under human control necessarily got there by the same route; therefore, all domesticated animals should not be treated as equivalent evolutionary entities, even though they may now fill equivalent cultural roles. There are serious objections to the idea that humans are responsible for deliberately initiating heterochronic changes in all domesticates, which Budiansky (1992) and O'Connor (1997) address in detail. An exhaustive discourse on the subject can be avoided, however, simply by acknowledging that the most significant objection to the traditional explanations of domestication is that they are not testable. Without refutability, none of these explanations for how domesticates were initially produced can ever be a scientific hypothesis. A more useful approach is to examine domestication in biological terms, as an evolutionary process compared and contrasted with speciation.

Domestication as Heterochronic Speciation

The model illustrating domestication as a biological process (Fig. 6.2) defines both incipient species and incipient domesticates as equivalent population entities that invade new habitats (stage 1). However, there are differences in the nature of the new environments they colonize. For incipient domesticates, the habitat is anthropogenic, a term that describes a localized set of environmental conditions created by the physical effects of permanent or semipermanent human settlements (Tchernov 1993). An anthropogenic environment, however, is also a habitat dominated by the continuous presence or proximity of people. In contrast, for incipient species the new environment may be either a previously unoccupied adjacent niche or a niche newly created by environmental or climatic change.

In both cases, the new habitat offers resources unavailable or scarce in the original or source territory (such as food and breeding sites), making it

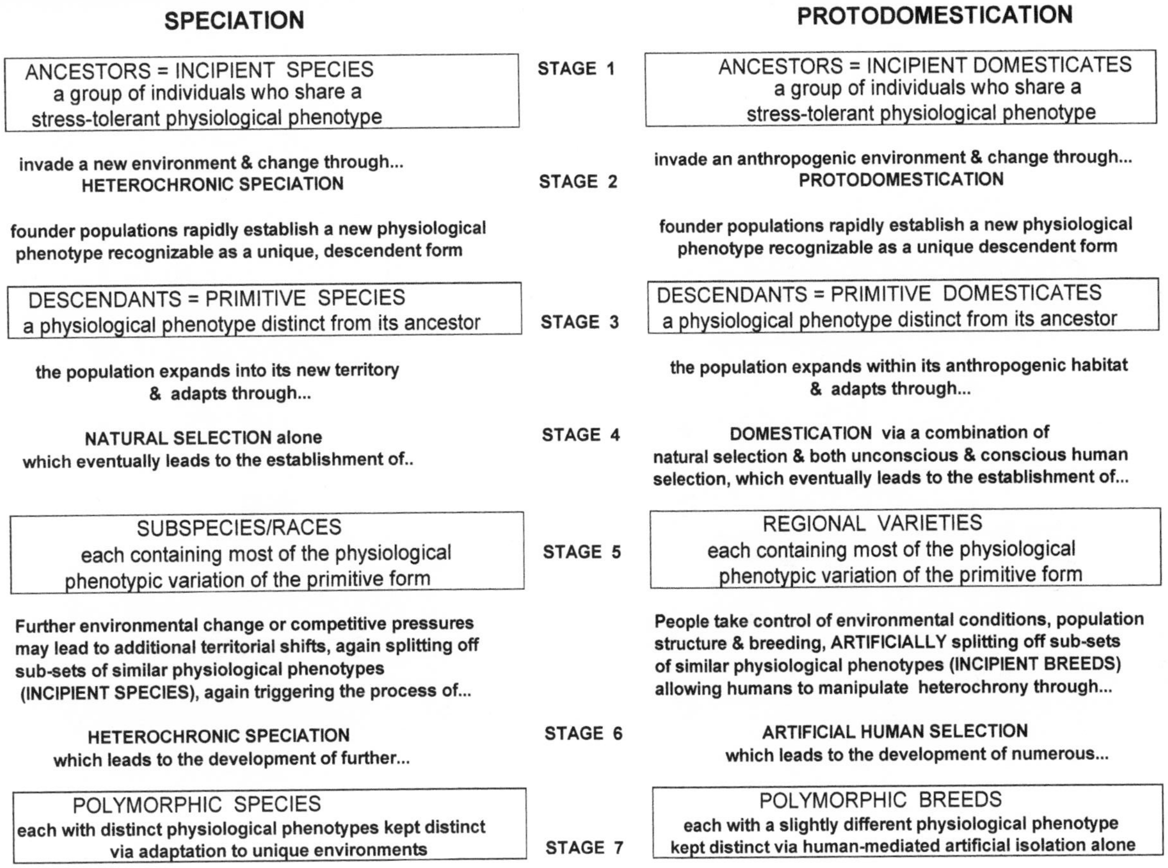

SPECIATION

STAGE 1

ANCESTORS = INCIPIENT SPECIES
a group of individuals who share a
stress-tolerant physiological phenotype

STAGE 2

invade a new environment & change through...
HETEROCHRONIC SPECIATION

founder populations rapidly establish a new physiological
phenotype recognizable as a unique, descendent form

STAGE 3

DESCENDANTS = PRIMITIVE SPECIES
a physiological phenotype distinct from its ancestor

the population expands into its new territory
& adapts through...

STAGE 4

NATURAL SELECTION alone
which eventually leads to the establishment of...

STAGE 5

SUBSPECIES/RACES
each containing most of the physiological
phenotypic variation of the primitive form

Further environmental change or competitive pressures
may lead to additional territorial shifts, again splitting off
sub-sets of similar physiological phenotypes
(INCIPIENT SPECIES), again triggering the process of...

STAGE 6

HETEROCHRONIC SPECIATION
which leads to the development of further...

STAGE 7

POLYMORPHIC SPECIES
each with distinct physiological phenotypes kept distinct
via adaptation to unique environments

PROTODOMESTICATION

STAGE 1

ANCESTORS = INCIPIENT DOMESTICATES
a group of individuals who share a
stress-tolerant physiological phenotype

STAGE 2

invade an anthropogenic environment & change through...
PROTODOMESTICATION

founder populations rapidly establish a new physiological
phenotype recognizable as a unique descendent form

STAGE 3

DESCENDANTS = PRIMITIVE DOMESTICATES
a physiological phenotype distinct from its ancestor

the population expands within its anthropogenic habitat
& adapts through...

STAGE 4

DOMESTICATION via a combination of
natural selection & both unconscious & conscious human
selection, which eventually leads to the establishment of...

STAGE 5

REGIONAL VARIETIES
each containing most of the physiological
phenotypic variation of the primitive form

People take control of environmental conditions, population
structure & breeding, ARTIFICIALLY splitting off sub-sets
of similar physiological phenotypes (INCIPIENT BREEDS)
allowing humans to manipulate heterochrony through...

STAGE 6

ARTIFICIAL HUMAN SELECTION
which leads to the development of numerous...

STAGE 7

POLYMORPHIC BREEDS
each with a slightly different physiological phenotype
kept distinct via human-mediated artificial isolation alone

Fig. 6.2.   The model: protodomestication as a speciation process. New environments of stage 1 may exist adjacent to occupied ones or may be created by climatic change or human activity. At all stages, *physiological phenotype* refers to all related expressions of a physiological genotype: morphological, behavioral, biochemical, reproductive, and so forth.

highly attractive. In both cases, individuals within the original population who have the highest physiological tolerance to stress (i.e., possess a particular physiological phenotype) are those most likely to invade a new territory. The division of the source population during colonization of a new habitat is thus distinctly nonrandom.

The effects of inbreeding and genetic drift within a small, isolated population of only stress-tolerant individuals would rapidly establish a new physiological equilibrium for the colonizing group. The establishment of a new physiological phenotype invariably precipitates changes in morphological, behavioral, and reproductive traits in descendant populations because of the intimate connection between the adrenal stress-response system and thyroid hormone metabolism. Thyroid hormone is also strongly implicated as an important control mechanism for developmental rates and timing. Descendants of physiologically similar population subsets that invade new territories are recognizable as different from their ancestors within relatively few generations because the new physiological phenotype established by the colonizing group has associated morphological, behavioral, and reproductive manifestations.

Viewed from this perspective, it is apparent that the initial populations of ancestors are equivalent subsets of wild populations (stage 1) and that the selection mechanism is the same in both heterochronic speciation and protodomestication (stage 2). The descendants of both processes are thus equivalent entities (stage 3) and must therefore be equally real species. This brings us to the second term that needs clarification.

Defining species is a topic that generates hot debate (Arnold 1997; Otte and Endler 1989). However, for the purposes of this discussion I have chosen to use Templeton's (1989, 12) cohesion species concept, which defines a species as "the most inclusive population of individuals having the potential for phenotypic cohesion through intrinsic cohesion mechanisms." Templeton's definition of a species differs somewhat from Mayr's widely accepted biological species concept, in that it emphasizes the mechanisms that induce reproductive cohesiveness within discrete populations rather than stressing mechanisms that prevent reproduction between them. Templeton's definition ascribes species status to members of syngameons and to genera composed of closely related species that can and do produce hybrids but that nevertheless each possess distinct morphological, ecological, genetic, and evolutionary characteristics. Such a situation exists for members of the genus *Canis,* for example (Fox 1978).

Returning to the comparison of speciation and protodomestication shown in Figure 6.2, it is apparent that stage 4 is the first point of major depar-ture between the two processes, the point at which adaptation via natural selection occurs. Although some might argue that deliberate human selec-tion played a significant role in the changes subsequent to protodomes-tication that affected primitive natural domesticates during stage 4, the prevailing view now concedes that natural selection was probably more sig-nificant. If human selection did occur, it was probably subconscious (Zo-hary et al. 1998). Subconscious selection refers to, among other examples, allowing individuals determined to escape to do so through inadequate fencing (thus removing them from the breeding pool), choosing to slaugh-ter individuals with intractable temperaments, and sparing from slaughter individuals that are good mothers or produce large litters.

Adaptation of primitive domesticates to the conditions within their an-thropogenic environment proceeds via the same mechanisms of natural selection that adapt wild primitive species to their new environments (Dar-win 1859). However, anthropogenic environments do possess some inher-ent properties that may not be present in the wild, including the constant physiological stress of close association with humans, the constant im-munological stress of very close association with conspecifics, the inbreed-ing stress imposed by limited choices of mates, and the nutritional stress imposed by severely limited food resources, especially for confined popu-lations (Hafez 1968). Therefore, the selective forces operating on primitive natural domesticates will be different from those acting on wild primitive species, even if the actual process of adaptation is the same.

It is not until stage 6 that the greatest difference in the history of the two types of descendant species occurs. This is the point at which humans as-sume absolute control over primitive domesticates and, through deliberate artificial selection, develop distinctly polymorphic breeds. Humans are able to discern and then artificially isolate subsets of individuals with phys-iologically similar phenotypes (incipient breeds) that share behavioral, morphological, and reproductive traits (Coppinger and Schneider 1995). Artificial isolation of specific physiological phenotypes and subsequent in-breeding and selection allow humans to initiate and perpetuate hetero-chronic processes decoupled from nature. This has resulted in a truly as-tonishing number of domestic variants of some animals, including more than four hundred recognized breeds of dogs and almost as many pig breeds (Alderson 1978; Darwin 1868; Wilcox and Walkowicz 1989). However, the polymorphic forms resulting from human manipulation stay distinct only if artificial isolation is maintained (stage 7). In contrast, the polymorphic wild forms of stage 7 are true species that stay distinct because they are uniquely adapted to specific habitats and have developed reproductive cohesion mechanisms.

As the model summarized in Figure 6.2 emphasizes, the heterochronic changes associated with protodomestication are significant consequences of the evolutionary process involved. Animals that do not manifest the distinctive biological changes associated with protodomestication, such as Asian elephants (Olivier 1984) and reindeer (Skjenneberg 1984), are more appropriately referred to as *managed species*. Managed species also include those classic domesticates for which historical records indicate that humans deliberately removed animals from the wild and subsequently maintained populations in captivity, such as golden hamsters (Clutton-Brock 1992b).

The morphological changes to animals produced through protodomestication include significant modification of the skeleton. Consequently, archeological remains are an essential source of evidence for unraveling the history of natural domestic species (Crabtree 1993). However, the archeological record of skeletal remains does not show the clear intermediate stages between wild and domestic forms we are led to expect from the cultural explanation of domestication, which assumes that changes are gradual (Clutton-Brock 1992b). Admittedly, there is controversy over some data (e.g., Dayan 1994). However, the absence of clear intermediate forms in all taxa suggests that the physical changes associated with protodomestication were not gradual but abrupt, implying that the process of protodomestication must have been relatively rapid. There are nondomestic populations in which rapid speciation rates seem to have occurred. For example, the distinct forms that make up "species flocks" of fish in some freshwater lakes could have developed in as little as two hundred years (Owen et al. 1990). The mechanisms that initiate and implement the process of such rapid speciation clearly must be identified if we are to understand how evolution actually works.

Small mutations within regulatory genes that operate during embryonic development have been suggested as the most probable mechanism by which large morphological changes could occur without substantial genetic change (Richardson 1995). A suite of regulatory genes known as the homeotic complex (or *Hox* genes) have been identified as the actual sites of developmental control. For example, *Hox* genes direct such early embryonic developmental functions as the diversification of vertebrae along the central body axis and digit formation in the limb of vertebrates (Krumlauf 1994).

Although control over large morphological changes (some of which might be heterochronic) might be attributed to *Hox*-type genes, we still do not know exactly how such change is either initiated or coordinated, although a hormonal mechanism has long been suspected (De Pablo 1993; McKinney 1998). Thyroid hormone has been implicated in heterochronic speciation in several taxa (Härlid and Arnason 1999; Jennings and Hanken 1998), and the adaptive nature of thyroid hormone's correlation to heterochrony has been demonstrated (Reilly 1994). In addition, thyroid hormone is essential for early embryonic development (e.g., Pavan et al. 1995). Surprisingly, the broader evolutionary significance of the association of heterochrony with thyroid hormone has not been actively pursued until now.

Experimental Domestication

An elegant example of the intricate biochemical, physiological, behavioral, and developmental interactions that are pertinent to this discussion is provided by a series of selection experiments on foxes that began in the 1950s at the Siberian Institute of Cytology and Genetics in Novosibirsk under D. K. Belyaev (Belyaev 1979) and continues under his colleague, L. N. Trut (Trut and Osadchuk 1997). The experiments began with a large population of farmed silver foxes (a naturally occurring black morph of the red fox, *Vulpes vulpes*) that retained all characteristics of the wild form; these foxes were seasonally monoestrous with an annual molt and were generally timid of people.

The researchers assessed foxes from several fur farms and selected individuals that demonstrated noticeably less "fearful" behavior toward people (about 20% of the total) as an experimental population. When approached, these animals reacted with limited curiosity rather than aggression or fear, although they still could not be handled. Females of the selected population of less fearful animals turned out to be the earliest breeders of the original total population, which suggests that there was an existing polymorphism for timing of estrus within the original population that was correlated with the selected behavior.

After several generations of breeding and selecting for nonfearful behavior, the estrous cycle and timing of the annual molt of many females had receded in the season by several months. As the experiment continued, estrous and molt receded further still, until several females were experiencing two estrous cycles annually. After twenty generations, some females were able to produce two litters per year. This diestrous pattern of reproduction (which occurs in most domestic dogs) was found to be inherited as an incompletely dominant trait. The really surprising result was that, after twenty generations, novel traits suddenly appeared: a curled tail, drooping ears, a coat color pattern of distinctive brown markings, and a classic white piebald pattern. All of these traits, once they had appeared, inherited in dominant fashion (in contrast, most rare colors and color patterns in mink are recessive traits that require intense inbreeding to generate consistent expression).

Physiologically, the animals from this last generation had smaller adrenal glands associated with lessened secretion of corticosteroid hormones and increased levels of serotonin. Females had higher levels of progesterone and estradiol in early pregnancy, accompanied by higher fertility than the original group. The pineal glands of these animals were also smaller. Subsequent osteometric analysis of selected foxes found changes in cranial conformation (Trut et al. 1991). Belyaev described the animals with these novel morphological and physiological traits as having remarkably "dog-like" behavior: they barked and were quite unafraid of people. Belyaev concluded that something in the selection for nonfearful behavior was not only causing paedomorphic changes in both morphology and behavior but also disrupting the normally constrained timing of sexual reproduction. Subsequent research indicated that the changes were not caused by selection for particular structural genes or by spontaneous mutations, leaving the investigators somewhat baffled by the results (Trut 1999).

Several similar experiments were carried out in other countries during this same period (Hemmer 1990; Keeler 1975). Although these researchers made valiant attempts to unravel the physiological basis for domestication changes, they were essentially doomed to fail because they asked their questions too soon. Genetic and biochemical knowledge essential for interpreting the results of their experiments have only become available, piece by piece, within the last ten years. Coppinger and Schneider recently (1995, 41) assessed Belyaev's experiments in their discussion of neoteny in relation to dog evolution. As they stated, "Belyaev thought that selection for tameness destabilized the genome in such a way as to create evolutionary novelties. As we learn more about gene action and biochemistry, we find that he was probably not very far from the truth."

Thyroid Hormone and Speciation

General Thyroid Metabolism

Complex interrelationships within the endocrine system (Fig. 6.3) control all of the physiological traits involved in protodomestication changes (see Fig. 6.1). The common denominator of these hormonal influences is thyroid hormone. Thyroid hormone is perhaps best known for its function in

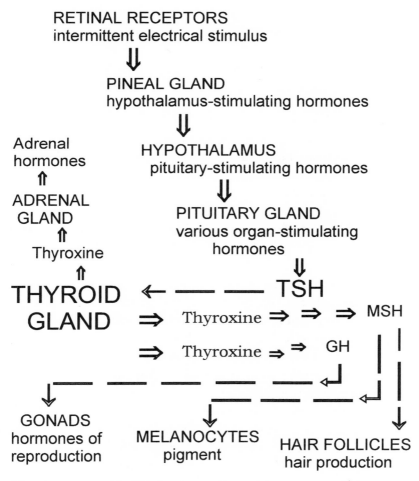

Fig. 6.3. A summary of endocrine interactions pertinent to protodomestication. *MSH,* melanocyte-stimulating hormone; *TSH,* thyroid-stimulating hormone; *GH,* growth hormone.

the maintenance of metabolism, for its role in the control of metamorphosis in amphibians and fish (Dickhoff 1993; Voss and Shaffer 1997), and for its effects on mammalian hibernation (Tomasi et al. 1998). However, the total range of influence of thyroid hormone on both embryonic development and adult physiology in all vertebrates is truly staggering. Thyroid hormone is essential (among other functions) for early embryonic cell migration and differentiation, both embryonic and postnatal growth and brain development, hair production, adrenal gland function, skin and hair pigment production, and the development and function of the gonads (Dawson et al. 1994; Hadley 1984; McNabb and King 1993). In essence, all of the traits that are known to change as a result of protodomestication are controlled by thyroid hormone.

The thyroid gland is ubiquitous among vertebrates; it arises early in the evolutionary sequence (in chordates) from endoderm of the cephalic portion of the alimentary canal of the embryo. The thyroid gland (or glands, where it has become a paired organ) is composed of numerous follicles that store the protein thyroglobulin, which acts as a substrate for tyrosine iodination. Iodine-bound tyrosine derivatives are stored in the thyroid gland until thyroid-stimulating hormone (TSH) triggers thyroglobulin to be hydrolyzed and released as thyroxine (T4). Some thyroxine is deiodinized to triiodothyronine (T3), which has a shorter half-life but is more metabolically active than T4. There is some indication that T3 may be more active in gene regulation and T4 more important in development (Brent et al. 1991), but as concentrations of each are clearly interdependent, they are often referred to collectively as thyroxine.

As a gene regulator, T3 binds to both nuclear and mitochondrial thyroid hormone receptors to form a ligand-receptor complex. This T3-receptor complex then binds to a specific DNA sequence located within the promoter region (the thyroid responsive element) of a number of genes, triggering the transcription of gene products within cell nuclei and mitochondria (Hadley 1984; Koibuchi et al. 1996). Thus, thyroid hormone, in its T3 state, is able to influence the transcription of a wide variety of different genes, including those involved in the synthesis of lung surfactants and pituitary growth hormone in rats (Oppenheimer 1979).

Thyroid hormones bind to several plasma proteins for circulation in the bloodstream. In humans and mice, these binding proteins are thyroxine-binding globulin (TBG), serum albumin, and transthyretin—a protein that also transports retinoic acid, a vitamin A derivative (Dulbecco 1991). The existence of a common carrier molecule for both thyroxine and retinoic acid may be significant to functions ascribed to both, since the molecular structure of thyroxine and retinoic acid is especially similar in the region of their nuclear receptor DNA binding sites (Morita et al. 1990).

Hormones released from the hypothalamus induce the production of TSH by the pituitary gland (Fig. 6.3), which initiates thyroxine production. TSH is produced incrementally because the stimulation that the pituitary gland receives from hypothalamic hormones is intermittent. Thus, in response to TSH stimulation, thyroxine is secreted into the bloodstream in a distinctly pulsatile manner. The precise frequency and amplitude of the thyroxine pulses change both seasonally and daily according to other physiological demands (e.g., Tomasi and Mitchell 1994). In most animals, thyroxine levels are highest during the middle of the day and lowest at night. Fluctuations also occur with age, reproductive stage (especially in females), psychological state, and general health. Since levels of thyroxine are known to fluctuate relative to the many variables described above, static (one-time) measurements of thyroxine and thyroid-binding protein concentrations often reported in the literature cannot be reliably compared.

Tests that measure thyroxine turnover rates are also commonly done, however, and these values are probably comparable. For example, the average half-life of T4 is 13 hours in dogs, 16.6 hours in cats, and 6.8 days in humans (Kaptein et al. 1994). Some differences between animal breeds have also been demonstrated. For example, the half-life of T4 in the beagle (a breed that is typically diestrous) is twice that in the basenji (a breed that is typically monoestrous) (Nunez et al. 1970). Although broad comparisons within and between species for measured values of T3 and T4 obtained by different methods are problematic, it is nevertheless apparent that significant differences do occur.

Few studies have sampled thyroxine or TSH levels often enough to determine the normal daily fluctuating pattern of thyroxine production for a species. Those studies that have been done suggest marked thyroid profile differences between species. Recently, for example, Gancedo et al. (1997) measured T3 and T4 levels four times a day in larvae of three anuran species with different phylogenetic origins, behaviors, and ecological habits (*Rana perezi, Xenopus laevis, Bufo calamita*). They found that the daily profiles of thyroid hormones differed significantly between species and during ontogenic development at both prometamorphosis and late climax stages for all three species. A similar situation has also been demonstrated in birds, where the pattern of posthatching rise in thyroxine levels differs significantly between neonates of species that produce precocious young, such as the European starling, and those with altricial young, such as the Japanese quail and the ostrich (Hrlid and Arnason 1999; Schew et al. 1996).

The pulsatile nature of TSH and thus thyroxine secretion exerts a strongly patterned influence throughout the endocrine system. For example, as production of growth hormone is induced by thyroxine (Harvey 1990) and reproductive hormones are subsequently stimulated by growth hormone (Ogilvy-Stuart and Shalet 1992; Thompson et al. 1992), levels of these critical hormones also fluctuate. The ultimate source of the pulsatile secretion of all of these hormones, as shown in Figure 6.3, is the pineal gland. The pineal is a small gland in the brain, unique in that it is stimulated electrically (by retinal receptors) but responds chemically with the intermittent release of hormones (melatonin, seratonin, and noradrenalin) (Korf 1994). The pineal gland is the organ most responsible for sensing changing environmental conditions of all kinds and initiating appropriate physiological responses (Hadley 1984). Thyroxine released as a consequence of pineal stimulation is the biochemical agent responsible for coordinating the body's total adaptive response to both short-term (daily) and long-term (seasonal) changes in environmental conditions. Thyroxine can thus be said to be the biological mechanism by which adaptation of individuals is achieved.

Individual Variation in Thyroid Physiology

The pulsatile pattern of thyroxine secretion is the property of thyroid hormone metabolism that seems to affect protodomestication and other types of heterochronic speciation. Precision in timing (frequency of pulses) and absolute amounts of thyroxine produced (amplitude of pulses) may be critical to certain target genes, cells, or organs during development (compare Nijhout 1999). If so, very slight variations in the fluctuating pattern of production could produce small physiological changes that would ultimately result in noticeable differences in morphological phenotype (principally variations in coat color and size) because of the pleiotropic influences of thyroxine on growth and development. Individual differences in thyroxine physiology may also underlie the slight variations in timing of ovulation, molt, and behavior among virtually all mammalian individuals (Banfield 1974) and recognizable individual ontogenies (Parker and McKinney 1999).

In addition to individual variation, particular intraspecific differences in thyroxine profiles are also expected within populations because of the intimate relationship between thyroid hormone and sex and growth hormones

(Fig. 6.3). The interaction between these hormones and thyroxine metabolism provides a physiological explanation for the apparent heterochronic nature of sexual dimorphism (Martin et al. 1994). A similar explanation may apply to other kinds of discrete within-species morphotypes, as in certain fish, amphibian, and reptile species (Dickhoff 1993; Voss and Shaffer 1997).

The importance of individual differences in physiology to evolution are becoming more clear as researchers attempt to understand what factors influence the success of populations over time (McNamara and Houston 1996). For example, Crowder et al. (1992) found that individual physiological differences among fish larvae are extremely critical to recruitment success; survivors were not "average individuals" but represented a specific fraction of the population that possessed particular physiological phenotypes. In other words, survivors were not a random subset of the original population.

Thyroid Hormone and Piebaldness

Although all of the biological traits associated with protodomestication are clearly under the control of thyroid hormone, the appearance of piebaldness (black or other solid color marked with white) is the hardest to explain in this context. The high incidence of piebaldness in domestic animals has always been somewhat of an enigma and is usually assumed to be a consequence of deliberate selection (Clutton-Brock 1992a). Piebald markings are rare in most wild mammals, including the ancestors (or their close extant relatives) of domestic taxa, although it is the norm in a small number of groups (e.g., skunks). Such a low natural incidence of piebaldness in wild populations would make increasing the proportion of individuals with this trait (via natural or human selection) very difficult. However, the early appearance of piebaldness in the silver foxes experimentally selected for behavioral phenotypes (Belyaev 1979) suggests that piebaldness could be an inevitable consequence of the heterochronic process of protodomestication. To explain why this is so, we need a short explanation of pattern formation and pigment production.

Pigment (technically melanin) is produced in special migratory cells called melanocytes that exist in skin, hair, and other body tissues; this is where the chemical conversion of tyrosine (derived from the ingested amino acid phenylalanine) to melanin takes place. Many different factors can affect melanocyte activity during pigment production (including hormonal influences from the pituitary [Fig. 6.4] and spontaneous mutations in hormone receptors and their cofactors in the follicles), resulting in different coat and skin colors (Kijas et al. 1998; Robbins et al. 1993; Searle 1968).

Patterning of color, however, is determined during fetal development. The hypothesis that, thus far, best explains color pattern anomalies suggests that, during early embryonic development, seven pairs of melanoblasts (undifferentiated melanocyte precursors) must migrate from their origins in the neural crest to areas in the body where pigment is required. Through controlled proliferation of these melanoblasts and their subsequent maturation to functional melanocytes, a normal pattern of pigmentation is produced (Silvers 1979). Areas that do not receive melanoblasts because of a disruption of these processes remain white, resulting in a coat color pattern that is variously spotted—the piebald coat patterns in affected silver foxes from Belyaev's selected population were ultimately determined to be caused by a one- to two-day delay in the migration rate of the melano-

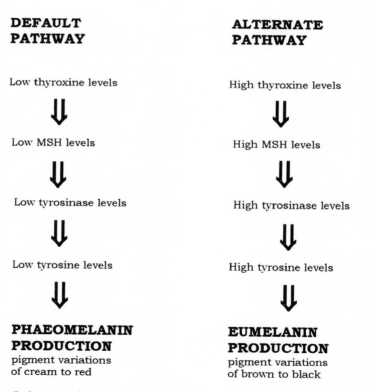

DEFAULT PATHWAY

ALTERNATE PATHWAY

Low thyroxine levels

⇓

Low MSH levels

⇓

Low tyrosinase levels

⇓

Low tyrosine levels

⇓

PHAEOMELANIN PRODUCTION
pigment variations
of cream to red

High thyroxine levels

⇓

High MSH levels

⇓

High tyrosinase levels

⇓

High tyrosine levels

⇓

EUMELANIN PRODUCTION
pigment variations
of brown to black

Fig. 6.4. Influences of thyroxine and melanocyte-stimulating hormone (*MSH*) levels on coat color production within hair follicles.

blasts from the neural crest in the embryo (Prasolova et al. 1997). Piebald coloration can vary from only a spot or two of white on a solid color background to an all-over white that is essentially one big spot.

Both thyroxine and retinoic acid have been identified as essential to the orderly movement of cells out of the neural crest during early development (Barres et al. 1994; Pavan et al. 1995). The thyroxine needed for neural crest cell migration is required by the embryo very early in development and, in mammals, the maternal thyroid gland is the source of the thyroxine utilized (Porterfield and Hendrich 1992). The fetal thyroid gland is not functional until fairly late into development (Pickard et al. 1993), at which time it begins to augment rather than replace the maternal contribution. Consequently, although a direct correlation between disruption of thyroxine production and piebaldness has not yet been demonstrated, the circumstantial evidence in favor of such a relationship is very strong.

Thyroid Hormone and Behavior

In addition to controlling the development of specific morphological traits, thyroxine also affects behavior. Principally, thyroid hormone mediates those behavioral responses to stress and stimuli that are influenced by sympathetic adrenal gland function. This is because the production of adrenal hormones (catecholamines: epinephrine, norepinephrine, and dopamine) is controlled by adrenergic receptors. The activity of these receptors is strongly affected by levels of available thyroid hormones (Schreibman et al. 1993). Thus, behavior relating to an animal's stress response is fundamentally under thyroid hormone control and, just as thyroxine profiles can show individual and intraspecific variation, so do stress responses (Boissy 1995). For example, differences in biochemical stress responses of both dog and rat breeds have been demonstrated (Arons and Shoemaker 1992; Windle et al. 1998).

Hadley (1984, 337) defines stress as "the state resulting from events (stressors) of external or internal origin, real or imagined, that tend to affect the homeostatic state . . . any condition tending to elevate plasma catecholamine levels in response to exogenous or endogenous stimuli." Therefore, the ability to respond to the varied exogenous stresses fundamental to survival should be affected by an individual's particular pattern of thyroid hormone metabolism. Individual differences in thyroid hormone production or utilization could therefore account in part for individual differences in stress response within species.

Thyroid Hormone and Development

As discussed earlier in relation to the development of piebald phenotypes, thyroxine is required for normal embryonic development very early in the process, supplied by the maternal system either directly (for mammals) or via reserves stored in egg yolk (for nonplacental vertebrates). This early direct role for thyroxine is not the only aspect of its effect on development, because of the way in which it interacts with the production of other hormones. Growth and sex hormones are also required for both embryonic and postnatal development but are dependent on thyroxine for their expression.

The role of thyroxine in somatic growth and embryonic development provides really significant insight into how species-specific developmental control can be achieved. Because of the critical role played by maternal thyroxine in early embryonic development (both directly and indirectly), it is clear that the precise endocrine physiology possessed by a mother (or passed along into egg yolk) must control the early development of her offspring and will continue to influence growth to some extent until they are born (Piosik et al. 1997; Wilson and McNabb 1997). Only from the point during development when the fetal thyroid gland becomes functional do genes controlling thyroxine function that were contributed by the sire to the offspring have an opportunity to be expressed. The distinct stages at which offspring are affected by each set of parental genes controlling thyroxine production (and thus growth and development) may provide a partial explanation as to why the phenotype of offspring from hybrid crosses differs depending on the species or phenotype of the dam (Wichman and Lynch 1991).

Thyroid Hormone and Nutrition

One other aspect of thyroid hormone metabolism that merits discussion is the relationship between thyroxine production and nutrition. There are several ways in which nutrition can affect thyroid hormone physiology in all animals, including humans. The first is through consumption of foods that contain the essential amino acid phenylalanine. Phenylalanine is converted to tyrosine, which is needed for both thyroxine and melanin production (Hadley 1984). Consumption of foods containing significantly inadequate or excessive amounts of phenylalanine may be capable of acting as a selective force on a portion of a population that is sensitive to available tyrosine levels. A similar situation might occur in relation to dietary sources of the iodine required for thyroxine production or through consumption of dietary flavonoids. Flavonoids naturally occur in many plants. Some flavonoids are goitrogenic: they are able to interfere with normal transport, synthesis, and action of thyroxine by competitive exclusion because they possess a similar three-dimensional molecular structure (Hadley 1984).

Last, and perhaps most drastically, the thyroid function of individuals may be affected by the direct consumption of thyroxine from prey animals and their eggs. Thyroxine is present in the flesh, blood, organ tissue, egg yolks, and thyroid glands of all vertebrates and, unique among hormones, can be easily absorbed directly through the digestive tract when consumed by predators (Hadley 1984). The consumption of thyroxine-rich foods must add considerably to an animal's daily thyroxine load, especially for carnivores. This may explain the large disparity in turnover rates of thyroxine between carnivores and humans mentioned previously (13–16.6 hours vs. 6.8 days). Carnivores obviously require an active metabolism capable of rapidly clearing the massive input of exogenous thyroxine, perhaps coupled with a relative insensitivity to temporarily high hormone levels or an ability to produce similarly high levels in the absence of exogenous hormone.

In all taxonomic groups the changes in diet that undoubtedly accompanied habitat shifts may have contributed to the precipitation of heterochronic speciation of all kinds (including protodomestication). Colonization of radically different environments may have occasionally necessitated quite dramatic dietary shifts that involved adding or deleting exogenous thyroxine sources or goitrogenic compounds, factors perhaps especially significant in human evolution.

Thyroid Hormone and Heterochronic Speciation

Individual variation for thyroxine metabolism in all animals is the key to proposing a scenario in which wild animals could be exposed to the selection pressures necessary for the biological changes associated with protodomestication and heterochronic speciation to occur. Slight variations in the pattern of thyroxine production or utilization among individuals are reflected in phenotypic differences of a physiological nature, creating polymorphisms (controlled by at least one or several genes) in thyroid physiology that are heritable traits.

These heritable phenotypes provide a mechanism for rapid adaptation of a portion of a population to new environments. The stresses and stimuli associated with attempts to invade and colonize a radically new territory (utilizing new food sources, adjusting to new predators and competitors, etc.) provide the selection pressures that divide a population. The colonizing subset of the population will thus be composed only of individuals who possess the physiological ability to tolerate increased stress and stimulus loads, while the subset remaining in the old habitat contains their less stress-tolerant cohorts. Individuals that can tolerate the stresses of a new habitat (including anthropogenic environments) constitute a similar non-random segment of any population. When interbreeding among these animals occurs within a restricted gene pool, offspring inherit a limited complement of alleles for the genes controlling thyroxine profiles. Without the full range of variability, descendants of this population are almost certain to be behaviorally, reproductively, and morphologically different from their ancestral population after relatively few generations because of the pleiotropic influence of thyroid hormone. Thus, a well-coordinated suite of traits in descendant populations would be recognizable relatively early as classic heterochronic changes of one kind or another.

Although this hypothesis does not preclude the possibility that protodomestication of any single taxon could have happened more than once (e.g., Bradley et al. 1998), it is doubtful whether it happened often. The attractions of the human-dominated habitat would need to have been very strong to encourage wild taxa to expand into an anthropogenic environment. In addition, the stress-tolerant behavioral types would need to have been present as natural variations in the ancestral population to start with (i.e., populations of animals who all tolerate this kind of stress well—or very poorly —do not lend themselves to protodomestication). This combination of necessary factors could explain why colonization of anthropogenic environments by wild taxa happened rarely overall in relation to the number of potential animal species available.

Perhaps the best example of protodomestication is that which has been proposed for the speciation of mithan from gaur (a type of wild Asian cattle). The method of handling modern domestic mithan stocks in Bangladesh and Burma described by Simoons (1984) seems to perpetuate the protodomestication process. Mithan, like their wild gaur ancestors, are attracted to salt. People today manage their mithan stocks in the most minimal fashion, simply encouraging the animals to return to habitation areas by providing salt. No attempts are made to handle or tame mithan or to confine them, since the regular proximity of the animals adequately facilitates the occasional culling of individuals for ritual use (they are not eaten or milked). The animals feed on available wild forage and breed entirely at will, and yet mithan are classic domesticates: they are smaller than wild gaur, relatively docile in behavior, and often piebald.

Although it is possible that previously domesticated mithan stocks are merely maintained by this system, there are compelling reasons to believe that protodomestication of wild guar occurred under these (or similar) conditions. This scenario provides the ecological components necessary for precipitating protodomestication changes: an attractive resource not readily available in the wild (salt), the constant proximity of humans to provide the selection pressure, and the freedom of any stress-intolerant individual (born in any generation) to leave at will.

An example of speciation mediated by heterochronic processes in a natural (as opposed to anthropogenic) environment may be the polar bear, a naturally occurring, extreme piebald animal. Molecular genetic research by Talbot and Shields (1996) on brown bear (*Ursus arctos*) populations in Alaska revealed surprising similarities between several mitochondrial genes of the polar bear (*Ursus maritimus*) and one particular coastal subspecies of brown bear. Polar bears had previously been proposed as descending from brown bear stock based on morphological criteria, but the results of this study indicate that at least some populations of these taxa are still exceedingly close genetically. Talbot and Shields themselves concluded that "the morphological features distinguishing polar bears from brown bears have evolved rapidly in response to selective pressures of adapting to a new environment, prior to the emergence of distinguishing molecular features."

If polar bears possess extreme piebaldness as a consequence of the rapid speciation changes precipitated by colonization of a radically new environment, then perhaps other white taxa (such as mountain goats) are white for this reason as well. Less extreme examples of naturally piebald taxa (such as skunks) may be equally significant, since piebaldness is normally far too rare in most wild populations for selection alone (especially in top predators like polar bears, which have no or few natural enemies) to have created a whole population of extreme piebald animals unless something occurred to increase the natural incidence of this anomaly substantially (Belyaev et al. 1981; Searle 1968). Piebaldness could only have become subject to selection pressure because heterochronic speciation changes made it an available phenotype.

Testing the Hypothesis

The hypothesis generated by this paradigm states that individual thyroid hormone phenotypes exist within populations and that these phenotypes are the real targets of natural selection in instances of adaptation to new environments. This implies that nonrandom subdivision of populations according to differences in thyroid hormone metabolism occur during speciation and that subsequent shifts in thyroid hormone profiles constitute the control mechanism by which heterochronic changes are achieved in descendant taxa.

I suggest that thyroid hormone profiles should therefore show hierarchical patterns. First, there should be slight variations between individuals that are contained within a pattern distinctive for discrete morphs and each sex. Ultimately, all of the aforementioned patterns should be contained within a distinctive, species-specific pattern. However, devising experiments that can reliably test the hypothesis will undoubtedly be difficult because of the dynamic and pleiotropic nature of the endocrine system and its inherent sensitivity to stress of any kind (Boissy 1995). Such sensitivity presents a unique challenge to the determination of normal thyroid hormone profiles within groups, since thyroid hormone levels clearly must be tested frequently (perhaps as often as every 15–20 minutes) under controlled conditions for many individuals.

Windle et al. (1998), however, demonstrated that automated sampling may circumvent many of the difficulties of testing the thyroxine hypothesis. For their study on corticosterone levels in rats, a surgically implanted cannula connected to an automated blood-sampling apparatus allowed minute quantities of blood (10–20 ml) to be collected every ten minutes over a twenty-four-hour period without disturbing the animals by repeated handling. The two strains showed significant differences in mean profiles of hormone production (as well as slight individual variations within strains) and significant differences in behavioral responses to a controllable stress (white noise). The success of these experiments in demonstrating the existence of fine-scale patterns of corticosterone production and in correlating these profiles to stress responses suggests that a similar method might be suitable for testing the thyroxine hypothesis.

In the meantime, ongoing research into the regulatory mechanisms of embryonic development should unravel some of the essential physiological interactions that involve thyroid hormones. Research on *Hox* genes has thus far revealed that they respond to retinoic acid and perhaps other molecules as well (Lawrence and Morata 1994). In light of the developmental regulation functions that retinoic acid shares with thyroxine, or in which their roles cannot be distinguished (Barres et al. 1994), it would perhaps be prudent to look at the response of some *Hox* genes to thyroxine pulses in combination with this molecularly similar vitamin A derivative. In addition, research into the physiological and genetic basis of natural piebaldness (rather than aberrant white spotting mutants) may also be illuminating. Piebaldness, if we can come to understand exactly what it signifies, could serve as an especially useful diagnostic marker.

I concede that the assumptions I have made about the role of thyroxine in protodomestication, heterochrony, and speciation may turn out to be too simplistic. However, if precisely controlled patterns of thyroxine production or utilization do not constitute the biological mechanism through which developmental control and change are achieved, we may have a better chance of finding out what that mechanism actually is by looking at the problem from this physiological perspective.

Conclusions

As Darwin (1859, 1868) seems to have suspected, domestic animals (or, rather, the process that produced them) may be the key to unlocking the mysteries of evolutionary change. I suggest that protodomestication, as I have defined it here, becomes an appropriate paradigm for describing the truly dynamic relationship between individual variation, adaptation, and speciation in all taxa, including humans. I have tentatively identified precisely controlled patterns of thyroid hormone metabolism as the essential factor in vertebrate heterochrony. This implicates physiologically determined behavioral phenotypes as the targets of selection in certain adaptation processes and nonrandom population subdivision as an essential component for the initiation of developmental change. Since thyroxine metabolism is responsible for keeping individuals adapted to environmental conditions that change on a daily and seasonal basis, this physiological system may well have the potential to implement the changes that allow species to adapt over longer periods.

The range of variation for thyroxine production or utilization by any species undoubtedly influences its evolutionary adaptability. We would therefore expect some species to be better able to adapt to specific kinds of environmental change because their populations are composed of individuals with more physiological, morphological, and behavioral variation. In contrast, some sets of quite distantly related species could all change in a similar fashion on exposure to identical shifts in environmental regimes over evolutionary time. Similar ontogenetic responses, such as the changing body sizes of mammalian taxa during the Pleistocene and Holocene (Kurtén 1988), could have occurred across taxonomic groups because both the stressor and the stress response system would have been the same for all taxa (in much the same way that protodomestication produces paedomorphic changes in every instance).

As a consequence of the interaction between thyroxine and sex and growth hormones, heterochrony can also generate both individual and intraspecific variants, as well as interpopulational and interspecific differences, and it can accomplish these results both slowly and rapidly. In addition, as development is a process that involves both embryonic and postnatal programmed growth that is dependent on thyroxine, heterochronic changes can be implemented at all ontogenic stages of embryonic and postnatal growth.

Is this paradigm applicable only to vertebrates? Taken in its broadest sense, I suspect not. Gibson and Hogness (1996), for example, have presented experimental evidence for *Drosophila* that repeated selection for a physiological phenotype combined with small population sizes affects developmental processes. In their study, flies selected for sensitivity to ether produced, over several generations, descendants with bithorax anomalies (four wings) indicative of developmental disruption. In addition, Nijhout's (1999) overview of polyphenic development in insects provides specific support for the suggestion that a physiological mechanism controlling development, similar to that proposed for vertebrates, also exists for invertebrates. The same may be true for plants: heterochrony has been shown to operate as a significant process in plants (Guerrant 1988), which also possess hormones whose interactions affect growth and development (Hoffmann and Parsons 1991).

Ever since the role of steroid hormones in development and gene regulation was clearly demonstrated (Evans 1988), the critical importance of hormonal interactions began to be more fully appreciated. However, real progress has been severely hampered by the lack of an appropriate theoretical framework that ties endocrine effects into an evolutionary context. Using protodomestication as a model for the speciation of all taxa—including humans—focuses a spotlight on the critical role played by thyroid hormone metabolism in vertebrate development and adaptation. Decoupling protodomestication from subsequent cultural processes provides a novel theoretical framework that has enormous explanatory potential. Such a paradigm has the power to demonstrate how the developmental changes associated with heterochrony could actually drive evolution in the mode described so eloquently by the concept of punctuated equilibrium.

Acknowledgments

For their intellectual support during the formulation of the ideas expressed in this chapter, I thank Cairn Crockford, Becky Wigen, Margot Wilson, David Moyer, and Nicholas Rolland, all at University of Victoria, British Columbia, and Gay Frederick at Malaspina University College, Nanaimo, British Columbia. Glenn Northcutt (Scripts Institute of Oceanography), Michael McKinney (University of Tennessee), I. Lehr Brisbin (Savannah River Ecology Laboratory), and Atholl Anderson (Australian National University) provided enthusiastic support when it was needed most. I extend particular thanks to several reviewers of earlier drafts, who provided severe but pertinent criticisms concerning some passages, which helped enormously in my effort to present a cogent argument—whether I succeeded in the end is my own responsibility. Finally, I acknowledge the significant economic support provided by my late mother, Barbara Crockford, when I was developing the ideas expressed in this chapter—she believed without a doubt that I had something important to say and, fortunately, lived long enough to see that others agreed.

References

Alberch, P. 1991. From genes to phenotype: dynamical systems evolvability. *Genetica* 84:5–11.

Alderson, L. 1978. *The Chance to Survive: Rare Breeds in a Changing World.* London: Cameron and Tayleur Books.

Arnold, M.L. 1997. *Natural Hybridization and Evolution.* Oxford: Oxford University Press.

Arons, C.D., and Shoemaker, W.J. 1992. The distribution of catecholamines and beta-endorphin in the brains of three behaviorally distinct breeds of dogs and their F1 hybrids. *Brain Research* 594:31–39.

Banfield, A.W.F. 1974. *Mammals of Canada.* Toronto: National Museum of Canada, University of Toronto Press.

Barres, B.A., Lazar, M.A., and Raff, M.C. 1994. A novel role for thyroid hormone, glucocorticoids and retinoic acid in timing oligodendrocyte development. *Development* 120:1097–1108.

Belyaev, D.K. 1979. Destabilizing selection as a factor in domestication. *Journal of Heredity* 70:301–308.

Belyaev, D.K., Ruvinsky, A.O., and Trut, L.N. 1981. Inherited activation-inactivation of the star gene in foxes: its bearing on the problem of domestication. *Journal of Heredity* 72:267–274.

Boissy, A. 1995. Fear and fearfulness in animals. *Quarterly Review of Biology* 70: 165–191.

Bradley, D.G., Loftus, R.T., Cunningham, P., and MacHugh, D.E. 1998. Genetics and domestic cattle origins. *Evolutionary Anthropology* 6:79–86.

Budiansky, S. 1992. *The Covenant of the Wild: Why Animals Chose Domestication.* London: Weidenfeld and Nicolson.

Brent, G.A., Moore, D.D., and Larsen, P.R. 1991. Thyroid hormone regulation of gene expression. *Annual Review of Physiology* 53:17–35.

Clutton-Brock, J. 1992a. Domestication of animals. In S. Jones, R. Martin, and D. Pilbeam (eds.), *The Cambridge Encyclopedia of Human Evolution,* 380–385. Cambridge: Cambridge University Press.

———. 1992b. The process of domestication. *Mammal Review* 22:79–85.

Coppinger, R., and R. Schneider. 1995. Evolution of working dogs. In J. Serpell (ed.), *The Domestic Dog: Its Evolution, Behaviour, and Interactions with People,* 21–47. Cambridge: Cambridge University Press.

Crabtree, P.J. 1993. Early animal domestication in the Middle East and Europe. In M.B. Schiffer (ed.), *Archaeological Method and Theory,* 201–245. Tucson: University of Arizona Press.

Crowder, L.B., Rice, J.A., Miller, T.J., and Marschall, E.A. 1992. Empirical and theoretical approaches to size-based interactions and recruitment variability in fishes. In D.L. DeAngelis and L.J. Gross (eds.), *Individual-based Models and Approaches in Ecology,* 237–255. New York: Chapman and Hall.

Darwin, C. 1859. *On the Origin of Species.* London: J. Murray.

———. 1868. *The Variation of Animals and Plants under Domestication.* London: J. Murray.

Dawson, A., Deeming, D.C., Dick, A.C.K., and Sharp, P.J. 1994. Plasma thyroxine concentrations in farmed ostriches in relation to age, body weight, and growth hormone. *General and Comparative Endocrinology* 103:308–315.

Dayan, T. 1994. Early domesticated dogs of the Near East. *Journal of Archaeological Science* 21:633–640.

De Pablo, F. 1993. Introduction. In M.P. Schreibman, C.G. Scanes, and P.K.T. Pang (eds.), *The Endocrinology of Growth, Development, and Metabolism of Vertebrates,* 1–11. New York: Academic Press.

Dickhoff, W.W. 1993. Hormones, metamorphosis and smolting. In M.P. Schreibman, C.G. Scanes, and P.K.T. Pang (eds.), *The Endocrinology of Growth, Development, and Metabolism of Vertebrates,* 519–540. New York: Academic Press.

Dulbecco, R. (ed.). 1991. *Encyclopedia of Human Biology.* New York: Academic Press.

Eldridge, N., and Gould, S.J. 1972. Punctuated equilibria: an alternative to phyletic gradualism. In T.J.M. Schopf (ed.), *Models in Paleobiology,* 82–115. San Francisco: Freeman and Cooper.

Evans, R.M. 1988. The steroid and thyroid hormone receptor superfamily. *Science* 240:889–895.

Fox, M.W. 1978. *The Dog: Its Domestication and Behavior.* New York: Garland STPM Press.

Gancedo, B., Alonso-Gomez, A.L., de Pedro, N., Delgado, M.J., and Alonso-Bedate, M. 1997. Changes in thyroid hormone concentrations and total contents through ontogeny in three anuran species: evidence for daily cycles. *General and Comparative Endocrinology* 107:240–250.

Gibson, G., and Hogness, D.S. 1996. Effect of polymorphism in the *Drosophila* regulatory gene Ultrabithorax on homeotic stability. *Science* 271:200–203.

Gould, S.J., and Eldridge, N. 1993. Punctuated equilibrium comes of age. *Nature* 366:223–227.

Guerrant, E.O. 1988. Heterochrony in plants: the intersection of evolution, ecology and ontogeny. In M.L. McKinney (ed.), *Heterochrony in Evolution: A Multidisciplinary Approach,* 111–133. New York: Plenum Press.

Hadley, M.E. 1984. *Endocrinology.* Englewood Cliffs, N.J.: Prentice Hall.

Hafez, E.S.E. (ed.). 1968. *Adaptation of Domesticated Animals.* Philadelphia: Lea and Febiger.

Härlid, A., and Arnason, U. 1999. Analysis of mitochondrial DNA nest ratite birds within the Neognathae: supporting a neotenous origin of ratite morphological characters. *Proceedings of the Royal Society of London* B 266:305–309.

Harvey, S. 1990. Thyroid inhibition of growth hormone secretion: negative feedback? In M. Wada, S. Ishii, and C.G. Scanes (eds.), *Endocrinology of Birds,* 111–127. Tokyo: Japan Scientific Society Press/Berlin: Springer-Verlag.

Hemmer, H. 1990. *Domestication: The Decline of Environmental Appreciation.* Cambridge: Cambridge University Press.

Hoffmann, A.A., and Parsons, P.A. 1991. *Evolutionary Genetics and Environmental Stress.* Oxford: Oxford University Press.

Isaac, E. 1970. *Geography of Domestication.* Englewood Cliffs, N.J.: Prentice Hall.

Jennings, D.H., and Hanken, J. 1998. Mechanistic basis of life history evolution in Anuran amphibians: thyroid gland development in the direct-developing frog, *Eleutherodactylus coqui. General and Comparative Endocrinology* 111: 225–232.

Jones, S., Martin, R., and Pilbeam, D. (eds.). 1992. *The Cambridge Encyclopedia of Human Evolution.* Cambridge: Cambridge University Press.

Kaptein, E.M., Hays, M.T., and Ferguson, D.C. 1994. Thyroid hormone metabolism: a comparative evaluation. In D.C. Ferguson (ed.), *Thyroid Disorders,* 431–463. Veterinary Clinics of North America: Small Animal Practice 24(3). Philadelphia: W. B. Saunders Co.

Keeler, C. 1975. Genetics of behaviour variations in colour phases of the red fox.

In M.W. Fox (ed.), *The Wild Canids: Their Systematics, Behavioural Ecology, and Evolution,* 399–415. New York: Van Nostrand Reinhold Co.

Kijas, J.M.H., Wales, R., Trnsten, A., Chardon, P., Moller, M., and Andersson, L. 1998. Melanocortin receptor I (MC1R) mutations and coat color in pigs. *Genetics* 150:1177–1185.

Koibuchi, N., Natsuzaki, S., Ichimura, K., Ohtake, H., and Yamaoka, S. 1996. Ontogenetic changes in the expression of cytochrome c oxidase subunit I gene in the cerebellar cortex of the perinatal hypothyroid rat. *Endocrinology* 137:5096–5108.

Korf, H-W. 1994. The pineal organ as a component of the biological clock: phylogenetic and ontogenetic considerations. In W. Pierpaoli, W. Regelson, and N. Fabris (eds.), *The Aging Clock: The Pineal Gland and Other Pacemakers in the Progression of Aging and Carcinogenesis,* 13–42. New York: Annals of the New York Academy of Science 719.

Krumlauf, R. 1994. Hox genes in vertebrate development. *Cell* 78:191–201.

Kurtén, B. 1988. *On Evolution and Fossil Mammals.* New York: Columbia University Press.

Lawrence, P.A., and G. Morata. 1994. Homeobox genes: their function in *Drosophila* segmentation and pattern formation. *Cell* 78:181–189.

Martin, R.D., Willner, L.A., and Dettling, A. 1994. The evolution of sexual size dimorphism in primates. In R.V. Short and E. Balaban (eds.), *The Differences between the Sexes,* 159–200. Cambridge: Cambridge University Press.

Mayr, E. 1994. Recapitulation reinterpreted: the somatic program. *Quarterly Review of Biology* 69:223–232.

McKinney, M.L. 1998. The juvenilized age myth: our overdeveloped brain. *BioScience* 48:109–116.

McKinney, M.L., and McNamara, K.J. 1991. *Heterochrony: The Evolution of Ontogeny.* New York: Plenum Press.

McNabb, A.F.M., and King, D.B. 1993. Thyroid hormone effects on growth, development, and metabolism. In M.P. Schreibman, C.G. Scanes, and P.K.T. Pang (eds.), *The Endocrinology of Growth, Development, and Metabolism of Vertebrates,* 393–417. New York: Academic Press.

McNamara, J.M., and Houston, A.I. 1996. State-dependent life histories. *Nature* 380:215–221.

Morey, D.F. 1992. Size, shape and development in the evolution of the domestic dog. *Journal of Archaeological Science* 13:119–145.

———. 1994. The early evolution of the domestic dog. *American Scientist* 82:336–347.

Morita, S., Matsuo, K., Tsuruta, M., Leng, S., Yamashita, S., Izumi, M., and Nagataki, S. 1990. Stimulatory effects of retinoic acid and tri-iodothryonine in rat pituitary cells. *Journal of Endocrinology* 125:251–256.

Nijhout, H.F. 1999. Control mechanisms of polyphenic development in insects. *BioScience* 49:181–192.

Nunez, E.A., Becker, D.V., Furth, E.D., Belshaw, B.E., and Scott, J.P. 1970. Breed differences and similarities in thyroid function in purebred dogs. *American Journal of Physiology* 218:1337–1341.

O'Connor, T.P. 1997. Working at relationships: another look at animal domestication. *Antiquity* 71:149–156.

Ogilvy-Stuart, A.L., and Shalet, S.M. 1992. Growth hormone and puberty. *Journal of Endocrinology* 135:405–406.

Olivier, R.C.D. 1984. Asian elephant. In I.L. Mason (ed.), *Evolution of Domesticated Animals*, 185–192. London: Longman Co.

Oppenheimer, J.H. 1979. Thyroid hormone action at the cellular level. *Science* 203:971–979.

Otte, D., and Endler, J.A. (eds.). 1989. *Speciation and Its Consequences*. Sunderland, Mass.: Sinauer.

Owen, R.B., Crossley, R., and Johnson. 1990. Major low levels of Lake Malawi and their implications for speciation rates in cichlid fishes. *Proceedings of the Royal Society of London*, Series B 240:519–53.

Parker, S., and McKinney, M.L. 1999. *Origins of Intelligence: The Evolution of Cognitive Development in Monkeys, Apes, and Humans*. Baltimore: Johns Hopkins University Press.

Pavan, W.J., Mac, S., Cheng, M., and Tilghman, S.M. 1995. Quantitative trait loci that modify the severity of spotting in piebald mice. *Genome Research* 5:29–41.

Pickard, M.R., Sinha, A.K., Ogilvie, L., and Ekins, R.P. 1993. The influence of the maternal thyroid hormone environment during pregnancy on the ontogenesis of brain and placental ornithine decarboxylase activity in the rat. *Journal of Endocrinology* 139:205–212.

Piosik, P.A., van Groenigen, M., van Doorn, J., Baas, F., and de Vijlder, J.J.M. 1997. Effects of maternal thyroid status on thyroid hormones and growth in congenitally hypothyroid goat fetuses during the second half of gestation. *Endocrinology* 138:5–11.

Porterfield, S.P., and Hendrich, C.E. 1992. Tissue iodothyronine levels in fetuses of control and hypothyroid rats at 13 and 16 days gestation. *Endocrinology* 131:195–200.

Prasolova, L.A., Trut, L.N., and Vsevolodov, E.B. 1997. Morphology of mottling hairs in domesticated silver foxes (*Vulpes vulpes*) and relation between the expression of the star and the mottling mutation. In L.N. Trut and L.V. Osadchuk (eds.), *Evolutionary-genetic and Genetic-physiological Aspects of Fur Animal Domestication: A Collection of Reports*, 31–40. Oslo: IFASA/Scientifur.

Reed, C.A. 1984. The beginnings of animal domestication. In I.L. Mason (ed.), *Evolution of Domesticated Animals*, 1–6. London: Longman Co.

Reilly, S.M. 1994. The ecological morphology of metamorphosis: heterochrony and the evolution of feeding mechanisms in salamanders. In P.C. Wainwright and S.M. Reilly (eds.), *Ecological Morphology: Integrative Organismal Biology*, 319–338. Chicago: University of Chicago Press.

Rice, W.R., and Hostert, E.E. 1993. Laboratory experiments on speciation: what have we learned in 40 years? *Evolution* 47:1637–1653.

Richardson, M.K. 1995. Heterochrony and the phylotypic period. *Developmental Biology* 172:412–421.

Robbins, L.S., Nadeau, J.H., Johnson, K.R., Kelly, M.A., Roseli-Rehfuss, L., Baack, E., Mountjoy, K.G., and Cone, R.D. 1993. Pigmentation phenotypes of variant extension locus alleles result from point mutations that alter MSH receptor function. *Cell* 72:827–834.

Schew, W.A., McNabb, F.M.A., and Scanes, C.G. 1996. Comparison of the ontogenesis of thyroid hormones, growth hormone, and insulin-like growth factor-I in ad libitum and food-restricted (altricial) European starlings and (precocial) Japanese quail. *General and Comparative Endocrinology* 101:304–316.

Schreibman, M.P., Scanes, C.G., and Pang, P.K.T. (eds.). 1993. *The Endocrinology of Growth, Development, and Metabolism of Vertebrates.* New York: Academic Press.

Searle, A.G. 1968. *Comparative Genetics of Coat Colour in Mammals.* London: Logos Press.

Silvers, W.K. 1979. *The Coat Colors of Mice: A Model for Mammalian Gene Action and Interaction.* New York: Springer-Verlag.

Simoons, F.J. 1984. Gayal or mithan. In I.L. Mason (ed.), *Evolution of Domesticated Animals,* 34–39. London: Longman Co.

Skjenneberg, S. 1984. Reindeer. In I.L. Mason (ed.), *Evolution of Domesticated Animals,* 128–137. London: Longman Co.

Talbot, S.L., and G.F. Shields. 1996. A phylogeny of the bears (Ursidae) inferred from complete sequences of three mitochondrial genes. *Molecular Phylogenetics and Evolution* 5:567–575.

Tchernov, E. 1993. From sedentism to domestication—a preliminary review for the southern Levant. In A. Clasen, S. Payne, and H.P. Uerpmann (eds.), *Skeletons in Her Cupboard: Festschrift for Juliet Clutton-Brock,* 189–233. Monograph 34. Oxford: Oxbow Books.

Templeton, A.R. 1989. The meaning of species and speciation: a genetic perspective. In D. Otte and J.A. Endler (eds.), *Speciation and Its Consequences,* 3–27. Sunderland, Mass.: Sinauer.

Thompson, D.L., Jr., Rahmanian, M.S., DePew, C.L., Burleigh, D.W., DeSouza, C.J., and Colborn, D.R. 1992. Growth hormone in mares and stallions: pulsatile secretion, response to growth hormone-releasing hormone, and effects of exercise, sexual stimulation, and pharmacological agents. *Journal of Animal Science* 70:1201–1207.

Tomasi, T.E., Hellgren, E.C., and Tucker, T.J. 1998. Thyroid hormone concentrations in black bears (*Ursus americanus*): hibernation and pregnancy effects. *General and Comparative Endocrinology* 109:192–199.

Tomasi, T.E., and Mitchell, D.A. 1994. Seasonal shifts in thyroid function in the cotton rat (*Sigmodon hispidus*). *Journal of Mammalogy* 75:520–528.

Trut, L.N. 1999. Early canid domestication: the fox-farm experiment. *American Scientist* 87:160–169.

Trut, L.N., Dzerzhinsky, F.Y., and Nikolsky, V.S. 1991. A principal component analysis of changes in cranial characteristics appearing in silver foxes (*Vulpes vulpes* Desm.) under domestication [in Russian]. *Genetika* 27:1440–1449.

Trut, L.N., and Osadchuk, L.V. (eds.). 1997. *Evolutionary-Genetic and Genetic-Physiological Aspects of Fur Animal Domestication: A Collection of Reports.* Oslo: IFASA/Scientifur.

Voss, S.R., and Shaffer, H.B. 1997. Adaptive evolution via a major gene effect: paedomorphosis in the Mexican axolotl. *Proceedings of the National Academy of Sciences, USA* 94:14185–14189.

Wayne, R.K. 1986. Cranial morphology of domestic and wild canids: the influence of development on morphological change. *Evolution* 40:243–261.

Wichman, H.A., and Lynch, C.B. 1991. Genetic variation for seasonal adaptation in *Peromyscus leucopus:* nonreciprocal breakdown in a population cross. *Journal of Heredity* 82:197–204.

Wilcox, B.W., and Walkowicz, C. 1989. *The Atlas of Dog Breeds of the World.* Neptune City, N.J.: TFH Publications.

Wilson, C.M., and McNabb, F.M.A. 1997. Maternal thyroid hormones in Japanese quail eggs and their influence on embryonic development. *General and Comparative Endocrinology* 107:153–165.

Windle, R.J., Wood, S.A., Lightman, S.L., and Ingram, C.D. 1998. The pulsatile characteristics of hypothalamo-pituitary-adrenal activity in female Lewis and Fischer 344 rats and its relationship to differential stress responses. *Endocrinology* 139:4044–4052.

Zohary, D., Tchernov, E., and Horwitz, L.K. 1998. The role of unconscious selection in the domestication of sheep and goats. *Journal of Zoology, London* 245:129–135.

Chapter 7

The Role of Heterochrony in Primate Brain Evolution

Sean H. Rice

Nearly everyone agrees that heterochrony has played some important role in human evolution. The exact nature of this role and, in particular, the type or types of heterochrony involved are not so well agreed upon. The notion of humans as neotenic apes has a certain poetic resonance (as depicted, e.g., in Aldous Huxley's *After Many a Summer Dies the Swan* [1939]) and makes sense of the physical resemblance between infant apes and humans. On the other hand, humans are larger and longer lived than our closest relatives, the chimpanzees, implying some sort of hypermorphosis (Shea 1988; McKinney and McNamara 1991). Distinguishing between these options has been made difficult by the fact that the concept of heterochrony is used alternately as a catch-all description for any morphological change that involves a change in rate or timing (i.e., any morphological change) and as a specific theory that identifies a well-defined set of mechanistic transformations that may underlie some, but not all, morphological evolution.

Theoretical concepts in science allow us to represent the world in ways that draw our attention to fundamental relationships between different phenomena and give us a way to bring rigorous logic or mathematics to bear so as to discover new phenomena and direct further empirical research. If the concept of heterochrony is to serve as a theory in this sense, and I think that it should, then we must be able to distinguish between what is and is not heterochrony and have some notion of what it means biologically to make such a distinction. I addressed this elsewhere (Rice 1997) by showing that, if we define heterochrony as a uniform change in the rate or timing of some developmental process, with no other internal change to that process, then the traditional categories of heterochrony correspond to meaningful biological transformations that we can test for by comparing ontogenetic trajectories.

In this chapter, I develop a statistical test for heterochrony based on this definition and then apply this test to the trajectories for brain growth in humans and some other primates. The differences between human and chimpanzee brain growth are largely a result of uniform changes in rate and timing, thus heterochrony. Compared to other primates, though, humans and chimps show a novel phase of brain growth that is not a simple heterochronic modification of an ancestral trajectory. I also compare the overall growth of the body in humans and chimpanzees and show that heterochrony seems to be a factor here also, but with different kinds of transformations acting at different stages of growth.

Analyzing Ontogenetic Trajectories

Figure 7.1 shows the transformations that correspond to different types of heterochrony (see Rice 1997 for justification and derivation). I refer to two trajectories as being *commensurate* if we can superimpose one on the other by applying some combination of these transformations. If two trajectories are commensurate, then we can infer that the difference between the two growth processes *could* be accounted for by a uniform change in rate or timing.

By contrast, if two trajectories cannot be related by some combination of these transformations, then we can infer that there must have been some change in the nature of the interactions underlying the growth process, not just a change in the rate or timing of that process. This definition of the types of heterochrony is compatible with that of Alberch et al. (1979), with one modification. In Alberch et al., the endpoint of the trajectory was held fixed in time unless there was progenesis or hypermorphosis. I am allowing the endpoint to shift if the entire growth process is slowed down (neoteny) or sped up (acceleration). This will be important in the discussion of whole-body growth.

Because this definition assigns so much significance to the superposition of trajectories and because data for actual trajectories are likely to be noisy, we seek a statistical test to compare trajectories and potentially reject a hypothesis of heterochrony. Often, ontogenetic trajectories must be inferred from clouds of points, each of which represents a separate individual (Fig.

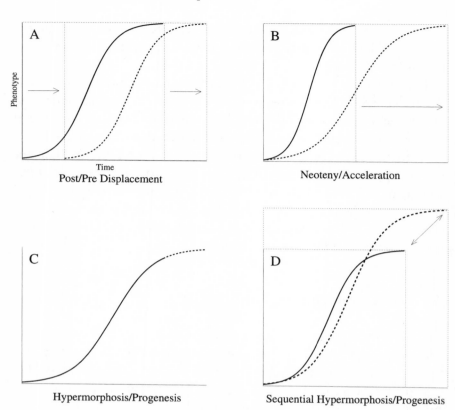

Fig. 7.1. Transformations of ontogenetic trajectories corresponding to different types of heterochrony. *A:* Shifting the entire trajectory to the *right* constitutes *postdisplacement;* the converse is *predisplacement. B:* A uniform stretching of the time axis, as if each value on that axis were multiplied by a constant greater than 1, constitutes *neoteny.* Contracting the time axis by multiplying each time value by a constant between 0 and 1 produces *acceleration. C:* Allowing the trajectory to continue beyond its ancestral termination is *hypermorphosis.* Stopping it before this point is *progenesis. D: Sequential hypermorphosis* involves stretching both the time and phenotype axes by the same amount.

7.2). After overlapping the trajectories by applying some combination of the transformations in Figure 7.1 to one of them, we need to decide whether they are commensurate. One way to do this is to fit a curve through one of the sets of points (species A), making sure that the points representing species A are symmetrically distributed around this curve, then to check the distribution of points representing species B around the same curve. If the two trajectories are commensurate, implying heterochrony, then each point representing species B should have a probability of 0.5 of lying above (or below) the line. If the points represent different individu-

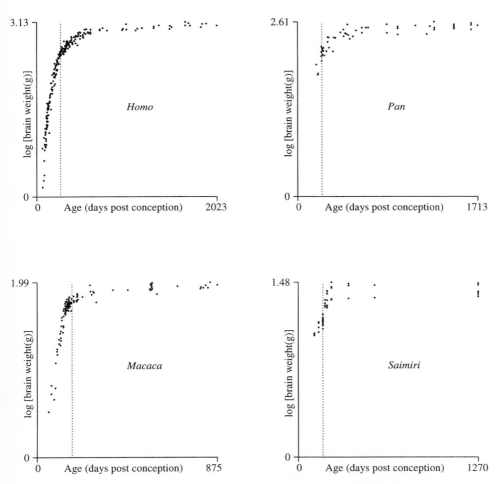

Fig. 7.2. Ontogenetic trajectories for log of brain weight (in grams) with time for human, chimpanzee (*Pan paniscus*), macaque (*Macaca mulatta*), and squirrel monkey (*Saimiri sciureus*). The bimodal adult trajectory for squirrel monkeys is a result of sexual dimorphism. The *dotted lines* represent age at birth. *Sources:* Human data are from Dobbing and Sands 1973 and Blinkov and Glezer 1968. Chimpanzee data are from Vrba 1998 and Shantha and Manocha 1969. Macaque data are from Cheek 1975. Squirrel monkey data are from Manocha 1978. Except for Dobbing and Sands 1973 and Manocha 1978, the data were calculated from digitized images of plotted points.

als, then each is an independent event, and we can think of the sequence of points lying above or below the line as analogous to a sequence of coin flips.

This suggests a test: define a run of length *x* as a set of *x* points for one species, adjacent to one another along the time axis, that all lie on the same side of the fitted curve for the other species. We seek the probability of encountering a run of length x or longer if the two trajectories are in fact commensurate.

Let $\Pi_{x,n}$ be the probability that there is *no* run of length x or longer in a set of n residuals. Then,

$$\Pi_{x,n} = 1 \text{ for } n < x, \Pi_{x,n} = 1 - \frac{1}{2^{x-1}}$$

Values of $\Pi_{x,n}$ can be found from these relations and the recursion:

$$\Pi_{x,n} = \Pi_{x,n-1} - \frac{1}{2^x}\Pi_{x,n-x}$$

(see the appendix for the derivation). Thus, if we observe a run of x in set of n points, then

$$p = 1 - \Pi_{x,n}$$

gives the probability of observing such a run (or a longer one, since a run of $>x$ contains a run of x) if the trajectories were commensurate.

The most problematic step in this process is that of fitting a curve to one of the datasets. The best way to do this depends on the nature of the data. In the analysis that follows, I first decide which dataset will be transformed and then find the distance between each point in this set and the regression line through the ten points in the other dataset that surround it (five on each side along the time axis); only those points for which this can be done are considered. This gives the necessary result that, when any of the datasets is compared with itself in this way, the residuals are evenly distributed around zero. I refer to this distance, from a point for one species to the local regression line for the other species, as the *between-species residual* for that point.

Brain Growth

Figure 7.3*A* shows growth trajectories for humans and chimpanzees (*Pan paniscus*) and, below these, a plot of the between-species residuals representing the distance between each chimpanzee data point and the mean of the human points at that age (calculated by taking the regression through ten

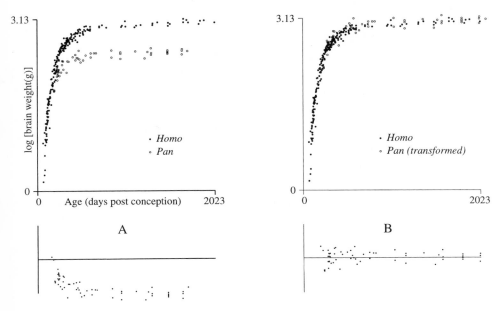

Fig. 7.3. Comparison of human and chimpanzee brain growth trajectories. The plots below the trajectories show the distribution of residuals representing the distance between each point on the chimpanzee curve and the local best-fit regression line through human data. *A:* Untransformed trajectories. *B:* The chimpanzee trajectory has been transformed by sequential hypermorphosis, multiplying both the phenotype and time axes by 1.22. With this transformation, the residuals come to be evenly distributed around the human curve. There are fifty-four residual points, and the longest run on the same side of the human line is four.

points of the human data, five on either side of the chimpanzee point being considered). Figure 7.3*B* presents the same comparison after the chimpanzee data have been transformed by sequential hypermorphosis. The between-species residuals are now evenly distributed around the human mean. Out of fifty-four points, the longest run of points on the same side of the regression line is four. Since the probability of getting a run of four or more if the transformed chimpanzee points were drawn from the same distribution as the human points is >0.9 and since the common ancestor had a brain closer in size to that of a chimpanzee, we conclude that the difference between these two trajectories is explained by sequential hypermorphosis along the human lineage. Vrba (1998) arrives at the same conclusion.

Showing that two ontogenetic trajectories are not commensurate is sufficient to show that there must have been some change in the nature of the developmental process underlying the characters under study, not just a uniform rate change. The converse, however, is not true. Commensurate trajectories *could* be related by some sort of heterochrony, but it is always possible that changes internal to the developmental process might not significantly alter the overall trajectory. This said, the match between the human and chimpanzee trajectories indicates that there have been no large-scale changes in growth of different parts of the brain since our mutual common ancestor.

The same analysis applied to humans and macaques (*Macaca mulatta*) tells a different story (Fig. 7.4). Though sequential hypermorphosis applied to the macaque data brings it into line with the fetal stage of human growth and the later stage after about one year after birth, the two curves are clearly not commensurate over the period from birth to about one year of age in humans. This is particularly apparent in the plot of between-species resid-

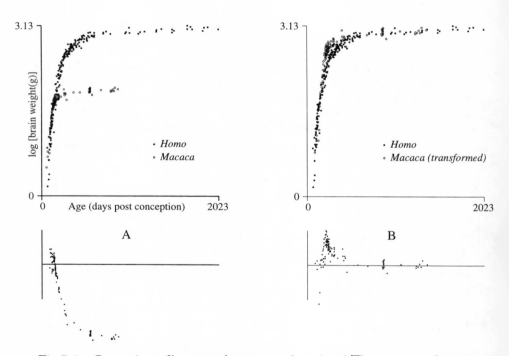

Fig. 7.4. Comparison of human and macaque trajectories. *A:* The macaque trajectory closely follows that for humans initially and then deviates at about the time of birth in macaques. *B:* Expanding the macaque curve by a factor of 1.56 produces a good match for the early and late parts of the trajectories but leads to an overshoot at around the time of birth. This is apparent in the plot of residuals, which has a run of 66 points (out of 101) above the human curve. The probability of a run of this length or longer arising by chance if the curves were actually the same is less than 10^{-6}.

uals, which shows a distinct bump during this stage. The longest run of residuals on the same side of the line is 66 (out of 101 points); the probability of such a run (or a longer one) is less than 10^{-6}. Further transforming the macaque data with neoteny improves the fit but not to the degree that we would accept heterochrony as an explanation of the differences between these two species.

Considering just three species, we cannot say whether the human/chimp brain growth trajectory is ancestral or derived within the primates. To resolve this, we need another outgroup (Maddison et al. 1984). The squirrel monkey (*Saimiri sciurius*) is a New World monkey that is an outgroup to the clade containing the Old World monkeys (including the macaque) and the apes. The ontogenetic trajectory of the squirrel monkey brain differs from all those considered thus far in that the growth spurt is delayed until after birth. When it occurs, though, the growth spurt and subsequent leveling off of brain size are commensurate with the trajectory for macaques (Fig. 7.5). Thus, the kind of trajectory exhibited by chimpanzees and humans seems to be derived within the primates.

It is generally argued that postnatal growth of the human brain is simply an extension of the fetal pattern of growth beyond our premature (relative to other primates) birth date (Deacon 1997; Martin 1983), either with a simple extension or through sequential hypermorphosis (McKinney and McNamara 1991). The data considered here suggest that this is not the case. When we plot brain size against time, the initial growth phase following birth clearly shows a slope different from that of the prenatal phase (Fig. 7.6*A*). When we plot brain size against body size (as is usually done), the difference in slopes is less pronounced but still significant (Fig. 7.6*B*; using an *F* test for equality of slopes [Sokal and Rohlf 1995]). Thus, the distinctive growth phase seen in humans and chimpanzees does not seem, in and of itself, to be a heterochronic modification of any phase in the growth of other primates. Count (1947) noted this pattern also, defining a "transitional" growth phase between the "fetal" and "post-infantile" phases (see Count's Figs. 1 and 2).

Putting all of these results together on a phylogeny (Fig. 7.7) suggests that, sometime between the common ancestor of Old World monkeys and apes, around 25 million years ago (Goodman et al. 1982), and that of chimpanzees and humans (around 5 million years ago), a novel phase of brain growth appeared in the hominoid line. This growth phase begins at around the time of birth and continues for approximately nine months in chimpanzees and one year in humans.

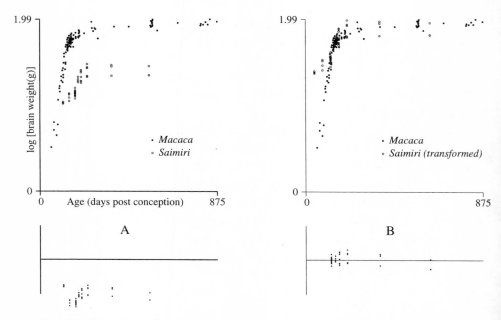

Fig. 7.5. Macaque and squirrel monkey trajectories. *A:* The growth spurt of the squirrel monkey brain occurs much later than that of the macaque. *B:* Transforming the squirrel monkey trajectory with a combination of sequential hypermorphosis ($\times$ 1.35) and predisplacement brings the growth spurt up to a fairly good match with that of the macaque. The maximum run of residuals is six in a total of forty-two points ($p = 0.477$).

This notion that the brains of humans and chimpanzees are developmentally similar to one another and different from those of other primates is consistent with the observation that chimpanzees show a humanlike pattern in the asymmetry of the planum temporale, a region of the brain associated with language in humans (Gannon et al. 1998). Thus, whatever internal modifications of brain growth set the stage for the evolution of the human brain, they seem to have been in place by the time of the most recent common ancestor of humans and chimpanzees.

Finally, a note on hypermorphosis. Although it seems to be one of the commoner types of heterochrony, hypermorphosis is, from the standpoint of analysis, the most problematic because we cannot know where a trajectory would go if "allowed" to continue. The test for sequential hypermorphosis discussed here is applicable only if the trajectories in question are composed of linear segments (which seems to be the case for log-transformed brain trajectories [Vrba 1998]).

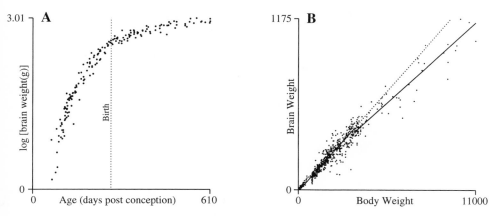

Fig. 7.6. *A:* Human brain growth trajectory up to the age of about one year after birth. The postnatal growth phase is not a hypermorphic extension of prenatal growth. *B:* Brain weight in relation to body weight for humans up to about one year after birth. The *solid lines* are the reduced major axis regression lines through the points up to 3,500 g body weight (around birth) and after this point. The *dotted line* is the extension of the fetal regression line. The longest run of points on one side of this extended fetal line is fifteen ($p < 0.01$ by the test described earlier). *Sources:* Data are from Dobbing and Sands 1973, Blinkov and Glezer 1968, Larroche 1967, and Burn et al. 1975. Estimates of age by weight are from Dobbing and Sands 1973.

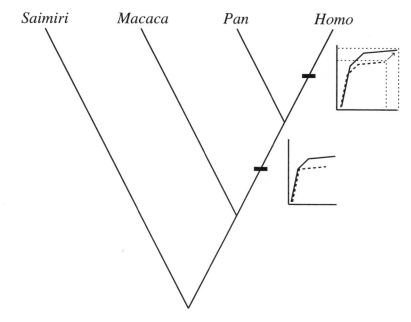

Fig. 7.7. Phylogeny of anthropoids showing changes in the ontogenetic trajectory for brain size.

Body Size

Figure 7.8 shows trajectories of body size for humans and chimpanzees (males in each case). A combination of neoteny and sequential hypermorphosis applied to the entire chimp trajectory yields a reasonable overlap (Fig. 7.8B; Rice 1997 considered only this comparison). If, however, we consider separately the curves before and after the onset of the adolescent growth spurt, we find that a combination of sequential hypermorphosis, neoteny, and predisplacement explains the differences between the trajectories up to the onset of the growth spurt (Fig. 7.8C), while the same amount of sequential hypermorphosis combined with slight acceleration yields a striking match for the trajectories after this point (Fig. 7.8D).

The need for some predisplacement to align the trajectories is probably an artifact caused by measuring age since birth, which in chimpanzees occurs after a shorter gestation than in humans. The apparent postdisplacement in Figure 7.8D is a consequence of anchoring the trajectories at the point at which they diverged in Figure 7.8C. Because some of these data are derived from following the growth of particular individuals, we cannot treat each point as an independent event and thus cannot apply the statistical test discussed above. The close match of the trajectories, though, especially over highly nonlinear regions, makes a strong case for heterochrony.

Both neoteny and hypermorphosis thus seem to have played roles in the evolution of human growth. More significantly, this analysis draws our attention to the onset of the adolescent growth spurt in humans as a point at which there has been a change in the nature of the growth process. This is consistent with the arguments of Bogin and Smith (1996) that the adolescent growth spurt in humans is a novel character distinct from the small increases in growth seen in some other primates. Here, this novelty is seen to be a consequence of slowing growth (neoteny) until the onset of puberty.

Partitioning an ontogenetic trajectory in this way and then looking for heterochrony in the different parts is dangerous. We can render any trajectory into a large number of approximately linear segments; if we are then allowed to apply any kind of heterochrony independently to each segment, we can make the resulting trajectory look like whatever we want it to. Breaking up the trajectory is appropriate in this case, since the two parts have distinct, nonlinear forms and the breakpoint corresponds to a developmentally distinguishable event, the onset of puberty.

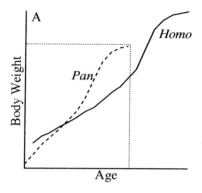
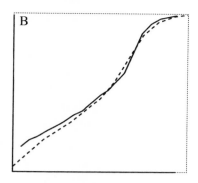
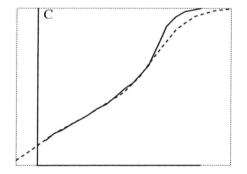
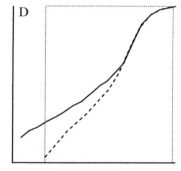

Fig. 7.8. Ontogenetic trajectories for body weight as a function of age for humans and chimpanzees (*Pan*). *A:* Untransformed trajectories. *B:* The best match for the entire trajectories obtained with a combination of sequential hypermorphosis and neoteny applied to the chimpanzee curve. *C:* Chimpanzee curve transformed by sequential hypermorphosis (both axes stretched by a factor of 1.28) and neoteny (age axis stretched by a factor of 1.6) combined with predisplacement. *D:* The same amount of sequential hypermorphosis combined with a slight amount of acceleration (untransformed trajectory compressed by a factor of 0.96). The chimp curve here is positioned so that the onset of the growth spurt occurs at the same point as in *C. Sources:* Human data are from Tanner 1978; chimpanzee data are from Grether and Yerkes 1940.

Conclusions

Even using a definition that would reject most changes in development, one can still conclude that heterochrony has played an important role in human evolution. In particular, the human brain follows a growth curve that is almost exactly what would be expected from sequential hypermorphosis applied to the chimpanzee brain. This said, the case in which heterochrony is rejected as an explanation (the comparison of human/chimp brain growth with that of other primates) is in some ways the most interesting. Rejecting heterochrony as a mechanism here highlights the appearance of a novel phase of brain growth that arose before the common ancestor of humans and chimpanzees. Along with the results of Gannon et al. (1988), who showed that chimpanzee brains are morphologically similar to human brains even in a character that (in humans) functions in language processing, the results presented here imply that much of the important rearranging of brain growth that led to the evolution of the modern human brain had already taken place before the split with our closest living relative.

The potential to reject a hypothesis of heterochrony is also important in the comparison of overall growth in body size. We can only accept or reject a hypothesis of heterochrony with respect to a particular trajectory or segment thereof. Rejecting heterochrony at a particular level may direct our attention to other levels. In this case, rejecting heterochrony at the level of the entire body size trajectory leads to the observation that different kinds of transformations are involved at different stages in the growth process. This, in turn, leads to insight into the evolution of another novel character, the human adolescent growth spurt. Thus, rather than rendering the concept too restrictive to be useful, narrowing and clarifying the definition of heterochrony makes it a more useful theoretical tool for the analysis of morphological evolution.

APPENDIX

Sequential Hypermorphosis

The transformations shown in Figure 7.1, *A* through *C,* have the property that the same transformation of the phenotype axis will linearize both trajectories (Rice 1997). They thus correspond to the traditional types of heterochrony discussed by Alberch et al. (1979). Figure 7.1*D* shows a transformation that does not meet this criterion but is still reasonably interpreted in terms of heterochrony. Because this particular transformation will play a major role in the discussion of primate evolution and because it strictly lies outside the classical categories of heterochrony, I discuss it in a little more detail.

Consider a phenotypic character, ϕ, present in both an ancestor and a descendant, that follows ontogenetic trajectories $\phi_A(t)$ and $\phi_D(t)$ in the ancestor and descendant, respectively. We can write the derivatives of these trajectories as

$$\frac{d\phi_A}{dt} = \omega\,(\phi,p) \text{ and } \frac{d\phi_D}{dt} = \omega'\,(\phi,p')$$

where p and p' are sets of parameters that include any other factors influencing the development of the character. The transformations shown in Figure 7.1A–C are cases in which, for a particular value of ϕ, call it ϕ^*, that is visited by both trajectories,

$$\omega(\phi^*,p) = C\omega'(\phi^*,p')$$

where C is a constant. In other words, the growth process has been sped up ($C > 1$) or slowed down ($C < 1$) or shifted as a unit, with no change in the structure of the equation describing growth. By contrast, the transformation in Figure 7.1D corresponds to

$$\omega(\phi^*,p) = \omega'(C\phi^*,p')$$

Here, the constant C is moved inside the function and multiplies ϕ wherever it occurs. This corresponds to a case in which, for each value of the phenotype, the developmental process of the descendant behaves as that of the ancestor would at an earlier ($C < 1$) or later ($C > 1$) point in development. Geometrically, if we think of the trajectory as being made up of many small linear segments, then this is the same as extending ($C < 1$) or contracting ($C > 1$), each segment by the same amount. This corresponds to what McNamara has called sequential hypermorphosis (or progenesis). However, sequential hypermorphosis of each segment of the trajectory will look like Figure 7.1D only if those segments are linear (Vrba 1998).

Statistics

Consider a string of points each designated either A (above the line) or B (below). The state of each point is independent of the others, and each has a probability of 0.5 of being A (or B). A run of length x is an uninterrupted sequence of x points with the same state.

To derive the recursion given in the text, we seek the probability that adding the nth point onto a string does not complete the first run of length x. This is equal to the probability that there was no run of x in the first $n - 1$ points, $\Pi_{x,n-1}$, multiplied by the probability that, given that there was no such run, the nth point does not complete one.

First, we calculate the probability that the nth point does complete the first run of length x. For this to be the case, the last $x - 1$ points [counting backward from the $(n - 1)$th point] must have been the same and the one just before these different. There are only two ways to achieve this, $(x - 1)$ As followed by a B or the reverse. The probability that the last x points have this property is thus $2(2^{-x}) = \frac{1}{2}^{x-1}$, and the number of possible sequences of $n - 1$ points with just the last $x - 1$ the same is $2^{n-1}/2^{x-1}$. The probability that there was no run of length x in the previous $(n - 1) - (x - 1)$ points leading up to these is, by definition, $\Pi_{x,n-x}$, so the total number of possible sequences with no run of length x but the last $x - 1$ constituting a run is

$$\frac{2^{n-1}}{2^{x-1}} \Pi_{x,n-x} \tag{7A.1}$$

The conditional probability, given that there was no previous run of x, of the last $x - 1$ points being the same is simply Equation 7A.1 divided by the total number of sequences of $n - 1$ with no run of x, which is

$$2^{n-1}\Pi_{x,n-1} \tag{7A.2}$$

There is a probability of $1/2$ that the next (nth) point has the same value as the previous $x - 1$ points, so the probability that the nth point completes a run of x, given that there is no such run in the first $n - 1$ points, is $\frac{1}{2}$ times Equation 7A.1 divided by Equation 7A.2, or

$$\frac{1}{2^x} \frac{\Pi_{x,n-x}}{\Pi_{x,n-1}} \tag{7A.3}$$

The probability that the nth point does not complete the first run of x is then $\Pi_{x,n-1}$ multiplied by $[1 - (7A.3)]$, which gives the result presented in the text.

References

Alberch, P., Gould, S.J., Oster, G.F., and Wake, D.B. 1979. Size and shape in ontogeny and phylogeny. *Paleobiology* 5:296–317.

Blinkov, S.M., and Glezer, I.I. 1968. *The Human Brain in Figures and Tables.* New York: Plenum Press.

Bogin, B., and Smith, B.H. 1996. Evolution of the human life cycle. *American Journal of Human Biology* 8:703–716.

Burn, J., Birkbeck, J.A., and Roberts, D.F. 1975. Early fetal brain growth. *Human Biology* 47:511–522.

Cheek, D.B. 1975. The fetus. In D.B. Cheek (ed.), *Fetal and Postnatal Cellular Growth,* 3–22. New York: John Wiley and Sons.

Count, E.W. 1947. Brain and body weight in man: their antecedents in growth and evolution. *Annals of the New York Academy of Sciences* 46:993–1122.

Deacon, T.W. 1997. *The Symbolic Species: The Coevolution of Language and the Brain.* New York: Norton.

Dobbing, J., and Sands, J. 1973. Quantitative growth and development of human brain. *Archives of Diseases in Childhood* 48:757–767.

Gannon, P.J., Holloway, R.L., Broadfield, D.C., and Braun, A.R. 1998. Asymmetry of chimpanzee planum temporale: humanlike pattern of Wernicke's brain language area homolog. *Science* 279:220–222.

Goodman, M., Romero-Herrera, A.E., Dene, H., Czelusniak, J., and Tashian, R.E. 1982. Amino acid sequence evidence on the phylogeny of primates and other eutherians. In M. Goodman (ed.), *Macromolecular Sequences in Systematics and Evolutionary Biology,* 115–191. New York: Plenum Press.

Grether, W.F., and Yerkes, R.H. 1940. Weight norms and relations for chimpanzees. *American Journal of Physical Anthropology* 27:181–197.

Huxley, A. 1939. *After Many a Summer Dies the Swan.* New York: Harper.

Larroche, J-C. 1967. Maturation morphologique du systeme nerveux central: ses rapports avec le developpement ponderal du foetus et son age gestationnel. In A. Minkowski (ed.), *Regional Development of the Brain in Early Life,* 247–256. Philadelphia: F. A. Davis.

Maddison, W.P., Donoghue, M.J., and Maddison, D.R. 1984. Outgroup analysis and parsimony. *Systematic Zoology* 33:83–103.

Manocha, S.L. 1978. Physical growth and brain development of captive-bred male and female squirrel monkeys, *Saimiri sciureus. Experientia* 35:96–98.

Martin, R.D. 1983. *Human Brain Evolution in an Ecological Context: 52nd James Arthur Lecture.* New York: American Museum of Natural History.

McKinney, M.L., and McNamara, K.J. 1991. *Heterochrony: The Evolution of Ontogeny.* New York: Plenum Press.

Rice, S.H. 1997. The analysis of ontogenetic trajectories: when a change in size or shape is not heterochrony. *Proceedings of the National Academy of Sciences, USA* 94:907–912.

Shantha, T.R., and Manocha, S.L. 1969. The brain of chimpanzee: I. External morphology. In G.H. Bourne (ed.), *The Chimpanzee,* 1:188–237. Basel: S. Karger.

Shea, B.T. 1988. Heterochrony in primates. In M.L. McKinney (ed.), *Heterochrony in Evolution: A Multidisciplinary Approach.* New York: Plenum.

Sokal, R.R., and Rohlf, F.J. 1995. *Biometry,* 3d ed. New York: Freeman.

Tanner, J.M. 1978. *Foetus into Man.* Cambridge: Harvard University Press.

Vrba, E.S. 1998. Multiphasic growth models and the evolution of prolonged growth exemplified by human brain evolution. *Journal of Theoretical Biology* 190:227–239.

Part II

The Evolution of Hominid Life-History Patterns

Chapter 8

Brain Evolution by Stretching the Global Mitotic Clock of Development

M ICHAEL L. M C K INNEY

> Neurologically, there is nothing paedomorphic about the human brain.
>
> Gibson 1991

> Mutation of a regulatory gene that controls the timing of cell divisions in the proliferative zone could create an expanded cortical plate.
>
> Rakic 1995

> The present multiphasic results suggest that a single kind of heterochrony affected whole-brain weight in a relatively simple and orderly way.
>
> Vrba 1998

In this chapter, I discuss evidence that mammalian and human brain evolution exemplifies a classic heterochronic pattern because brain development is highly constrained compared with most other organs. The particular type of heterochronic pattern that dominates human brain evolution is sequential delay of developmental stages to produce an overdeveloped brain (McKinney 1998). McNamara (in Chap. 5) provides an overview of the role of sequential timing change in human evolution. My goal here is to review some of the recent findings on the genetic, cellular, and developmental mechanisms that have influenced human brain evolution. Such mechanistic explanations are essential if we are to understand the manifold causes of heterochronic patterns and lift the study of heterochrony beyond a taxonomy of words (McKinney 1999), as suggested in many past critiques of heterochronic studies (e.g., Raff 1996).

The accumulating evidence strongly disagrees with the recurring argument that humans have a juvenilized brain (reviewed in Gibson 1991; Bogin 1997; McKinney 1998). Instead, the opposite process, of overdevelopment by sequential extension of all stages of brain growth and development, is a far more enlightening way of explaining the evolution of human brain morphology and function. In general, human evolution by developmental timing changes can be ultimately attributed to a few relatively simple changes that are first evident in very early ontogeny: (1) allotment of relatively more brain stem cells and (2) a global mitotic clock that permits sequential delays in cell differentiation, causing more mitotic cell cycles, which produce a larger, more complex brain and a longer duration of all developmental stages.

Data support a model that sequential delays in the global mitotic clock produce disproportionately longer delays in later developmental events. This occurs because the rate of cell cycling decreases with developmental age so that extending each developmental stage by a consistent number of cell cycles will produce progressively longer delays when measured by absolute time.

Why Is Brain Evolution So Constrained?

Constraints on the evolution of brain shape are seen in the extremely high correlation among brain structures during evolutionary size increase across a wide size range of mammal species (Finlay and Darlington 1995). This is also seen in the tendency of body and brain growth to follow the same ontogenetic trajectory in many mammal species (Deacon 1990). Deacon (1997) noted the interesting implication that the evolution of brain/body ratios has most often involved changes in body size growth rather than brain growth.

Upon reflection, it seems apparent why brain shape is so evolutionarily conservative. Compared with most other morphological traits, shape in the brain is much less closely related to function. Unlike muscular organs such as limbs, where kinetic functions affect shape, or intestines, where digestive functions affect shape, the function of the brain for storing and processing information is largely independent of shape, so that there has been no selective demands to alter brain shape. Instead, new environmental demands have been met by alterations in the neural networks at fine microscopic scales. Deacon (1997), for example, reviewed evidence for evolutionary changes in network allometries among neurons and dendrites that occur with brain size change.

This process has been promoted by the extreme functional plasticity of cortical neurons. This plasticity allows them to be coopted for a variety of different purposes in both development and evolution. For instance, much of our prefrontal area, which is essential for synthesis and planning, has apparently been coopted from other parts of the primate brain that are no longer used for other functions (Deacon 1997). Cooption of function among body parts is a common theme in the evolution of morphological development (Raff 1996). It is therefore not surprising to see it occurring in the evolution of brain development, especially given the brain's intrinsic ability to adjust to environmental changes during development.

Evidence of the cellular mechanisms causing such cooptive adjustments during development indicates that growing body parts actually compete among themselves for resources (Nijhout and Emlen 1998). In the case of the brain, the growth of neurons and their axonal, dendritic, and synaptic extensions clearly follows a similar competitive process whereby increasing use promotes the survival of some neural pathways over others (Deacon 1997). This kind of competition among developing cells and parts produces a developmentally buffered system that can adjust to environmental change. The result is a correlated growth process that produces the mathematically elegant allometric patterns that characterize almost all developmental (and evolutionary) trajectories (McKinney and McNamara 1991). In the brain, this allows the organ to alter the size and even function of certain parts without substantial alteration of the overall shape of the organ itself.

Besides shape, brain evolution is conservative in other ways. Genetically, there is often surprisingly little difference among brains of greatly varying morphology. Even among distant phyla, the homeotic genes that influence segmental divisions of the embryonic nervous system are remarkably conserved (Holland et al. 1992; Finkelstein and Boncinelli 1994). Again, this is probably produced by the uniquely plastic nature of the developing brain. The number of genes required to specify all of the information coded into a large brain would be prohibitively huge (Purves 1988). Instead, large brains rely on the production of a nervous system that is strongly influenced by the environment to record and process information, by such processes as competition among neurons and synaptic pathways (Edelman 1987). In terms of natural selection, this has the effect of buffering genes influencing the brain from many direct effects of environmental changes.

Rates of neuron mitosis in the mammalian fetus are also constrained. Neural mitotic rates are very similar in all mammal species, perhaps representing the maximum rate in the prenatal environment (Sacher and Staffeldt 1974). The crucial result is that the rate of human brain growth does not distinguish us from other mammals (Deacon 1997, 173). Instead, large mammalian brains are produced by extending the period of brain growth (Finlay and Darlington 1995). Equally important, humans also slow down the rate of body growth to produce our very high brain/body ratio. Deacon (1997) estimates that our brains would grow about as fast if they were in an ape body that would otherwise attain an adult body size exceeding 1,000 pounds.

In addition to extending the fetal period of neuron mitosis, all phases of human brain development seem to be extended (reviewed in McKinney 1998): glial cell mitosis, dendritic growth, synaptogenesis, dendritic pruning, and nerve myelination are examples of processes that are prolonged in humans relative to all other primates. Because the rate of growth is not reduced (as it is with body size), these delays produce an overdeveloped brain that is the opposite in almost every way of that expected from a juvenilized brain model (McKinney 1998; see also Gibson 1991; Deacon 1997). This process of developmental prolongation to produce more cells is another common theme in evolution. Developmental evolution often proceeds via repetitive duplication of preexisting units (Raff 1996), so we see that, as with cooption, brain evolution exhibits patterns seen in other morphological units.

Mechanisms of Sequential Brain Development

The observation that general aspects of the human brain, such as its size, evolved via delayed development is not new (Gould 1977). Only recently, however, have the genetic and developmental details of this process begun to emerge. Of special note is the discovery of global mitotic clocks in brain tissues indicated by xenotransplants of brain tissue grafts from the fetuses of one species into another. When fetal pig (Deacon et al. 1994) or fetal human (Wictorin et al. 1992) neuroblasts are transplanted into adult rat host brains, they mature far slower than do cells from the corresponding rat brain structures. This, plus other lines of evidence (reviewed in Deacon

1997), support the presence of a global mitotic clock in mammalian brains, as advocated by Finlay and Darlington (1995): all cells in a zygote share a common mitotic setting that influences the length of time until cells differentiate to their terminal state and therefore cease mitotic multiplication.

The xenotransplant experiments show pig and human neuroblast prolongation in both (1) the time to cell differentiation and (2) the duration of cell differentiation itself. Because cell differentiation terminates mitosis, prolonging the time to differentiation will increase the total number of mitotic cell cycles and thereby increase the total number of cells. Prolonging differentiation will increase the degree of morphological differentiation of the terminal cells themselves; in this case, neurons. Clearly, both of these play key roles in determining the size and complexity of the human brain.

The biochemical mechanisms that cause these relative prolongations of cell differentiation apparently involve the accumulation of signal molecules in the cytoplasmic material of daughter cells during mitosis (Raff 1996). Given that these signal molecules eventually initiate cell differentiation, slower accumulation (or a smaller initial amount) of these molecules during mitotic cycles would permit more cycles to occur before differentiation.

Regional modifications of the embryo are superimposed on this global mitotic clock. Deacon (1997) presents evidence that these regional modifications originate very early in development by the action of homeotic genes that apportion segments within the developing embryo. Organs within segments that receive relatively larger portions of initial stem cells than in the descendant will grow to larger sizes than the same organs in the ancestor, given that the same number of cell cycles occurs in both ancestor and descendant. Conversely, organs in segments with smaller onset cell populations will produce smaller organs than in the ancestor. Primates, for example, generally have larger brain/body ratios than other mammals because primate embryos receive a smaller initial population of somatic (body) cells and a relatively larger initial population of brain cells (Deacon 1997). This process can be shown experimentally in vertebrates by removing a substantial number of undifferentiated cells from the developing blastula to produce a dwarfed but otherwise normal body that develops at a normal rate (Deacon 1997).

The role of onset cell population size is evident in the evolution of the large human neocortex (now often called the isocortex), which is the basis of human cognitive abilities. Of special note is the surface area of the neocortex, which has ratios of about 1:100:1000 in mouse:monkey:human, while the thickness of the neocortex varies only by a factor about 2 across these species (Blinkov and Glezer 1968). Rakic (1995) proposed a widely cited radial unit hypothesis that the much greater surface area of human neocortex compared to monkeys (macaque) largely arises from a roughly three-day prolongation of neocortical neuronogenesis when founder cells are being produced.

Specifically, study of the macaque embryo shows that there are two basic phases of neocortical neuranogenesis: the phase before E40 (fortieth day after fertilization), when most of the founder cells of the prospective cerebral cortex are formed, and the phase of ontogenetic column formation, which begins around E40 and continues until the completion of corticogenesis in a given region. The duration of the first phase determines the number of radial units in the cortex of a given species and, indirectly, the size of the cortical surface, whereas the duration of the second phase regulates the number of neurons within each ontogenetic column (Rakic 1995, 386). The tenfold larger human cortex arises because the first phase is prolonged by about three days, to E43, allowing an extra three or four rounds of founder cell mitosis (Rakic 1995). In contrast, the twenty-day delay in termination of phase 2 that occurs in humans (E120 for humans vs. E100 for the monkey) produces an increase of only about 10 percent in human cortical thickness.

The subsequent growth of this enlarged cortical area in humans is, of course, not a simple matter of homogeneous expansion. Different areas of cortex are themselves allotted more or fewer neurons. In part, this can be related to different patterns in the origin of cortical founder cells. For example, in humans, some cortical areas begin to terminate founder cell proliferation before E43 and others continue after E43 (Rakic 1995). Such local differences in cortex size and maturation may be ultimately attributable to homeotic gene patterning. Finlay and Darlington (1995) show that the later an area of brain cortex develops, the larger is the relative size of that area in the developing brain. For example, prefrontal cortex, which is the crucial area for many higher brain functions, is the last-developing area of cortex and is expanded over 200 percent compared to the same area in apes (Deacon 1990).

It is therefore notable that these enlarged regions of the human brain also correspond to adjacent domains of expression in homeotic genes (Deacon 1995). The enlargement of brain structures therefore seems to be produced, once again, by changes in the apportionment of initial cell populations in the very early embryo, as divided up by homeotic genes. Specifically, there are distinct temporal gradients in the expression of many segment-specific homeotic gene families (e.g., Holland et al. 1992). Thus, the *Hox, Otx,* and *Emx* gene families are expressed in a rostral to caudal temporal gradient across segments that affect brain development in the same sequence that those segments are expressed in the developing brain (Simeone et al. 1992).

Apportionment of cortical neurons to specific regions may thus be determined very early in the embryo. After this, growth of and within brain regions is a highly integrated process that involves the proliferation of supportive glial cells and of increasing neural interconnections via dendritic and synaptogenetic expansion. The great complexity of these neural orchestrations requires that they are under epigenetic controls involving cell-cell interactions. This has the crucial implication that increased numbers of cortical neurons will enter into developmental cascades that allow them to be epigenetically integrated with one another as functional parts of a larger brain. Neural pathways that are used are strengthened while those not used will atrophy. Such processes may produce allometric patterns among neural networks (Deacon 1997) and the parcellation of these networks into local and regional patterns. Once larger numbers of cortical neurons are produced, both globally and regionally, much of the rest will take care of itself because evolution has endowed each mammalian brain with the ability to assemble itself.

A Model of Increasingly Delayed Developmental Stages

Table 8.1 summarizes the main developmental changes that apparently produced modern humans. I focus on cellular-level changes because they provide more direct insight into the mechanisms that produced our overdeveloped brain. The less ($-$) or more ($+$) values reflect the relationship of those parameters to our ancestors, where ($+$) causes an increase in size of the cell population in the adult and ($-$) causes a decrease (McKinney and McNamara 1991). Our high brain/body ratio occurs because the brain is allotted relatively greater numbers of embryonic stem cells, with fewer cells being allotted to the body (Table 8.1). Body growth is slowed (reduced rate of mitotic cycling) while differentiation of body cells is delayed (see also McNamara, Chap. 5; Rice, Chap. 7). The result of these counteracting processes ($2-$, $1+$) is to produce an adult that is somewhat larger than our ancestors but grows much more slowly. In contrast, brain growth (mitotic) rate is not relatively slowed and brain cells generally have both a larger on-

Table 8.1 Main Developmental Changes That Produced Modern Humans

	Body	Brain	Traditional Heterochronic Terms
Cell population onset size	less ($-$)	more ($+$)	predisplacement of brain
Cell population growth rate	less ($-$)	same (0)	neoteny of body growth
Cell differentiation delay	more ($+$)	more ($+$)	hypermorphosis of both

set population size and delayed stages of growth and maturation to produce a larger, more complex structure (Table 8.1).

Aside from my focus on cell-level processes, none of the above paragraph is especially new. Even proponents of juvenilization in human brain evolution have emphasized that humans delay developmental events (e.g., Gould 1977; reviewed in Bogin 1997 and McKinney 1998). The problem is that these delays have been explained by rather vague references to timing or regulatory genes without any detailed, quantitative descriptions of the mechanisms that cause them (McKinney 1998). I have already reviewed some mechanisms relevant to human evolution that have emerged from recent research, including the fact that the initial stem cell can be adjusted to modify adult organ size. Initial stem cells are determined very early by the action of homeotic genes. Also, the rate of organ growth can be altered to modify adult organ size. Mechanisms that control growth rates include global biochemicals (e.g., hormones) and cell-cell interactions (Raff 1996).

However, perhaps the most central mechanism behind our evolutionary uniqueness is that the timing of cellular differentiation can be altered to either truncate or prolong the number of mitotic cycles during each stage of development. I will therefore test, with data, a specific quantitative model that I call the *increasing mitotic delay model*. This model focuses on the delay in cellular differentiation as a key mechanistic explanation for sequential hypermorphosis (see McNamara, Chap. 5) (i.e., the delay in all phases of brain and body development in humans). As reviewed above, there is evidence that all cells in a zygote share a common mitotic setting that influences the length of time until cells differentiate to their terminal state and therefore cease mitotic multiplication (Finlay and Darlington 1995; Deacon 1997). Prolonging the time to differentiation (perhaps by slower accumulation of signal molecules) will increase the total number of mitotic cell cycles and thereby increase the total number of cells produced during that phase (Raff 1996; Deacon 1997).

A basic prediction of a mitotic delay model is that sequential developmental delays may be strongly correlated and probably show mathematical regularities of an internal mitotic clock. Perhaps the simplest example is that of constant proportional change: each stage in human development would be truncated or prolonged by a constant proportion. In this case, if we regress the onset time of developmental events (e.g., birth, puberty, death) in descendant species Y versus the onset time of the same event in its ancestral species X, then

$$Y = bX \qquad\qquad (8.1)$$

Here $b > 1$ if developmental events in species Y are prolonged by some constant proportion, and $b < 1$ if events are truncated.

However, this constant proportion model is probably an inaccurate description of most actual sequential timing developmental evolutionary changes. The reason is that the rate of cell cycling (i.e., mitosis) steadily (monotonically) decreases from conception to death. Conversely, cell cycle (e.g., mitosis) duration undergoes a monotonic increase from conception to death. For example, a complete neuron mitotic cycle in macaques takes about 21.5 hours at E40 (fortieth day after conception), 46.5 hours at E60, and 91 hours at E95 (Caviness et al. 1995). This has the important potential effect of causing progressively increasing delays in developmental events because the global internal clock will be progressively slowed as cell cycles are slowed. The addition of three additional cycles of neuron mitosis in macaques, for example, at E40 would extend that phase for about 65 ($= 3 \times 21.5$) hours. In contrast, adding three cycles at E95 would add 273 hours. Similar patterns exist for all cell cycles throughout the life span (Murray and Hunt 1993).

This increasing delay should be especially pronounced when comparing delays in a phase that occurs very early in the first few months of primate development versus one that occurs much later in life (e.g., puberty). This is because the rate of cell cycling decreases rapidly during the first few months of life compared with later years, when the decrease is much slower (Murray and Hunt 1993). Thus, we would expect puberty to be delayed for a relatively longer period of absolute (external) time than neurogenesis, even if both were extended by the same number of cell cycles.

If these inferences about progressively increasing delays are true, then when we regress the onset time of developmental events (e.g., birth, puberty, death) in descendant species Y versus the onset time of the same event in its ancestral species X, a better model will be a power law equation:

$$Y = bX^k \tag{8.2}$$

The exponent k measures any relative change in a developmental event in species Y compared with the time of the same developmental event in species X: $k = (dY/Y)/(dX/X)$. Therefore, $k > 1$ indicates that there is a consistent increase in the delay of events in species Y. That is, $k > 1$ if events in later life are prolonged longer (in absolute time) compared with species X than are events in early development. This is what would be predicted if slower cell cycling increases developmental delays in later life. If $k < 1$, then the opposite, a consistent truncation in species Y, is indicated.

This model was tested using data on developmental events (landmarks) to compare humans with chimpanzees (Table 8.2) and humans with macaques (Table 8.3). Regression of the timing of human developmental events to the same events in chimpanzees is shown in Figure 8.1. The regression is log-transformed so that the slope of the regression = k in Equation 8.2. That is, $\log Y = k(\log X) + \log b$. The regression indicates a very strong relationship:

$$\log(\text{human onset age}) = 1.06 (\log[\text{chimp onset age}]) \, 0.30$$

Table 8.2 Time of Major Events in the Development of Chimpanzees and Humans

	Days after Conception		
Developmental Landmark	Chimpanzee	Human	Source
Brain transition 1	105	132	Vrba 1998
Gestation	232	280	Nowak and Paradiso 1991
Brain transition 3	423	546	Vrba 1998
End brain spurt	475	620	Rice, Chapter 7
Brain transition 4	2,040	3,420	Vrba 1998
Puberty	2,700	4,500	Nowak and Paradiso 1991
Menopause	14,000	18,250	Nowak and Paradiso 1991
Life span	15,000	27,000	Nowak and Paradiso 1991
Molar (*Australopithecus*)	1,500	2,500	Smith 1991
Life span (*Australopithecus*)	15,000	27,000	Smith 1991

Note: Two estimates of early australopithecines are shown for comparison.

Table 8.3 Time of Major Events in the Development of Macaques and Humans

| | Days after Conception | | |
Developmental Landmark	Macaque	Human	Source
Founder neurogenesis ends	40	43	Rakic 1995
Neurogenesis ends	100	120	Rakic 1995
Brain myelination ends	1,443	5,000	Gibson 1991
Cortical histogenesis ends	1,825	7,300	Caviness et al. 1996
Gestation	166	280	Nowak and Paradiso 1991
Puberty	1,000	4,000	Nowak and Paradiso 1991
Menopause	9,125	18,250	Nowak and Paradiso 1991
Life span	11,000	27,000	Nowak and Paradiso 1991

The r^2 of 0.996 indicates a nearly perfect fit, and the regression is highly significant ($p < 0.01$). There is a substantial standard error with all these parameters because there are only eight points in the regression. The 95 percent confidence interval for the slope is 1.01–1.11. Interestingly, inclusion of two data point estimates for early *Australopithecus* results in no

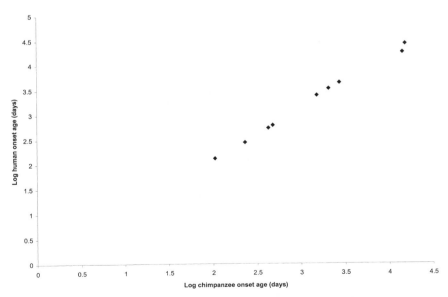

Fig. 8.1. Regression of log(human onset age) versus log(chimpanzee onset age) for developmental landmarks in those two species. Data are given in Table 8.2. Also included in the table are two estimated landmarks for australopithecines.

change in the regression results. This supports other suggestions of developmental similarities between chimps and early australopithecines (e.g., Vrba 1998; Rice, Chap. 7, this volume).

When regressing human onset age versus macaque onset age, a similarly strong relationship is obtained, though with predictably different parameters:

$$\log(\text{human onset age}) = 1.15 \, (\log[\text{macaque onset age}]) \, 0.07$$

(In this case, $r^2 = 0.976$, $p < 0.01$.) With 95 percent standard error, the slope estimate ranges from 0.98 to 1.31.

Discussion: Disproportionately Stretching the Global Mitotic Clock

It is well established that developmental events in humans are delayed relative to our ancestors (see McNamara, Chap. 5, and Rice, Chap. 7). The results here refine this pattern to describe that these delays act in a pattern that disproportionately stretches the delays, so that increasingly later developmental events are increasingly delayed. For example, compared with macaques, human founder cell neurogenesis is prolonged only by a factor of 8 percent (43 days/40 days = 1.08) compared with our delay in puberty, which is prolonged by a much greater 400 percent (= 4,000 days/1,000 days). This stretching is greater, as expected, when we are compared with macaques (k = 1.15) than when we are compared with chimpanzees (k = 1.06). The surprisingly simple mathematical progression that apparently underlies the progressive delays in human development supports the model of a global mitotic clock shared by all cells in each individual organism. The internal timekeeping in this clock seems to be at least partly (entirely?) based on the cell cycle, so that a single change that delays all events by a certain number of cycles could cascade to cause progressively longer delays as the individual develops.

The history of evolutionary biology is replete with examples of simple explanations for developmental changes, such as proposed here, that were wrong. Human developmental evolution was clearly more complicated than can be explained by a mechanism as simple as delaying differentiation to increase the number of cell cycles in each developmental stage by a constant amount. However, the patterns described here conform to a quantitative, testable null model. In addition, evidence from other studies supports the model. Caviness et al. (1996) analyzed human brain development by taking developmental data from macaques and stretching the macaque time line. They reviewed evidence that early human gestation equals early macaque gestation multiplied by a factor of two. Prepubertal years are multiplied by three, and puberty and beyond are multiplied by a factor of four. Also, Vrba (1998) used a multiphasic analysis to show that four brain growth transitions in humans are delayed relative to chimpanzees. Her data also show that the last transition is delayed the longest in humans. The last transition for the chimpanzee occurs at 68 months, and the same transition is 114 months for humans. The last delay is therefore about 67 percent (114/68) in humans relative to chimpanzees. This is far longer than the delays of the first three brain transitions, which are each delayed by no more than 30 percent in humans relative to chimpanzees.

Summary and Implications

The large human brain and cortex is a product of delayed brain growth and maturation. Homeotic genes carve up the early embryo to produce a very high brain/body ratio by allotting relatively large numbers of neural stem cells and fewer somatic (body) stem cells. Subsequently, at about E40 (the fortieth day after conception) there is a delay in humans of about three days, relative to monkeys, in the production of founder cortical cells. This results in a tenfold increase in the overall area of modern human neocortex relative to the monkey brain.

Some areas of this enlarged neocortex are themselves disproportionately enlarged, especially those areas that mature late, such as the prefrontal cortex. These areas are enlarged from a combination of (1) a larger initial cell population size (predisplacement) and (2) a prolonged period of neuron mitosis caused by a delay in cell differentiation to the terminal state (hypermorphosis). Importantly, the latter process of prolonged mitosis is not accompanied by a reduction in mitotic rates. This leads to enormous increase in the number of neurons in the affected brain regions, most notably the prefrontal cortex. The large brain size increase is not replicated with human body size because, while our body growth is prolonged by delayed development, the rate of growth is also reduced (Rice 1997; Chap. 7, this volume). Prolonged neuron mitosis is followed by a similar prolongation in the growth of associated neural structures (e.g., dendrites, synapses).

I also reviewed evidence for a global mitotic clock that is progressively stretched in human evolution. Internal timekeeping by this clock is at least partly based on the number of cell cycles. Because the rate of cell cycling decreases with age, a relatively consistent delay in the termination of cell cycling in each developmental stage will cause progressively longer delays in later developmental events. As a result, our most prolonged stages of life are the later ones, middle and older adulthood. This has interesting implications for the extent to which neurobiological development constrains natural selection to produce human evolution. Is our prolonged old age a consequence of natural selection favoring the accumulated knowledge of older individuals, or is it a by-product of the way developmental processes produce larger brains?

The neurobiological mechanisms sketched in this chapter also provide more specific reasons why the juvenilized ape myth of human origins is invalid. These mechanisms trace how our enhanced cognitive abilities are produced, without recourse to vague (and untestable) notions of retained juvenilized traits. Further insight on this is gained by comparing human evolution to cases where behavioral evolution actually is produced via juvenilization. Domestication of animals, for example, has been generally accompanied by juvenilized behaviors as humans have selected for such traits as reduced aggression and increased playfulness (in pets). Yet, in contrast to human evolution, many authors cite considerable neurobiological evidence that such juvenilized domestication often produces reduced cognitive potential, as seen in learning abilities, brain/body ratios, and general neural complexity (Price 1984). This illustrates the importance of distinguishing cognitive evolution caused by increased neocortex development, as outlined here, from the evolution of aggression, playfulness, and many other behaviors caused by changes in emotional development (i.e., limbic structures). Evolutionary changes in the latter, such as retention of juvenilized temperamental states, clearly involve brain alterations besides changes in the information-processing abilities of the neocortex.

References

Blinkov, S., and Glezer, I. 1968. *The Human Brain in Figures and Tables*. New York: Plenum.

Bogin, B. 1997. Evolutionary hypotheses for human childhood. *Yearbook of Physical Anthropology* 40:1–27.

Caviness, V.S., Kennedy, D.N., Bates, J.F., and Makris, N. 1996. The developing brain: a morphometric profile. In R.W. Thatcher, G. Lyon, J. Rumsey, and N. Krasnegor (eds.), *Developmental Neuroimaging*, 3–14. San Diego: Academic Press.

Caviness, V.S., Takahashi, T., and Nowakowski, R.S. 1995. Numbers, time and neocortical neurogenesis: a general developmental and evolutionary model. *Trends in Neuroscience* 18:379–383.

Deacon, T.W. 1990. Rethinking mammalian brain evolution. *American Zoologist* 30:629–705.

———. 1995. On telling growth from parcellation in brain evolution. In E. Alleva, A. Fasola, H. Lipp, L. Nadel, and L. Ricceri (eds.), *Behavioural Brain Research in Naturalistic and Semi-naturalistic Settings*, 237–268. Dordrecht: Kluwer.

———. 1997. *The Symbolic Species*. New York: W. W. Norton.

Deacon, T.W., Pakzaban, P., Burns, L.H., Dinsmore, J., and Isacson, O. 1994. Cytoarchitectonic development, axon-glia relationships and long-distance axon growth of porcine striatal xenografts in rats. *Experimental Neurology* 130:151–167.

Edelman, G. 1987. *Neural Darwinism*. New York: Basic Books.

Finkelstein, R., and Boncinelli, E. 1994. From fly head to mammalian forebrain: the story of otd and Otx. *Trends in Genetics* 10:310–315.

Finlay, B.L., and Darlingon, R.B. 1995. Linked regularities in the development and evolution of mammalian brains. *Science* 268:1578–1584.

Gibson, K.R. 1991. Myelination and behavioral development: a comparative perspective on questions of neoteny, altriciality and intelligence. In K.R. Gibson and A.C. Petersen (eds.), *Brain Maturation and Cognitive Development*, 29–64. New York: Aldine de Gruyter.

Gould, S.J. 1977. *Ontogeny and Phylogeny*. Cambridge: Belknap Press of Harvard University.

Holland, P., Ingham, P., and Krauss, S. 1992. Development and evolution: mice and flies head to head. *Nature* 358:627–628.

McKinney, M.L. 1998. The juvenilized ape myth: our overdeveloped brain. *BioScience* 48:109–116.

———. 1999. Heterochrony: beyond words. *Paleobiology* 25:149–153.

McKinney, M.L., and McNamara, K.J. 1991. *Heterochrony: The Evolution of Ontogeny*. New York: Plenum.

Murray, A., and Hunt, T. 1993. *The Cell Cycle*. New York: Freeman.

Nijhout, H.F., and Emlen, D.J. 1998. Competition among body parts in the development and evolution of insect morphology. *Proceedings of the National Academy of Sciences, USA* 95:3685–3689.

Nowak, R.M., and Paradiso, J.L. 1991. *Walker's Mammals of the World*, 5th ed. Baltimore: Johns Hopkins University Press.

Price, E.O. 1984. Behavioral aspects of animal domestication. *Quarterly Review of Biology* 59:1–32.

Purves, D. 1988. *Body and Brain: A Trophic Theory of Neural Connections*. Cambridge: Harvard University Press.

Raff, R. A. 1996. *The Shape of Life*. Chicago: University of Chicago Press.

Rakic, P. 1995. A small step for the cell, a giant leap for mankind: a hypothesis of neocortical expansion during evolution. *Trends in Neuroscience* 18:383–388.

Rice, S.H. 1997. The analysis of ontogenetic trajectories: when a change in size or shape is not heterochrony. *Proceedings of the National Academy of Sciences, USA* 94:907–912.

Sacher, G.A., and Staffeldt, E.F. 1974. Relation of gestation time to brain weight for placental mammals. *American Naturalist* 108:593–615.

Simeone, A., Acampora, D., Gulisano, M., Stornaiuolo, A., and Boncinelli, E. 1992. Nested expression domains of four homeobox genes in developing rostral brain. *Nature* 358:687–690.

Smith, B.H. 1991. Dental development and the evolution of life history in Hominidae. *American Journal of Physical Anthropology* 86:157–174.

Vrba, E.S. 1998. Multiphasic growth models and the evolution of prolonged growth exemplified by human brain evolution. *Journal of Theoretical Biology* 190:227–239.

Wictorin, K., Brundin, P., Sauer, H., Lindvall, O., and Bjorklund, A. 1992. Long distance directed axonal growth from human dopaminergic mesencephalic neuroblasts. *Journal of Comparative Neurology* 323:475–494.

Chapter 9

Natural Selection and the Evolution of Hominid Patterns of Growth and Development

NINA G. JABLONSKI, GEORGE CHAPLIN, AND KENNETH J. MCNAMARA

One of the few uncontested facts of paleoanthropology is that the establishment of habitual bipedalism was the key innovation that created the hominid clade. All other signal characteristics or trends of the hominid lineage evolved subsequent to this innovation. Among the most important of these were sequential, escalating increases in body size in members of the monophyletic lineage leading to *Homo erectus* and through to *Homo sapiens* and prolongation of all phases of ontogeny (see McNamara, Chap. 5). Because of their importance in determining fitness, age and size at maturity are two of the most frequently studied parameters of organismal life history (Stearns 1992). These parameters are known to have changed remarkably through hominid history, particularly in the phase leading up to the appearance of *H. erectus* (McHenry 1992, 1994; Smith 1991, 1993). Other recognized trends in hominid evolution were the establishment of unique patterns of sexual dimorphism in overall body dimensions and the canine teeth and the development of distinctive allometries of the limbs, jaws, teeth, and, eventually, the brain.

Morphological and morphometric documentation of the anatomical changes occurring during hominid evolution is now thorough and sophisticated, but an explication of the selective factors that operated to promote bipedalism and increased body and brain sizes and an explanation of the constraints that had previously prevented the expression of these characteristics are still lacking. Exactly what happened in the late Miocene and early Pliocene to release the apparent constraints limiting hominid ancestors to small body and brain sizes? More specifically, what was the selective advantage of the dominant pattern of heterochrony that is observed in the hominid clade, and what factors had previously prevented the expression of this pattern?

In this chapter, we present evidence supporting the hypothesis that the evolution of bipedalism was inaugurated by a process that reduced the morbidity and mortality from intra- and intergroup aggression among prehominid individuals. This effect, which amounted to a reduction in predation pressure on prehominids, relaxed specific constraints on individual life histories and made possible profound changes in hominid growth trajectories and increases in the duration of all phases of hominid life history.

Some influential authors (e.g., Gould 1977; Montagu 1981) have argued that hominids have evolved largely through paedomorphosis—that is, that the adult characteristics of modern hominids are retentions of the juvenile characteristics of ancestral forms. More recently, it has been argued that, on the contrary, many aspects of hominid evolution reflect the operation of peramorphic processes (McNamara 1997 and many chapters in this book).

The major heterochronic pattern observed in the course of hominid evolution has been the prolongation of all developmental stages, from the embryonic period through adulthood. Extension of growth periods is termed *hypermorphosis,* and extension of the preadult phase (i.e., a delay in the onset of sexual maturity) is referred to as *terminal hypermorphosis.* Prolongation of successive developmental stages is known as *sequential hypermorphosis* (McNamara 1983) and, as argued in Chapter 5, is the process that seems to have been dominant in hominid evolution (McKinney and McNamara 1991; McNamara 1997). In pure sequential hypermorphosis, there will be no differences in growth rates and ancestral allometries will be extended at each growth stage, whether they are positive or negative. This is not to say that sequential hypermorphosis has been the only pattern of heterochrony observed in hominid evolution. One of the main problems in unraveling the role of heterochrony in human evolution has been confronting the strong preconception that *H. sapiens* is the product of either paedomorphosis or peramorphosis. As in the evolution of other organisms, human evolution reflects the operation of both paedomorphic and peramorphic processes (see McNamara, Chap. 5).

The questions are: Which process has been the most dominant, and which has had the most significant effect in the evolution of the lineage?

One of the most important and earliest recognized trends in hominid evolution was the reduction of tooth and jaw size. This is an example of a paedomorphosis occurring *pari passu* with peramorphosis. Such a phenomenon is termed *dissociated heterochrony* and is the dominant pattern observed in the history of evolving lineages. The most extreme example of dissociation in hominid evolution is the very early descent of the testes relative to the onset of reproduction in males (Martin 1990). This seems to have evolved because of the importance of the testes in hormone production connected with somatic growth and differentiation but is not connected to the organs' role in reproduction (McNamara 1995). Despite dissociation of particular heterochronies, the dominant trend in hominids has been one of sequential hypermorphosis.

One of the consequences for hominids of progressive delay in the onset of each growth stage, including the last, was the evolution of descendants that were larger than their ancestors. Larger bodies and commensurately larger brains were not, however, ends in themselves and would not have evolved unless they conferred on individuals enhanced fitness. Increases in size and complexity carried with them inevitable costs, but that these features evolved at all meant, in cost-benefit terms, that the benefits outweighed the costs. The trade-off between the costs of somatic cell division and reproductive cell division were extensively investigated by Stearns (1992), Stearns and Koella (1986), and Charnov (1991, 1993). Their work showed that somatic growth ceases when further growth increments do not add to increased evolutionary fitness. When having a larger body size does not add to future overall fitness, it behooves the individual to engage in reproduction (Charnov 1991, 1993). In evolving hominids the costs of selection for larger individuals were longer generation times and relatively more altricial young. These two factors worked together to constrain the degree of hypermorphosis that was possible.

Predation exerts strong selective pressures on life-history traits and is one of the most potent factors affecting size and age at maturity (Abrams and Rowe 1996; Endler 1986). Through its direct link with the evolution of these life-history traits, heterochrony is strongly influenced by predation (McKinney and McNamara 1991; McNamara 1997). The theory that hominid bipedalism evolved as an extension of the bipedal threat display (Jablonski and Chaplin 1992, 1993) predicts that hominid life-history parameters would change as a result of the reduction of intra- and intergroup morbidity and mortality, which amounts to a reduction in functional predation pressure. In this chapter we identify intraspecific predation as the primary constraint that was relaxed to bring sequential and terminal hypermorphosis into operation in hominid evolution. This proposal is conceived within the theoretical framework of optimization of life-history invariants (Charnov 1991, 1993), which allows predictions concerning the allocation of resources from somatic growth to reproductive effort.

Causes of Morbidity and Mortality in Living Apes

Examination of the causes of morbidity and mortality in the living African apes helps to shed light on the nature of the constraints on growth and development that may have existed in prehominid populations. Despite the fact that predation on chimpanzees by carnivores is rare (Cheney and Wrangham 1987; Tsukahara 1993), few chimpanzees live out their life span; indeed, only 62.5 percent of the animals at Gombe studied by Goodall (1986) made it to independence. At Gombe, disease accounted for the largest percentage of deaths (28%), but many cases could be attributed to the human introduction of polio and pneumonia (Goodall 1986). Without these epidemics, deaths from disease would have equaled those from trauma. Apart from disease, intraspecific aggression is the most common cause of morbidity and mortality in chimpanzees and gorillas (Bygott 1972; Fossey 1981, 1983; Goodall 1986; van Lawick-Goodall 1971; Lovell 1990). In immature gorillas, the largest single cause of death is infanticide by adult males (Fossey 1983).

Cheney and Wrangham (1987) noted that intraspecific killing was more important to apes than "conventional" predation because of its behavioral and demographic effects and because it often accounted for more deaths than predation. Some injuries and disabilities are not the obvious sequelae of aggressive intraspecific interactions but are ultimately traceable to such causes. Fractures of long bones resulting from falls from trees are common in both gorillas and chimpanzees (Lovell 1990), but the largest single cause of falls, at least among chimps, is aggressive interaction (Goodall 1986). Trauma to the anterior teeth has been observed by several workers (Cave and Jones 1952; Fossey 1983; Lovell 1990) and would seem to be caused by biting during aggressive intraspecific interactions (Fossey 1983). Fractures to the incisors or canines lead to dental infections, which can be disabling or fatal (Fossey 1983, Lovell 1990). Lacerations and fractures of the hands or feet resulting from biting are also common (Goodall 1986; Lovell 1990). Direct morbidity and mortality are not the only results of aggression that affect fitness. Orphans die, infants get injured "in the cross-fire," females miscarry, and stress caused by aggression may cause decreases in fertility and irregularities in cycling. Among chimpanzees, physical attacks directed to the external genitalia are common and have obvious effects on individual reproductive potential (Goodall 1986).

Modeling Changes in Patterns of Morbidity and Mortality

To examine how changes in patterns of morbidity and mortality might have affected populations of living apes or hypothetical prehominids, workers used the RAMAS/age program to perform a series of demographic simulations based on the well-studied population of common chimpanzees at Gombe (Ferson and Akcakaya 1993). This program allows population models to be built using specified parameters of starting population size, age structure, survival, and fecundity (with their respective variances) and is widely applied in ecological studies. RAMAS/age predicts how many individuals will be in an age class in future years and estimates the probability and time that the population will exceed or fall below a particular size. RAMAS/age uses Leslie matrices to predict population changes in light of the fact that individuals of different ages have different demographic rates (Ferson and Akcakaya 1993). In a Leslie matrix, the fecundities of the various age classes are arrayed along the top row and their survivals are along the subdiagonal. The entries in the matrix reflect how many individuals each member of a given age class is likely to contribute to another age class in the next year (Ferson and Akcakaya 1993). The long-term result expected from a Leslie matrix model is population growth or decline; population equilibrium is rarely observed in the long term because a change to even a single element of the matrix will destroy an equilibrium situation.

For purposes of these simulations, the population of Gombe chimpanzees was divided into eight age classes: 0–4, 5–9, 10–14, 15–19, 20–24, 25–29, 30–34, and 35+ years. The survival and fecundity of each class based on the known ages, survivorship, and fecundity of individuals was then calculated. In each of the following simulations, the population was followed for 50 generations and the simulation was repeated 500 times.

In the first simulation, the parameters of the actual, present-day Gombe population were followed, and the population can be seen to be slowly increasing in size (Fig. 9.1, *top*). If a decrease in the fecundity in the first fertile period for females is simulated (i.e., if we delay the onset of reproduction

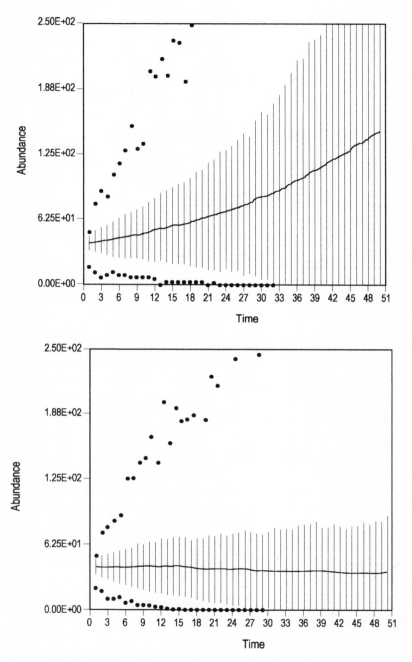

Fig. 9.1. *Top:* RAMAS/age simulation of the estimated trajectory of growth in a population of common chimpanzees over the course of fifty generations. In this simulation, the same demographic profile and the same patterns of morbidity and mor-

by half of the first eligible age class so that females commence reproduction in the 15–19 age class instead of later in the usual 10–14 class), the effect on the population's future is marked (Fig. 9.1, *bottom*). These conditions mimic those known to occur during extended periods of increased seasonality in chimps as well as those inferred to have been incorporated into the life-history evolution of early hominids. Rate of population increase is most sensitive to changes in age at maturity (Stearns and Koella 1986; Stearns 1992), and age of first reproduction in females is the life-history parameter that most influences the intrinsic rate of increase in slowly reproducing species (Charnov 1991, 1993). For primates, age at first reproduction is the most important single life-history parameter influencing the number of offspring realized per lifetime and, thus, the intrinsic rate of increase of population (Stearns 1992). Delay of onset of first reproduction thus carries with it significant costs to individuals and populations. In this simulation, population growth was halted through the delay of onset of female reproduction.

If normal onset of female reproduction is resumed but survivorship in all age classes is reduced by 10 percent, a slower intrinsic rate of increase than in our original simulation is observed, but the population is not in stasis or decline (Fig. 9.2, *top*). These conditions simulate those that might be observed under conditions of slightly increasing mortality.

If the parameters of the previous two simulations are combined—that is, if the onset of female reproduction is delayed and survivorship of all age classes is reduced by 10 percent—the population is sent into frank decline (Fig. 9.2, *bottom*). If conditions are simulated in which onset of female reproduction occurs at its normal time but in a population with a 10 percent across-the-board increase in survivorship, a dramatic rise in the intrinsic rate of population increase is observed (Fig. 9.3, *top*). These conditions simulate those proposed for protohominids by Jablonski and Chaplin (1992, 1993), in which morbidity and mortality have been reduced because of the introduction of a more efficacious method of settling potentially violent

---

tality as the population studied by Goodall (1986) at Gombe were used. The *vertical lines* represent the standard deviation of the estimated population size, based on 500 iterations of the simulation. *Bottom:* RAMAS/age simulation of growth in a population of common chimpanzees over the course of fifty generations in which the onset of reproduction in females has been delayed by half of the first eligible age class (i.e., so that females commence reproduction in the 15–19 age class instead of the 0–14 age class). The slow decline in population growth observed under such conditions, even with rates of morbidity and mortality held constant, demonstrates the great sensitivity of slow-reproducing mammals to delays in the onset of female reproduction.

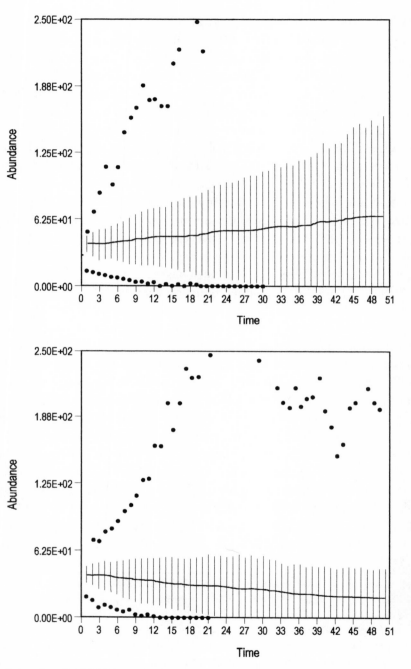

Fig. 9.2. *Top:* RAMAS/age simulation of growth in a population of common chimpanzees over the course of fifty generations in which survivorship in all age classes has been reduced by 10 percent. *Bottom:* RAMAS/age simulation of growth in a popula-

disputes over resources and mates. With a simulated 10 percent increase in survivorship in all age classes, we can, thus, demonstrate the relaxation of the heterochronic constraint. When onset of reproduction is offset by one age class, a rise, albeit slight, in the intrinsic rate of increase is still observed (Fig. 9.3, *bottom*), as compared with the decline seen in Figure 9.1, *bottom*.

The Effects of Increased Survivorship on the Evolution of Hominid Life Histories

In previous studies, two of us argued that the origin of habitual bipedalism in prehominids was intrinsically connected to the use of bipedal postures in the maintenance of social order, a relatively ancient behavior that was traced, using Lauder's (1981, 1982) method of historical (cladistic) reconstruction of morphological change, to the common origin of humans and modern African hominoids (Jablonski and Chaplin 1992, 1993). In contrast to most other theories of the origin of hominid bipedalism, this one proposes that the increased use of bipedal behaviors by both sexes in the context of social control had direct effects on individual reproductive success by making possible the less violent resolution of disputes over resources and mates, leading to lowered individual morbidity and mortality combined with a relaxation of intraspecific predation pressure.

It is clear that a level of increased survivorship as a result of reduced mortality would have a dramatic effect on the intrinsic rate of increase of the population of living chimps at Gombe or that of a putative pre- or protohominid population with broadly comparable life-history parameters (Fig. 9.3). What would have been the effects of such a change in prehominids? We reason that such a change would have relieved the pressure on females to reproduce early (Figs. 9.1, *bottom,* and 9.3, *bottom*). With decreased mortality due to intraspecific killing, the periods during which somatic growth could occur and before which reproduction must commence in females were extended, as per Stearns and Koella (1986) and Charnov (1991, 1993). Early phases of development could "afford" to be prolonged without leading to extinction.

---

tion of common chimpanzees over the course of fifty generations in which the onset of reproduction in females has been delayed by half of the first eligible age class and in which survivorship in all age classes has been reduced by 10 percent. Reduced survivorship combined with delayed onset of first female reproduction has sent the population into obvious decline.

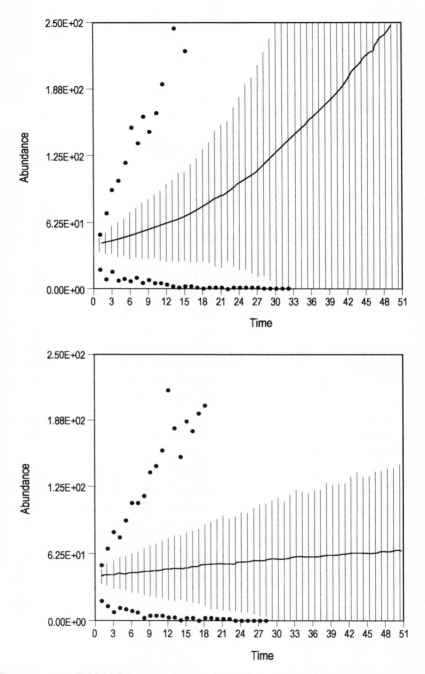

Fig. 9.3. *Top:* RAMAS/age simulation of growth in a population of common chimpanzees over the course of fifty generations in which onset of female reproduction has occurred at its normal time, but the population is showing a 10 percent across-the-

Optimization of life-history stages requires that somatic growth be channeled to reproductive effort when a further increment in body size does not increase future fecundity (Berrigan et al. 1993; Charnov 1991, 1993). In other words, when there is no further fitness to be gained by getting larger, growth stops. Larger bodies and brains are not ends in themselves and will not be selected for if they do not increase Darwinian fitness. Even if natural selection favors such increases, they cannot occur if life histories are constrained or if resources are not available because attaining large size is a cost in itself. At the point where another increment in resource use translates only to a corresponding increment in size without an enhancement of fitness, it becomes more advantageous to convert the next resource increment to reproductive effort instead (Charnov 1993).

The detailed and wide-ranging allometric studies of Shea (e.g., 1988, 1989) have shown that growth rates in humans represent an extension of those observed in chimpanzees rather than accelerations, as observed in the gorilla. The larger size of hominids has been achieved by sequential hypermorphosis: extension of the ancestral growth pattern at each successive stage, with the terminal extension—the adolescent growth spurt—leading to much larger individuals. The fossil record attests that this process occurred slowly in the putative lineage leading to *H. sapiens*. Evidence relating to dental development and patterns of dental incremental growth lines indicates that early fossil hominids displayed patterns and timing of dental development more similar to those of modern apes than those of humans (Beynon and Dean 1988; Conroy and Vannier 1987; Smith 1991; Anemone, Chap. 12, this volume).

The modern human pattern of greatly prolonged infant and child dependency seems unique to our species but is presaged by a shift to a more modern pattern by the early *H. erectus* stage, as evidenced by the "more human" standards of dental development seen by Smith (1993) in her study of the "Nariokotome Boy" *Homo ergaster* skeleton (KNM-WT 15000; referred to by some as an early form of *H. erectus*). Significantly, early fossil hominid species show similar patterns of body size increase, with only small

---

board increase in survivorship. A dramatic rise in the intrinsic rate of population increase is observed. *Bottom:* RAMAS/age simulation of growth in a population of common chimpanzees over the course of fifty generations in which a 10 percent increase in survivorship in all age classes has been combined with a delay of female reproduction by one age class. Even with a pronounced delay in the onset of female reproduction, a modest rise in the intrinsic rate of increase is observed.

shifts away from the small body size of the putative prehominid in *Australopithecus afarensis* and *Homo habilis,* followed by more significant shift to a larger body size in *H. ergaster* (Table 9.1; McHenry 1994).

When these figures are combined with those from other hominids (all estimated by McHenry 1994) and then compared with those of modern common chimpanzees (Fig. 9.4), it can be seen that in *A. afarensis* and *H. habilis* estimated body weights for both sexes are close to those of the common chimpanzee. The male *A. afarensis* was somewhat larger than and the female smaller than modern chimps. With the earliest *H. erectus* types, however, estimated body weights of both hominid sexes relative to chimps are considerably greater, with females showing the largest relative increase. Sexual dimorphism in body size is also seen to have changed profoundly over time. The index of sexual dimorphism shown in Figure 9.4 allows one to see that the smallest, earliest hominids had the greatest dimorphism (McHenry 1994), exceeding that observed in *Pan,* while the larger, later hominids had considerably less dimorphism, with males averaging 120 percent of female body weight for roughly the last 2 million years.

The results depicted in Figure 9.4 are in accord with the first of the morphological predictions of the theory of bipedalism advanced by Jablonski and Chaplin (1992, 1993). That is, high levels of sexual dimorphism in body size were observed in early hominids because increased size and stature would have been influential in determining the outcomes of encounters where bipedal displays were used to settle disputes between males. The second morphological hypothesis of that theory also warrants examination. If there was a trend toward lessened use of the canine teeth in agonistic encounters, reduced sexual dimorphism in canine tooth size in the earliest hominids should be observed. This prediction is also borne out by available fossil evidence (Table 9.2). Compared with modern apes and the early

Table 9.1 Estimated Body Weights of Selected Hominids

	Body weight (kg)	
Species	*Male*	*Female*
Australopithecus afarensis	44.6	29.3
Homo habilis	51.6	31.5
Homo erectus	63.0	52.3

Source: Data from McHenry (1994).

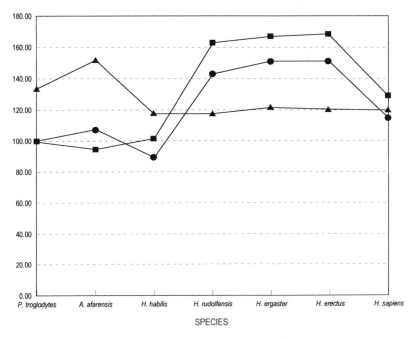

Fig. 9.4. Estimated body weights for males (*circles*) and females (*squares*) of selected fossil hominid species (McHenry 1994), shown as percentages relative to those in common chimpanzees (*Pan troglodytes*). The simple index of sexual dimorphism shown for the same species (*triangles*), calculated as estimated male size divided by estimated female size times 100, indicates that somatic sexual dimorphism in *Australopithecus afarensis* was greater than that in *Pan* or in later hominid species.

Table 9.2 Indices of Sexual Dimorphism in Upper Canine Tooth Dimensions for Living Apes Compared with the Early Miocene Hominoid *Proconsul nyanzae* and the Pliocene Hominid *Australopithecus afarensis*

Species	Height	Height/Length	Length
Hylobates spp.	110.7	101.6	108.9
Pongo pygmaeus	166.3	125.2	132.8
Pan troglodytes	149.4	118.4	126.2
Pan paniscus	142.6	117.0	121.9
Gorilla gorilla	189.0	130.8	144.5
Proconsul nyanzae	178.8	109.8	162.8
Australopithecus afarensis	117.4	105.8	111.0

Source: Data from Kelley (1995), Andrews (1978), and Pickford (1986).

Note: The indices were calculated as the male value divided by the female value times 100.

Miocene hominoid *Proconsul nyanzae, A. afarensis* shows less sexual dimorphism in both canine height and length.

The Evolution of Hominid Bipedalism and Life Histories

What are the implications of these findings for the evolution of early hominid life histories? As mentioned above, it is possible to apply the theoretical framework of optimization of life-history invariants (Charnov 1991, 1993) to the prediction of resource allocation from somatic growth to reproductive effort in early hominids. In a context in which morbidity and mortality due to intraspecific aggression has been reduced, predation pressure has been effectively reduced and the constraint on females to mature and commence reproduction earlier has been released. In early hominids, the relief from the pressure to reproduce sooner rather than later was translated first into body size increases by sequential hypermorphosis. Important in terms of the known environmental context of early hominid evolution is that an extended, rather than an accelerated, period of ontogeny did not require an improvement in the quality of food resources.

Body size increases were more marked in males during the earliest phases of hominid history because larger body size and stature enhanced the chances of favorable outcomes in encounters using bipedal displays. Increases in female body size followed a slower, catch-up trajectory until a standard differential of about 20 percent in body size was reached nearly 2 million years ago. This 20 percent differential, which could be termed *residual sexual dimorphism,* has persisted because of the necessity for females to curtail somatic growth before males so they can channel resources into reproductive effort instead of somatic growth. Because reproductive females must survive pregnancy, lactation, weaning, and a period of infant dependency before they have produced an independent offspring, they must suspend somatic growth and commence reproduction that much earlier than males.

Implications for Increases in Hominid Brain Size and the Evolution of Behavior

Expansion of the brain, in particular, the cerebral cortex, in hominid evolution is known to have been a protracted process (see Chap. 8). Increase in the size of a complex organ like the brain cannot occur as a result of accelerated development. The earliest and relatively minor phases of brain expansion, which occurred in australopithecine history, involved mostly areas of the brain related to the coordination of a new form of locomotion (Eccles 1989). Slow increases in brain expansion, related to increases in body mass, followed in the early history of the genus *Homo*. How, then, can we account for the steady and profound increases in brain size that occurred in fossil hominids after body size had reached its roughly modern level, around 2 million years ago? In a putative early hominid society with increased survivorship and reduced levels of physical aggression, increased group sizes and enhanced levels of sociality were inevitable. Effective social communication and individual social position strongly affect individual fitness in primates, and an organ that can enhance these attributes will be favored by natural selection. Human brains are those of highly evolved social communicators, which expanded slowly but steadily in relation to body size because of the tremendous enhancement in individual survival that could be realized by increased sociality, increased behavioral flexibility, and ever more sophisticated communication.

The evolution of behavioral plasticity facilitates the exploitation of a greater range of resources and the extraction of lower cost alternative resources. Consequently, behavioral plasticity enables "intelligent" guesses to be made with respect to the outcomes of various behaviors, as in the design of a foraging strategy relative to the spatiotemporal arrangement of food resources. Enhanced behavioral flexibility also makes possible faster responses to ecological change than do genetically dictated changes in phenotype because the latter can potentially be altered only at generational intervals. Finally, once behavioral flexibility has arisen, it is not constrained to the same degree as is morphological change.

Relaxation of the constraint to complete growth and engage in relatively early reproduction as a result of reduction in intraspecific predation pressure made possible sequential hypermorphosis in hominids. This profound heterochronic shift had marked morphological and behavioral consequences. Prolongation of the periods of infant and juvenile development increased the scope for individual learning from the environment and conspecifics and for the development of increasingly sophisticated social interactions. The longer the learning period, the more complex the cultural behaviors that were possible (see Parker, Chap. 14). Thus, the operation of heterochronic processes that extended the periods of most active brain growth and learning enhanced the potential for extending complex cultural behaviors.

Conclusions

Students of human evolution must be able to account, in terms of Darwinian fitness, for two important and highly unlikely phenomena: the morphologically difficult transition to habitual terrestrial bipedalism and the establishment in the hominid lineage of a dominant pattern of sequential hypermorphosis. It has been argued previously that the first of these phenomena could be accounted for through the elaboration of the preexisting bipedal threat display (Jablonski and Chaplin 1992, 1993). We present evidence here indicating that the second important phenomenon was facilitated by the first (i.e., that the evolution of bipedalism by elaboration of a behavior that significantly reduced morbidity and mortality, so effectively reducing predation pressure, led to relaxation of constraints on early maturation and reproduction). Relaxation of these constraints made possible the prolongation of all phases of ontogeny and various manifestations of sequential hypermorphosis.

Hominid evolution began with the giving up of "nature red in tooth and claw" for a society based on bluff and ritualized display. The ramifications of that "original shift" for individual ontogenies were profound, leading first to larger bodies and eventually to larger brains. Among mammals, primates are distinguished by their behavioral flexibility and their enhanced social abilities to mitigate aggression and avoid physically injurious confrontations. Hominids have not reversed this trend; they have exaggerated it.

Acknowledgment

We thank Rebecca German for her detailed and constructive comments on this chapter.

References

Abrams, P.A., and Rowe, L. 1996. The effects of predation on the age and size of maturity of prey. *Evolution* 50:1052–1061.

Andrews, P.J. 1978. A revision of the Miocene Hominoidea of East Africa. *Bulletin of the British Museum (Natural History) Geology* 30:85–224.

Berrigan, D., Charnov, E.L., Purvis, A., and Harvey, P.H. 1993. Phylogenetic contrasts and the evolution of mammalian life histories. *Evolutionary Ecology* 7: 270–278.

Beynon, A.D., and Dean, M.C. 1988. Distinct dental development patterns in early fossil hominids. *Nature* 335:509–514.

Bygott, J.D. 1972. Cannibalism among wild chimpanzees. *Nature* 232:410–411.

Cave, A.J.E., and Jones, T.S. 1952. Canine tooth fracture in two Congolese gorillas. *Proceedings of the Zoological Journal, London* 227:685–690.

Charnov, E.L. 1991. Evolution of life history variation among female mammals. *Proceedings of the National Academy of Sciences USA* 88:1134–1137.

———. 1993. *Life History Invariants: Some Explorations of Symmetry in Evolutionary Ecology.* Oxford: Oxford University Press.

Cheney, D.L., and Wrangham, R.W. 1987. Predation. In B.B. Smuts, D.L. Cheney, R.M. Seyfarth, R.W. Wrangham, and T.T. Struhsaker (eds.), *Primate Societies,* 227–239. Chicago: University of Chicago Press.

Conroy, G.C., and Vannier, M.W. 1987. Dental development of the Taung skull from computerized tomography. *Nature* 329:625–627.

Eccles, J.C. 1989. *Evolution of the Brain.* London: Routledge.

Endler, J.A. 1986. *Natural Selection in the Wild.* Princeton: Princeton University Press.

Ferson, S., and Akcakaya, H.R. 1993. *RAMAS/age User Manual: Modeling Fluctuations in Age-structured Populations. Applied Biomathematics.* New York: Setauket.

Fossey, D. 1981. The imperiled mountain gorilla. *National Geographic* 259:500–523.

———. 1983. *Gorillas in the Mist.* Boston: Houghton Mifflin.

Goodall, J. 1986. *The Chimpanzees of Gombe.* Cambridge: Belknap Press of Harvard University Press.

Gould, S.J. 1977. *Ontogeny and Phylogeny.* Cambridge: Belknap Press of Harvard University Press.

Jablonski, N.G., and Chaplin, G. 1992. The origin of hominid bipedalism reexamined. *Archaeology in Oceania* 29:115–125.

———. 1993. The origin of habitual terrestrial bipedalism in the ancestor of the Hominidae. *Journal of Human Evolution* 24:259–280.

Kelley, J. 1995. Sexual dimorphism in canine shape among extant great apes. *American Journal of Physical Anthropology* 96:365–389.

Lauder, G.V. 1981. Form and function: structural analysis in evolutionary morphology. *Paleobiology* 7:430–442.

———. 1982. Historical biology and the problem of design. *Journal of Theoretical Biology* 97:57–67.

Lovell, N.C. 1990. *Patterns of Injury and Illness in Great Apes.* Washington: Smithsonian Institution Press.

Martin, R.D. 1990. *Primate Origins and Evolution: A Phylogenetic Reconstruction.* Princeton: Princeton University Press.

McHenry, H.M. 1992. Body size and proportions in early hominids. *American Journal of Physical Anthropology* 87:407–431.

———. 1994. Behavioral ecological implications of early hominid body size. *Journal of Human Evolution* 27:77–87.

McKinney, M.L., and McNamara, K.J. 1991. *Heterochrony: The Evolution of Ontogeny.* New York: Plenum Press.

McNamara, K.J. 1983. Progenesis in trilobites. *Special Papers in Palaeontology* 30: 59–68.

———. 1995. Sexual dimorphism: the role of heterochrony. In K.J. McNamara (ed.), *Evolutionary Change and Heterochrony,* 65–89. Chichester, England: John Wiley and Sons.

———. 1997. *Shapes of Time: The Evolution of Growth and Development.* Baltimore: Johns Hopkins University Press.

Montagu, M.F.A. 1981. *Growing Young.* New York: McGraw Hill.

Pickford, M. 1986. Sexual dimorphism in *Proconsul.* In M. Pickford and A.B. Chiarelli (eds.), *Sexual Dimorphism in Living and Fossil Primates,* 133–170. Florence: Editrice Ill Sedicesimo.

Shea, B.T. 1988. Heterochrony in primates. In M.L. McKinney (ed.), *Heterochrony in Evolution: A Multidisciplinary Approach,* 237–266. New York: Plenum Press.

———. 1989. Heterochrony in human evolution: the case for neoteny reconsidered. *Yearbook of Physical Anthropology* 32:69–101.

Smith, B.H. 1991. Dental development and the evolution of life history in Hominidae. *American Journal of Physical Anthropology* 86:157–174.

———. 1993. Physiological age of KNM-WT 15000 and its significance for growth and development of an extinct species. In A.C. Walker and R.E.F. Leakey (eds.), *The Nariokotome* Homo erectus *Skeleton,* 195–220. Cambridge: Belknap Press of Harvard University Press.

Stearns, S.C. 1992. *The Evolution of Life Histories.* Oxford: Oxford University Press.

Stearns, S.C., and Koella, J.C. 1986. The evolution of phenotypic plasticity in life-history traits: predictions of reaction norms for age and size at maturity. *Evolution* 40:893–913.

Tsukahara, T. 1993. Lions eat chimpanzees: the first evidence of predation by lions on wild chimpanzees. *American Journal of Primatology* 29:1–11.

van Lawick-Goodall, J. 1971. *In the Shadow of Man.* Boston: Houghton Mifflin.

Chapter 10

Sexual Dimorphism and Ontogeny in Primates

REBECCA Z. GERMAN AND SCOTT A. STEWART

Understanding the developmental trajectories that generate any aspect of adult morphology, function, or behavior can ultimately elucidate the evolution of those traits. The biological basis for differences between adult males and females rests on differences that begin to accrue even before birth. Thus, this chapter might be called "The Ontogeny of Sexual Dimorphism" because the goal of this work on heterochrony is to enhance our understanding of the differences between the sexes. However, there are equally defensible questions concerning the developmental variation between the sexes that exist in mammals in general, primates and humans in specific. This suggests another title, "The Sexual Dimorphism of Ontogeny."

Many facets of sexual dimorphism in primates have been extensively studied, usually in adults. The sexual dimorphism that occurs during growth and development and the differences in development between males and females have received significantly less attention. One quite reasonable explanation of this lesser interest is that many of the significant morphological and behavioral differences between the sexes are not manifest until puberty. The more subtle differences that occur prenatally and prepubertally are less compelling as topics of scientific study (Mittwoch 1996). Still, several brave and intrepid workers have looked at sexual dimorphism during growth, often in the context of comparative studies answering questions about human evolution.

The kinds of questions that an ontogenetic or heterochronic study can address are, in fact, fundamental to understanding the nature of sexual dimorphism. The ontogenetic trajectory of a single morphological character, a more general behavior, or an entire organism is governed by several factors. Selection, the major force behind evolutionary change; epigenesis, the interactions and correlations that occur during growth; and the limits and constraints that evolutionary history put on evolution and epigenesis are all major contributors to the vast variation in sexual dimorphism that exists in the primates. One of the most potent scientific controversies surrounding sexual dimorphism concerns the relative significance of selective versus nonselective forces throughout the evolution of primate sexual differences.

Explanations for the evolution of sexual dimorphism rest on two major foundations: either sexual dimorphism evolves because of unequal selective pressures on the two sexes, or sexual dimorphism evolves because of phylogenetic constraints or phyletic inertia. Darwin's (1859, 1871) theory of sexual selection holds that selection operates within one sex on specific characters that confer an advantage to the individual in gaining access to mates. An alternative selective theory suggests that differences between males and females can also result as a means of reducing competition between the sexes for resources, as proposed by Selander (1972).

In contrast to selective theories, nonselective models argue that the evolution of sexual dimorphism is more closely associated with the interaction between ancestral life histories and subsequent selection on body size. In these models, descendant species are sexually dimorphic because of their ancestors' sexual dimorphism (Leutenegger 1978; Leutenegger and Cheverud 1982). As a species evolves a larger or smaller body size, the sexual size dimorphism that was present in the ancestral species is carried along with the evolution of body size. As several works make apparent (Clutton-Brock et al. 1977; Leutenegger 1978; Leutenegger and Cheverud 1982; Cheverud et al. 1985), the lack of phylogenetic corrections has hampered many studies of sexual size dimorphism in primates because it is not possible to estimate the independent amount of evolution due to sexual selection.

When Ernst Haeckel first proposed the term *heterochrony* (Haeckel 1866), referring to the change in the rates or timing of developmental events, he was referring to these differences in comparing ancestor to descent relationships. However, a broader definition of the heterochronic process can explain differences between any two groups, such as differences between sexes. Overtly or covertly, a heterochronic framework has been used to explain and understand the ontogeny of sexual dimorphism in many species of mammals. Evolutionary biologists and anthropologists interested in the genesis of sexual differences (Shea 1983; Cheverud et al. 1985; Leigh 1992;

Richtsmeier et al. 1993) have tended to use the formal terminology proposed by Haeckel, modified first by Gould (1977) and then by Albrech et al. (1979). The concepts, often without the terminology, have been used by workers more interested in growth as a process to describe the differences between sexes (Tanner 1981; Watts 1986).

The Sexual Dimorphism of Ontogeny

The sexual differences during growth that generate adult dimorphism are clearly best known for human beings (Tanner 1981, 1989; Marshall and Tanner 1986). Beyond people's inherent interest in their own biology, the clinical and societal implications and utility of basic growth processes have given a priority and urgency to collecting such data (e.g., Eveleth and Tanner 1990; Mascie-Taylor and Bogin 1995). Thus, the long history of data collection (see Tanner 1981) has resulted in enormous databases consisting of detailed and fine-grained data describing human growth. The detailed standards for human beings and data on biological and social variation in those standards far exceed anything that exists for other species of mammals, including standard laboratory models. In data with this degree of accuracy and precision, the differences between males and females are quite distinctive. One of the major uses for such data is clinical assessment of growth, and in such analyses males and females are nearly always separated. Additionally, discreet events, such as menarche, are significant markers not only for assessments of a single individual's growth but also for comparisons due to external factors, such as nutrition or degree of inbreeding (Marshall and Tanner 1986; Eveleth and Tanner 1990; Mascie-Taylor and Bogin 1995). Thus, aspects of the human growth curve that would be of interest in an evolutionary context not only are available but also exist in detail.

When considering the sexual dimorphism of ontogeny (i.e., the growth perspective), there are two distinctive aspects of human growth curves. It is debatable as to whether other primates share these ontogenetic traits: the distinctive adolescent growth spurt and the relatively lengthened duration of growth (Watts 1986). These two features, present in several different measurements (particularly height and weight) and said to characterize the human growth curve (Sinclair 1973; Tanner 1981; Bogin 1988), are not necessarily independent. It is the intervening low rates of growth between infancy and puberty that give a specific identity to the adolescent growth spurt.

In most mammals, growth rates of body weight as well as linear measurements of organs and body are highest immediately after birth (McCance and Widdowson 1986; Watts 1986). Even with detailed longitudinal data, rates of growth of weight and the skeletal system monotonically decrease with time (Fiorello and German 1997). In animals where sexual maturity is relatively close in time to the growth spurt occurring at birth, it is impossible to distinguish whether there are one or two peaks in growth rate. Given that the timing of puberty is a feature that is sexually dimorphic in humans, it becomes important to understand how the evolution of these two features in humans relates to their variation between the sexes (Fig. 10.1).

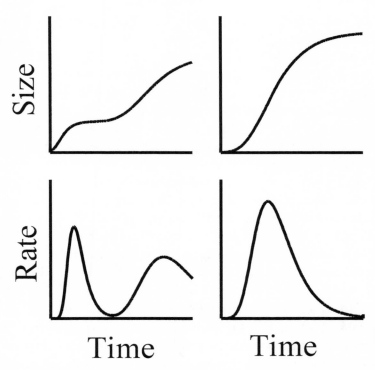

Fig. 10.1. The effect of delaying an adolescent growth spurt on size and rate curves. In the first pair of these hypothetical curves (*left*), the adolescent curve occurs significantly after the initial growth spurt, and they are separated in time, as is true for humans. The second pair (*right*) depicts the situation most often found in other mammals, where the two growth spurts occur so close in time that they are not discernible as separate peaks on the rate curves.

The Ontogeny of Sexual Dimorphism

It is widely accepted that heterochrony has played a major role in the evolution of modern humans, particularly with respect to the ontogeny of sexual dimorphism. That the two specific aspects of human growth curves identified above are the critical human characters has been debated. An earlier view was that human evolution is best explained by a heterochronic process that primarily consists of neoteny (retarded growth rate), while more recent work suggests that human evolution is in large part a result of hypermorphosis (prolonged growth). Neoteny, popular from the late nineteenth century through the 1970s, became the standard assumption, that a retarded growth rate was responsible for creating the juvenilized look of modern humans compared to our ancestors. Once ingrained into both scientific and popular views, the idea of human evolution by neoteny attained a resilience that even today is evident in the popular literature and general public ideology.

The more widely accepted view today is that hypermorphosis (in particular, sequential hypermorphosis—the increased duration of different stages of growth) is the defining heterochronic feature of humans (McKinney and McNamara 1991; see also McNamara, Chap. 5; McKinney, Chap. 8; and Jablonski et al., Chap. 9, this volume). If hypermorphosis is indeed significant to human evolution, then three questions arise: (1) does differential hypermorphosis result in sexual dimorphism in humans, as well as other primates; (2) is this the only dimorphism in humans; and (3) how does this relate to growth differences documented in other primates?

When considering comparative studies as a basis for understanding human evolution, it is relevant, if not necessary, to ask several questions. Do other primates show the same degree of extended duration between the infantile growth spurt and the adolescent growth spurt for any measurements, and if so which? Is the variation between sexes in the onset of an adolescent growth spurt consistent across primates, or does the existence and intensity of this feature vary among species? To differentiate between the selective and nonselective evolutionary models for explaining dimorphism, we need such comparative data.

Watts (1986) strongly supported the idea that both the increased duration and the adolescent spurt are major factors of sexual dimorphism in primates that would explain the evolution of the human growth curve. That is, males, relative to females, not only have a larger growth spurt at adolescence but also delay the onset of that spurt, giving them a longer prepubertal duration of growth. Watts pointed out that earlier work (e.g., Shea 1983) found only time hypermorphosis between the sexes, such that the rates of growth were similar in males and females, even though rate variation explains evolutionary differences among taxa. The results of her detailed analyses of longitudinal growth in linear body dimensions for rhesus macaques and chimpanzees differ from these earlier works as follows.

The conditions for the existence of rate variation between the sexes had three parts. First, there was variation in the degree of sexual dimorphism in the three species. Rhesus macaques are more dimorphic than humans, who in turn are more dimorphic than chimpanzees. Second, in regard to the relative delay in the onset of the male adolescent growth spurt, all males have equal delays, proportional to their total duration of growth. The third part is the conclusion that, given the first and second conditions, rate differences must explain the differences in sexual dimorphism among the species. The longitudinal data for macaques and chimpanzees permit measurements of the exact time delay to the adolescent growth spurt, as well as the rates of growth for five linear measurements of size: trunk, arm, forearm, thigh, and leg lengths. These data have consistent patterns of delay to the adolescent growth spurt between males and females across the different morphological parameters, indicating distinct patterns of sexual dimorphism of growth for the two species.

In the macaques studied by Watts (1986), size (body weight and linear body measurements) was nearly equal in males and females at birth, but sexual dimorphism increased as a function of age. When females hit an earlier puberty, males were already significantly larger than females. Male size as a function of female size is a constantly increasing function, always greater than 100 percent, in this species. In *Pan*, however (which were nearly equal in size at birth), males tended to grow slightly slower than females. In fact, males were smaller than females at various prepubertal ages. Males used the delayed onset of puberty to catch up with female size and ended up only slightly larger than females as adults. Male size in *Pan* as a function of female size starts out slightly higher than 100 percent, but drops below 100 percent from two to nine years and then rises over 100 percent only after eleven years. From these results Watts concluded that natural selection must have been operating on growth rates, not the delay in onset of the pubertal growth spurt.

One point that Watts emphasized is that differences could be seen because she had longitudinal data. It has long been clear to scientists working on human growth (Healy 1986; Tanner 1986, 1989) that cross-sectional data have significant limitations in the determination of the timing and extent of growth spurts. These problems result from significant interindividual variation, within sex or species, in the exact timing of a growth spurt. If variation occurs as a spike in a narrow range of time, cross-sectional data will tend to smooth out such spikes, shorten interspike duration, and flatten spikes that are present. The timing of growth spurts can always be more accurately determined from velocity curves than from growth or distance curves (Tanner 1986). Even with excellent cross-sectional data, both the time of onset of a spurt and the degree or intensity of the spurt will be significantly underestimated. Cross-sectional data will sample the variation among individuals at each age. Thus, while each individual may have a sharply defined growth spurt that would be a spike on a time-velocity graph, these spurts will be occurring at different ages (Fig. 10.2). One reason for the statistical significance of Watts's data is that her data are precisely the right kind to measure such events (Watts and Gavan 1982) as the onset and velocity of a growth spurt.

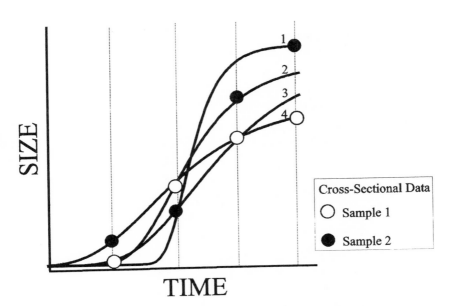

Fig. 10.2. The difference between cross-sectional and longitudinal data. Four sigmoidal growth curves are numbered 1 to 4. Two cross-sectional samples, each taking one point at the four dotted lines, produce very different-looking growth curves. Sample 1, the *open circles*, has almost no growth spurt. Sample 2, the *closed circles*, has no asymptote or shoulder.

Watts pointed out the dearth of longitudinal growth data for species other than humans and the necessity of this data for enhanced understanding of the origin of primate and, specifically, human sexual dimorphism. In the years since Watts's work, few additional longitudinal data have been accrued for primates. Two longitudinal samples for other species have been analyzed. Coehlo and collaborators worked (Glassman et al. 1984; Coehlo 1985) approximately simultaneously with Watts, measuring weight growth in savanna baboons. These authors suggested that both males and females have a significant adolescent growth spurt, which is much larger in males than in females. Furthermore, there is a delay in the onset of this spurt in males. The rates of growth before adolescence are equal in males and females, in a pattern that resembles what Watts found for chimpanzees more than what she described for macaques. Another significant longitudinal dataset on *Macaca nemistrina* was collected by Sirianni and Swindler (1985), with the weight data subsequently analyzed by German et al. (1994). In a nonlinear analysis they found evidence for two growth spurts, separated in time. Both the rates and the timing of these growth spurts were sexually dimorphic. However, because these data are often censored (i.e., missing data from either the beginning or the end of the growth period) or otherwise incomplete, this analysis was of a mixed cross-sectional design. Thus, the problem of averaging data compromises the precise identification of the timing of growth spurts in this cross-sectional analysis.

Beyond the studies mentioned, there have been cross-sectional studies of skeletal material for several primate species. One of the most significant early studies of primate heterochrony was Shea's (1983) work on African apes. His design included examining both within-species (intersexual) and between-species heterochrony and allometry. Shea's results have become one of the mainstays for understanding patterns of heterochrony and are cited to support a variety of different perspectives (e.g., Watts 1986; McKinney and McNamara 1991). Shea found that, although differences among species were due to rate hypermorphosis (i.e., ancestors and descendants having different rates of growth to affect a size change), intersexual differences were due entirely to time hypermorphosis where the duration varied, with males (designated as descendants) growing for a longer period.

Richtsmeier and coworkers have described ontogenetic sexual dimorphism in species of *Macaca, Ateles, Saimiri,* and *Cebus* (Corner and Richtsmeier 1991, 1992, 1993; Richtsmeier et al. 1993). The questions addressed in these studies concerned anatomically detailed analyses of various aspects of the craniofacial skeleton. Because they applied similar and consistent analyses to several species, this work has great comparative value for understanding growth differences among the face, cranial base, and neurocranium. One important advantage of the cross-sectional studies that use museum specimens is the number of different measurements and the accuracy of those measurements. Because these studies could use actual skulls and not radiographs, much of the studies of Richtsmeier and coworkers include three-dimensional data. Because they are measuring skeletal material, they are not restricted to either external body measurements, which are not as accurate as skeletal measurements, or radiographs, which impose their own methodological and logistic limitations. One of the most valuable aspects, though, is the number of different measurements that can be taken, so that differences in the ontogeny of sexual dimorphism for different parts of the body or skull can be assessed. Although the heterochronic conclusions that can be drawn from these studies are limited by the nature of the cross-sectional sample and the coarse time grain of the data, they are still useful when putting these results in a framework such as proposed by Watts (1986).

The spider monkey, *Ateles geoffroyi*, is generally considered to be monomorphic in adults. Corner and Richtsmeier (1993) found differences in ontogenetic trajectories between males and females that lead to adult similarity. In this cross-sectional study of craniofacial skeletal growth, they found evidence for earlier onset of female maturity. Answering valid questions about skeletal growth is feasible, even when using only museum material for which there is limited, if any, independent data on age. Thus, the ontogenetic basis of this work is broad-based dental age categories, which the authors point out are not necessarily of equal duration either over time or between the two sexes. Such data are not sufficiently detailed to repeat Watts's analyses and are exactly the form of data about which Watts expressed concern. Growth spurts would be nearly impossible to detect. This type of cross-sectional data is biased against finding delays in onset of growth spurts. Not only is there more variation around a growth or velocity curve, but the time basis is not continuous for the five dental age groups. Any evidence for sexual differences in timing, or bimaturation, must be strong indeed to be significant in these data.

Corner and Richtsmeier's studies of other primate species are of similar value to our understanding of comparative differences in growth patterns. They found in *Saimiri*, the relatively monomorphic squirrel monkey, that the sexes tend to grow in parallel until late in ontogeny, with male duration being longer (Corner and Richtsmeier 1992). This is in contrast to the results in *Cebus apella,* a markedly dimorphic capuchin, where rate hypermorphosis was detected, with males growing faster than females but for an identical duration (Corner and Richtsmeier 1991). Equal duration in age category data merely means that differences in timing cannot be ruled out. Finally, in their study of *Macaca fascicularis,* the long-tailed macaque, they again found evidence of rate hypermorphosis between sexes (Richtsmeier et al. 1993). It is difficult to classify these results into Watts's framework. What is admirably demonstrated, though, is that there are clearly very different patterns of ontogeny with respect to sexual dimorphism in this group of species.

Data on another species of macaque, *Macaca sinica,* the toque macaque, is mixed cross-sectional data (Cheverud et al. 1992). While these data did contain some repeated measurements on individuals, the data are pooled by sex, and single growth curves for each sex are used in the analysis. This work is also distinguished by an attempt to transcend linear regression, which does require an a priori estimate of the cessation of growth. By fitting splines, Cheverud et al. (1992) found that males and females had different numbers of distinct growth phases or growth spurts, two in females and three in males, although it is not clear whether they believe that males have an extra growth spurt. These results did enable them to identify an adolescent growth spurt in the males and several differences in timing between the sexes. They found little evidence of early differences between the infant sexes. It seems that growth in this species resembles Watts's pattern for rhesus macaques and provides more evidence for a shared primate growth pattern.

Time hypermorphosis, or bimaturism, also characterized Masterson and Leutenegger's (1992) results for *Pongo,* the orangutan. These results were again based on dental age categories and therefore are subject to the same caveats as other cross-sectional data. Masterson and Leutenegger also highlight another problem inherent in the application of heterochronic analysis to studies of sexual dimorphism. They overtly state, something assumed in many other studies, that females represent the "ancestral condition" and males are the modified descendants. This requires the use of hypermorphosis as opposed to progenesis. It has been pointed out, in studies with organisms that have different sex determination schemes, that this is not always a valid assumption (Brooks 1991; McNamara 1995). The possibility exists that both sexes are evolving away from a primitive condition.

While it may seem that the pubertal growth spurt in males is delayed, it could also be true that the onset of this growth is accelerated only in females or that both are happening simultaneously (see Jablonski et al., Chap. 9). Sorting out these alternatives would require longitudinal data for the ancestral taxon as well as the species under consideration.

Because none of the other examples mentioned are identical to either of the two patterns that Watts described, it seems unlikely that her data on macaques and chimpanzees represent two distinct categories of sexual dimorphism in primate ontogeny. This conclusion is supported by three other very distinct pieces of work. Leigh's (1992) comparative study of more than 40 species of dimorphic primates is different from the other studies of primate dimorphism. Rather than investigating detailed differences for different measurements in one species, he searched for broad patterns of ontogenetic differences in weight growth between sexes across many species. He devised a quantitative scheme separating out weight dimorphisms due to rate differences versus those due to time or duration differences between the sexes. Assuming that females grew for a shorter period, he looked at the relative proportion of size difference that accrued by the time the females stopped growing and attributed this difference to rate. The remainder of the size difference that accrued after females stopped growing was time difference. Thus, he had two simple measures of the proportions of variation due to the two factors. He applied these measures to cross-sectional data obtained from a variety of captive sources, including zoos and primate colonies.

Leigh's results suggest that rate and time hypermorphosis—the latter he calls *bimaturation*—are independent. In most species males grow longer than females, but in some there were rate differences and in other species differences in rate are the sole explanation of size differences. Leigh concluded that a difference in duration always leads to some degree of sexual dimorphism. This result is complicated by the fact that he excluded monomorphic species from his analysis.

Leigh emphasized two almost contradictory aspects of dimorphism. First, the processes that ontogenetically generate dimorphism are complex and variable. This is certainly consistent with the variation found in the more detailed studies of individual species described above. Second, however, he chose to classify species as either duration-differentiated or rate-dimorphic species. Whether these cross-sectional data are sufficiently fine-grained to be able to determine the extent of male lag to the onset of a pubertal growth spurt is not clear. Although he identified male spurts, his quantification was only visual separation of two aspects of growth. Although he acknowledged that growth rates vary over time, Leigh dismissed this as an insignificant problem to his essentially linear models. In fact, visual determination of the cessation of growth is both extremely difficult and statistically unreliable in detailed longitudinal data in mammalian species with far simpler growth curves than those of primates (Fiorello and German 1997). That these results differ from Watts's (1986) is not surprising. Leigh was measuring only the overall difference in growth duration, not more specifically the differences in prepubertal duration. A final difficulty in drawing more general conclusions about the evolutionary patterns in Leigh's results is that there is no attempt at phylogenetic correction. If several closely related species show the same pattern, it is possible, even likely, that the pattern evolved once, then to be shared by the descendent taxa.

More significant support for the variation in heterochronic mechanisms that generate sexual dimorphism comes from studies of purely adult dimorphism. Oxnard's (1987, 1988) multivariate examination of sexual dimorphism, repeated for several different anatomical regions, was admittedly serendipitous. However, he realized that he had been dealing with other adaptations without thinking about sexual dimorphism. In comparative studies of living primates, "sex is a nuisance. Almost all investigators attempt to ignore or eliminate sexual differences" (Oxnard 1988). This might also hold for growth studies, even after such differences are acknowledged as existing in adults. His results were that no single axis of sexual dimorphism existed in several large studies with many species. Thus, patterns of dimorphism for limb proportions were different from those for teeth.

One difficulty that dampens enthusiasm for these results is the lack of any phylogenetic corrections (Harvey and Pagel 1991). There is no assessment of the independence of the taxonomic units (i.e., species) used. These traits could have evolved once, at the base of the group, and the strong correlations that support these relationships are not real. Even so, Oxnard's view that sexual dimorphism in primates is a complex array of differences and not a single axis of "sexual dimorphism" is born out by examining what is known about sexual dimorphism in various ontogenetic studies. Richtsmeier and colleagues repeatedly found that the specific heterochronies of one part of the skull (the cranial base, for example) differed from those in another (e.g., the face).

Finally, variation in sexual dimorphism among neonates of different species suggests that prenatal growth also exhibits large interspecific variation. Generally, little attention has been given to the biological differences before or at birth. Mittwoch's (1996) work is one tantalizing exception in suggesting early chromosomal mechanisms that produce later dimorphisms. Most comparative studies of neonate body mass in primates have combined data for the sexes (e.g., Harvey and Clutton-Brock 1985; Hartwig 1996; Treves 1996). In the most comprehensive study of sexual dimorphism in primate neonates, Smith and Leigh (1998) found positive correlation with adult body mass and dimorphism in their study of twenty-three species of primates. Although these data do provide information on possible mechanisms, it is important to consider that there are significant differences in dimorphism variation in primates at the starting point (birth).

It seems that the patterns of heterochrony and sexual dimorphism vary, on one hand, among traits within a single species and then, on the other, among traits of many species of primates. Leigh's (1992) study, in particular, is to be commended for its breadth. The difficulties with these analyses, including cross-sectional data that in turn imply constant rates of growth, and simplified duration calculations suggest that these methods cannot provide answers for the fundamental questions about variation in the ontogeny of sexual dimorphism among primates. Further studies are necessary to obtain a more complete view of the role of heterochrony in producing sex differences in primates. Primate data with chronological ages, either longitudinal or cross-sectional, are difficult to obtain, even for the most cursory linear measurements of size. A definitive determination of the evolutionary patterns of sexual dimorphism in the distinctive traits of human or primate growth curves rests on such data.

Acknowledgments

This work was supported by a Fulbright Senior Fellowship to R.Z.G. We thank Andrew Lammers for help with the figures and one anonymous reviewer for insightful comments.

References

Alberch, P., Gould, S.J., Oster, G.F., and Wake, D.B. 1979. Size and shape in ontogeny and phylogeny. *Paleobiology* 5:296–317.

Bogin, B. 1988. *Patterns of Human Growth.* Cambridge: Cambridge University Press.

Brooks, M.J. 1991. The ontogeny of sexual dimorphism: quantitative models and a case study in labrisomid blennies (Telostei:Palaclinus). *Systematic Zoology* 40:271–283.

Cheverud, J.M., Dow, M.M., and Leutenegger, W. 1985. The quantitative assessment of phylogenetic constraints in comparative analyses: sexual dimorphism in body weight among primates. *Evolution* 39:1335–1351.

Cheverud, J.M., Wilson, P., and Dittus, W.P.J. 1992. Primate population studies at Polonnaruwa: III. Somatometric growth in a natural population of toque macaques (*Macaca sinica*). *Journal of Human Evolution* 23:51–77.

Clutton-Brock, T.H., Harvey, P.H., and Rudder, B. 1977. Sexual dimorphism, socionomic sex ratio and body weight in primates. *Nature* 269:797–800.

Coelho, A.M. 1985. Baboon dimorphism: growth in weight, length and adiposity from birth to 8 years of age. In E. Watts (ed.), *Nonhuman Primate Models for Human Growth and Development*, 125–159. New York: Alan R. Liss.

Corner, B.D., and Richtsmeier, J.T. 1991. Morphometric analysis of craniofacial growth in Cebus apella. *American Journal of Physical Anthropology* 84:323–342.

———. 1992. Cranial growth in the squirrel monkey *Saimiri sciureus:* a quantitative analysis using three dimensional coordinate data. *American Journal of Physical Anthropology* 87:67–82.

———. 1993. Cranial growth and growth dimorphism in *Ateles geoffroyi*. *American Journal of Physical Anthropology* 92:371–394.

Darwin, C. 1859. *The Origin of Species by Means of Natural Selection*. London: Murray.

———. 1871. *The Descent of Man, and Selection in Relation to Sex*. London: Murray.

Eveleth, P.B., and Tanner, J.M. 1990. *Worldwide Variation in Human Growth*, 2nd ed. Cambridge: Cambridge University Press.

Fiorello, C.V., and German, R.Z. 1997. Heterochrony within species: craniofacial growth in giant, standard, and dwarf rabbits. *Evolution* 51:250–261.

German, R.Z., Hertweck, D.W., Sirianni, J.E., and Swindler, D.R. 1994. Heterochrony and sexual dimorphism in the pigtailed macaque (*Macaca nemestrina*). *American Journal of Physical Anthropology* 93:373–380.

Glassman, D.M., Coelho, A.M., Carey, K.D., and Bramblett, C.A. 1984. Weight growth in savannah baboons: a longitudinal study from birth adulthood. *Growth* 48:425–433.

Gould, S.J. 1977. *Ontogeny and Phylogeny*. Cambridge: Belknap Press of Harvard University Press.

Haeckel, E. 1866. *Generelle Morphologie der Organismen: Allgemeine Grundzüge der organischen Formen-Wissenschaft, mechanisch begründet durch die von Charles Darwin reformirte Descendenz-Theorie*. Berlin: Riemer.

Hartwig, W.C. 1996. Perinatal life history traits in New World monkeys. *American Journal of Primatology* 40:99–130.

Harvey, P.H., and Clutton-Brock, T.H. 1985. Life history variation in primates. *Evolution* 39:559–581.

Harvey, P.H., and Pagel, M.P. 1991. *The Comparative Method in Evolutionary Biology.* Oxford: Oxford Press.

Healy, M.J.R. 1986. Statistics of growth standards. In F. Falkner and J.M. Tanner (eds.), *Human Growth,* 2nd ed., 3:47–58. New York: Plenum Press.

Leigh, S.R. 1992. Patterns of variation in the ontogeny of primate body size dimorphism. *Journal of Human Evolution* 23:27–50.

Leutenegger, W. 1978. Scaling of sexual dimorphism in body size and breeding system in primates. *Nature* 272:610–611.

Leutenegger, W., and Cheverud, J. 1982. Correlates of sexual dimorphism in primates: ecological and size variables. *International Journal of Primatology* 3: 387–402.

Marshall, W.A., and Tanner, J.M. 1986. Puberty. In F. Falkner and J.M. Tanner (eds.), *Human Growth,* 2nd ed., 2:171–209. New York: Plenum Press.

Mascie-Taylor, C.G.N., and Bogin, B. 1995. *Human Variability and Plasticity.* Cambridge: Cambridge University Press.

Masterson, T.J., and Leutenegger, W. 1992. Ontogenetic patterns of sexual dimorphism in the cranium of Bornean orangutans *Pongo pygmaeus pygmaeus. Journal of Human Evolution* 23:3–26.

McCance, R.A., and Widdowson, E.M. 1986. Glimpses of comparative growth and development. In F. Falkner and J.M. Tanner (eds.), *Human Growth,* 2nd ed., 1:133–151. New York: Plenum Press.

McKinney, M.L., and McNamara, K.J. 1991. *Heterochrony: The Evolution of Ontogeny.* New York: Plenum Press.

McNamara, K.J. 1995. Sexual dimorphism: the role of heterochrony. In K.J. McNamara (ed.), *Evolutionary Change and Heterochrony,* 65–89. Chichester, England: John Wiley and Sons.

Mittwoch, U. 1996. Sex-determining mechanisms in animals. *Trends in Ecology and Evolution* 11:63–67.

Oxnard, C.E. 1987. *Fossils, Teeth, and Sex: New Perspectives on Human Evolution.* Seattle: University of Washington Press.

———. 1988. Fossils, teeth and sex: new perspectives in human evolution. In N.W. Bruce, L. Freedman, and W.F.C. Blumer (eds.), *Perspectives in Human Biology,* 23–74. Nedlands: Australasian Society for Human Biology, University of Western Australia.

Richtsmeier, J.T., Cheverud, J.M., Danahey, S.E., Corner, B.D., and Lele, S. 1993. Sexual dimorphism of ontogeny in the crab-eating macaque (*Macaca fascicularis*). *Journal of Human Evolution* 25:1–30.

Selander, R.K. 1972. Sexual selection and dimorphism birds. In B. Campbell (ed.), *Sexual Selection and the Descent of Man,* 180–230. Chicago: Aldine Publishing.

Shea, B.T. 1983. Allometry and heterochrony in African apes. *American Journal of Physical Anthropology* 62:275–289.

Sinclair, D. 1973. *Human Growth after Birth,* 2nd ed. Oxford: Oxford University Press.

Sirianni, J.E., and Swindler, D.R. 1985. *Growth and Development of the Pigtailed Macaque.* Boca Raton: CRC Press.

Smith, R.J., and Leigh, S.R. 1998. Sexual dimorphism in primate neonatal body mass. *Journal of Human Evolution* 34:173–201.

Tanner, J.M. 1981. *A History of the Study of Human Growth*. Cambridge: Cambridge University Press.

———. 1986. Use and abuse of growth standards. In F. Falkner and J.M. Tanner (eds.), *Human Growth*, 2nd ed., 3:95–109. New York: Plenum Press.

———. 1989. *Foetus into Man*. Ware, U.K.: Castlemead Publications.

Treves, A. 1996. A preliminary analysis of the timing of infant exploration in relation to social structure in 17 primate species. *Folia Primatologica* 67:152–156.

Watts, E.S. 1986. Evolution of the human growth curve. In F. Falkner and J.M. Tanner (eds.), *Human Growth*, 2nd ed., 1:153–166. New York: Plenum Press.

Watts, E.S., and Gavan, J.A. 1982. Postnatal growth of nonhuman primates: the problem of the adolescent spurt. *Human Biology* 54:53–70.

Chapter 11

Life-History Evolution in Miocene and Extant Apes

Jay Kelley

One of the more profound changes in the evolution of the human lineage has been prolongation of the maturational or life-history profile. *Life history* refers to the timing or scheduling of life stages. It encompasses prenatal development and postnatal maturation, as well as the timing and frequency of reproduction. Collectively, therefore, life-history traits represent one of the most fundamental aspects of a species biology. There has been lively debate about when in the course of the human lineage the major part of this change took place and in which species, although this debate has not always been framed in terms of life history per se. However, there is no doubt when one compares the life histories of modern apes and humans that substantial change has taken place.

Prolongation of life history within the human lineage is associated with the emergence of attributes that distinguish modern humans from apes. Some of these, such as behavioral issues relating to a lengthened period of infancy, with implications for maternal behavior and division of labor, and the adolescent learning period, with its connection to increased cognitive abilities, have been the subjects of intense discussion and debate. Thus, understanding both the history and the process of life-history evolution in the human lineage is of fundamental importance.

Given the implications of life-history evolution to human evolution as a whole, one of the most meaningful questions we can ask is: What was the driving force behind life-history change in the human lineage? There has been a tendency for those who focus primarily on the human lineage to have somewhat different perceptions about life history than those whose focus is nonhuman primates or mammals more generally. This primarily concerns issues such as: Upon what is selection acting to prolong life history in the human lineage? Is the target of selection the lengthening of a particular life stage, as has often been suggested with respect to the adolescent period, or is selection acting within a different sphere entirely? Was life-history change in the human lineage simply a continuation of trends that had shaped ape life-history evolution, or were entirely different selective forces at work? To investigate these questions requires that we examine life-history evolution in apes, especially the selective forces driving this evolution.

The aim here is to explore these and other questions about life history from the perspective of nonhuman primates and other mammals and then to look at what is known about life-history evolution in the apes. Is it possible to trace a trajectory of life-history evolution in apes that might explain further life-history evolution in at least some portion of the human lineage?

Life History in Mammals

What Is Life History?

Life history is often described almost exclusively in terms of rates of maturation. This has been particularly true in discussions of life-history evolution in the human lineage, where debate has tended to focus on how quickly various fossil species grew up. As a consequence, life history is frequently conceptualized in much the same way. Such characterizations, however, have often led to misinterpretation of the adaptive significance of life history, as well as unsupported ideas about the causal mechanisms responsible for life-history variation. For example, limitations imposed by basal metabolic rate on rates of growth, including the special case of brain growth (Martin 1983), have often been considered to be a primary determinant of variation in life history (see references in Read and Harvey 1989). However, when adjustments are made for body size differences among a wide range of mammals, there is no correlation between basal metabolic rate and various life-history parameters (Harvey et al. 1991; Read and Harvey 1989).

In contrast, a substantial body of life-history theory characterizes life history primarily in terms of the scheduling of reproduction, particularly when to begin reproducing, and the allocation of resources to reproduction both pre- and postnatally (e.g., Charlesworth 1980; Charnov 1993; Stearns 1992). Viewed in reproductive terms, key life-history variables include age at first reproduction, litter size, neonatal mass relative to maternal mass, and interbirth interval. Selection for the scheduling of reproduction clearly will affect the entire ontogenetic process, which is reflected in the high degree of correlation among not only these variables but others as well, such as gestation length, age at weaning, and age at sexual maturity. Changes in life history, therefore, are primarily about developmental change as it relates to the timing and frequency of reproduction.

Life-History Variation and Its Causation

The high degree of correlation among life-history variables means that life history as a whole is expressed as distinct suites or packages. For example, a species with a short gestation period will also tend to have an early age at weaning and an early age at first reproduction. This makes intuitive sense if selection is acting to produce either earlier and more frequent or later and less frequent reproduction. As a result, organisms are arrayed along a continuum of general life-history schedules (Harvey et al. 1989; Promislow and Harvey 1990; Read and Harvey 1989). This has been referred to as the *fast-slow life-history continuum.*

The positions of different taxa along this continuum are generally highly correlated with body size. This is in part a consequence of simple growth rates; it takes a long time to produce a very large animal. In fact, this allometric reality has frequently been offered as the explanation for observed life-history variation. However, in the absence of hypotheses about causation, the allometric relationship between life history and body size is nothing more than a restatement of the facts that it is supposed to explain (Boyce 1988; Harvey et al. 1989). Further, there is some portion of life-history variation that is not correlated with body size. This suggests that (1) there are factors influencing life history other than body size and (2) the correlation between life history and body size might be due in part to the correlation of each to one or more of these other factors.

Numerous other such factors have been suggested to account for observed life-history variation. However, as noted above for basal metabolic rate, when the effects of body size are removed, the correlations between variation in life-history variables and these purported explanatory variables either disappear or become insignificant (Harvey et al. 1989, 1991; Read and Harvey 1989).

A successful causal explanation for life-history variation must be able to account for that portion of the variation that is correlated with body size, as well as that portion that is not. Life-history theory, with its emphasis on demographic variables related to reproduction, suggests such a causative factor. Since fecundity must be balanced by mortality, rates of mortality would seem to offer promise as an explanation for observed life-history variation. In a series of papers, Paul Harvey and coworkers examined the relationship between mortality and life-history variation both across and within the mammalian orders (see Promislow and Harvey 1991 for review). They found that, with the effects of body size removed, there are significant correlations between mortality and the residual variation in several life-history variables. As would be expected given the influence of body size on predator-prey relations, mortality is also significantly correlated with body size. Thus far, mortality rate seems to be the only proposed causative factor that accounts for the variation in life history that is independent of body size and the variation that is not.

In these same studies, it was found that consideration of age-specific mortality is also important for explaining the observed patterns of life-history variation. While the pattern of correlations between life-history traits and age-specific mortality is complex (Promislow and Harvey 1990), in general taxa that suffer relatively high adult mortality in relation to infant/juvenile mortality tend to have relatively fast life histories. Those that suffer relatively low rates of adult mortality tend to have slower life histories (also Horn 1978). The explanation for this phenomenon lies with the trade-off nature of life history. It might be expected that the most successful life-history strategy for all animals would be to reproduce as early and as prolifically as possible. However, there are very high costs associated with early reproduction, expressed as relatively high rates of reproductive failure and depressed future reproduction. The latter includes both delays in subsequent reproduction and a higher probability of death for a female that assumes the stress of reproduction before full adult body mass is achieved (Promislow and Harvey 1990). The trade-off, then, for the presumed advantage gained by early reproduction is the higher potential for reproductive failure and the cost incurred on future reproduction. If the probability of death as an adult is low, greater lifetime reproductive output will result from delaying reproduction and minimizing the deleterious consequences of early reproduction. If, on the other hand, the probability of death as an adult is high, greater lifetime reproductive output will result from early reproduction in spite of the costs.

Life History and Ecology

Life-history variation is certainly linked to ecology, but the relationship is complex. One reason for the complexity is that, for animals of very different body sizes and therefore very different mass/energy relationships, many ecological variables are not absolute. An ecological perturbation that might be catastrophic for one species might be nothing more than a nuisance for another. For this reason, there has been greater success in identifying apparent ecological correlates of life-history variation in studies of closely related species (e.g., Smith 1988) than there has been in identifying ecological variables that are universally applicable across a wide range of taxa (Partridge and Harvey 1988).

Another approach to identifying ecological correlates of life-history variation, one that is intermediate between narrowly focused comparative studies and broad syntheses, is to focus on a single higher taxonomic group that expresses a range of body sizes but a more or less common underlying biology. A series of such analyses of different primate higher-level taxa and the order as a whole consistently found a correlation between the intrinsic rate of natural increase (r)—which can substitute as an approximate measure of overall life history—and a bipartite division of habitats into rain forest and other kinds of habitats (see Ross 1998). There is a strong tendency for species living in rain-forest habitats to have a lower r (slower life histories) than those living in nonforest habitats.

Ross initially interpreted these results in terms of traditional r and K selection theory and the notion of r and K selecting habitats. However, the validity of these concepts to an understanding of life history is being increasingly criticized because their application frequently ignores many of the underlying assumptions of the theory (Charlesworth 1980; Harvey et al. 1989; Partridge and Harvey 1988; Promislow and Harvey 1991; Stearns 1992). In fact, Ross (1998) acknowledged these criticisms and attempted to reinterpret her results in terms of the possible influences of different habitat types on adult mortality.

Life-History Variation and Evolution in Catarrhine Primates

Primates have size-adjusted life histories that are slow compared to those of most other mammals. Only bats occupy a similar position on the fast-slow life-history continuum (Read and Harvey 1989; Harvey et al. 1989). Within primates, however, there is substantial variation, even when body size is taken into account. Apes, including both gibbons and great apes, have relatively prolonged life histories. Old World monkeys, on the other hand, have relatively accelerated life histories for their size (Kelley 1997). Thus, there seems to have been an actual divergence in life histories between the two groups of catarrhines, not just a slowing of life history in apes. The differences between monkeys and apes are most evident in the comparison of gibbons and monkeys, which overlap in size and therefore obviate the need for the statistical size adjustments necessary when comparing monkeys and great apes (Fig. 11.1).

Data relating to mortality in apes and monkeys are consistent with their positions on a primate fast-slow life-history continuum. Once past infancy, mortality from all causes is lower in apes than it is in monkeys (Gage and Dyke 1988; Goodall 1986; Nishida et al. 1990). This is also reflected in the

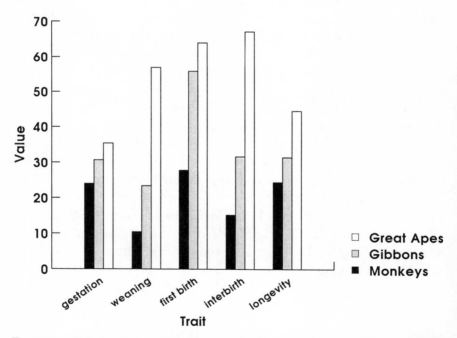

Fig. 11.1.   Relative timing of life-history traits in extant great apes, gibbons, and Old World monkeys. The absolute time scale varies among traits (e.g., gestation expressed in weeks, longevity in years) but is the same among taxa within traits. Values are means of included genus means from data in Harvey et al. (1987), with some adjustments to the great ape values (see Kelley 1997).

data on predation alone; the incidence of observed as well as presumed adult mortality from predation is far higher among species of Old World monkeys than it is in apes (Boesch 1991; Cheney and Wrangham 1987; Isbell 1994; Stanford et al. 1994; Tsukahara 1993). It is relevant to the question of life-history evolution in the human lineage to know when this divergence in ape and Old World monkey life histories began, the subsequent trajectory of life-history evolution in the apes, and the selective forces that might have been at work.

Life-History Evolution in Apes

Life-History Inference from Dental Development

The timing and overall duration of dental development are tightly linked to rates of maturation in organisms (Smith 1989); dental development is, in a sense, just another life-history trait (Smith and Tompkins 1995). Because of the strong correlations among life-history variables, rates of dental development can be used to infer the general life histories of extinct animals. There is nothing special about dental development with regard to life-history inference. It is simply one of the few life-history traits that is preserved in the fossil record. In fact, adult brain size is seemingly as reliable as dental development for the purpose of life-history inference (Smith et al. 1995), but there are simply a great many more jaws and teeth preserved in the fossil record than there are neurocrania that are sufficiently complete for reliable estimates of brain size. This is far more true for fossil apes than it is for fossil humans, a point to which I will return below.

Among living primates, it has been demonstrated that age at first molar emergence is a particularly good correlate of various life-history traits (Smith 1989). Since the analyses upon which this relationship is based include representatives of all living primate higher taxa, we can presume that all fossil primates within the radiation encompassed by the extant species will express the same relationship. Thus, if the average age at first molar emergence can be established for a fossil species, we can, within certain limits, infer its general life-history profile. The precision with which such inferences can be made is important and determines how questions about life-history evolution are framed.

The most straightforward approach to estimating age at first molar emergence in fossil species is to determine the age at death for individuals that died while in the process of erupting their first molars. Ages at death can be determined with a high degree of precision using the record of incremental growth lines that are preserved in all teeth. The use of dental histological growth records for aging individuals is described in Beynon et al. (1991), Dean (1989), Dean et al. (1993), and Kelley (1997) and will only be summarized here.

Several regular growth lines are preserved in dental enamel and dentine, but only two of these are of importance here: enamel cross-striations, which are daily increments of enamel deposition, and brown striae of Retzius, or Retzius lines, which are brief disruptions in enamel secretion that occur with a regular periodicity of between six and ten days in apes and humans (Bromage 1991; Dean 1987). Retzius lines are also expressed on the surface of teeth as regular undulations or ridges known as *perikymata*. There is some variation in Retzius line periodicity within species as a whole, but within individuals there is a constant number of daily increments between adjacent Retzius lines of all teeth (Beynon et al. 1991; Dean 1987; Dean and Beynon 1991). Thus, much the same as tree rings but with an approximately weekly rather than annual period, Retzius lines and perikymata precisely record the duration of tooth crown formation subsequent to the formation of the cusp apex (see below).

In apes and monkeys, the incisor crowns are still forming during first molar eruption. Therefore, the growth record of the incisors can be used to age individuals that died while their first molars were erupting. To do so, however, requires two pieces of information in addition to perikymata counts. First, a portion of crown formation time, consisting of the enamel formed at the cusp apex, does not leave a record of perikymata because the Retzius lines are buried under subsequently formed enamel and do not reach the surface. Second, incisors begin to mineralize anywhere from a few weeks to a few months after birth (Beynon et al. 1991; Dean 1987; Dean and Beynon 1991), so the time between birth and the beginning of mineralization (the postnatal delay) must be added to crown formation time for the total elapsed time between birth and death. This value must be estimated in most circumstances because it can be directly measured only by sectioning both the incisor and the first molar, registering growth lines in the incisor to those in the molar, and then counting the elapsed time between the neonatal line in the molar (produced at birth) and the inception of mineralization in the incisor (Beynon et al. 1991). Therefore, the complete formula for calculating age at death from the incisor is the sum of (1) the time between birth and the beginning of crown mineralization,

(2) the duration of cuspal enamel formation, and (3) the duration of enamel formation represented by surface perikymata, determined by multiplying the number of perikymata by the number of daily cross-striations between perikymata.

Life Histories of Miocene Apes

To date, estimates of age at first molar emergence using dental development have been made for only two fossil apes. The first was an individual of *Sivapithecus parvada,* from a 10 million-year-old site in the Siwalik Sequence of Pakistan (Kelley 1997). The second, reported here, is an individual of *Afropithecus turkanensis* from the approximately 17 million-year-old site of Moruorot in Kenya. *Sivapithecus* is generally considered to be on the orangutan lineage (Andrews and Cronin 1982; Ward and Brown 1986), whereas *Afropithecus* is a more primitive species not obviously linked to any living ape (Leakey and Walker 1997).

Sivapithecus. The mandible of the *Sivapithecus* juvenile for which age at first molar emergence was estimated is shown in Figure 11.2*A.* The mandible preserves both deciduous premolars and the permanent anterior teeth and premolars embedded within their crypts. The matrix-filled alveolus of the first molar reveals that this tooth had erupted but fallen out before fossilization, an indication that it was still relatively early in the eruption process. Also belonging to this individual is a fragment of premaxilla with the upper central and lateral incisors exposed within their crypts. Perikymata are well expressed on the central incisor from near the crown apex to the enamel that was forming at the advancing cervical margin at the time of death (Fig. 11.2*B*).

In calculating the age at death of this individual (Kelley 1997), the number of daily increments between perikymata, the duration of cuspal enamel formation, and the length of the postnatal delay in incisor mineralization all had to be estimated from the literature on extant primates, since it was decided not to section the incisor of this individual and there were virtually no data for these growth parameters for any fossil species. Thus, maximum, modal, and minimum values for each growth parameter were obtained from the literature on extant humans, apes, and large monkeys to produce corresponding maximum, modal, and minimum ages at death. The results are shown in Table 11.1.

Since the first molar had emerged from its crypt and had almost certainly undergone gingival emergence, it was necessary to estimate the time

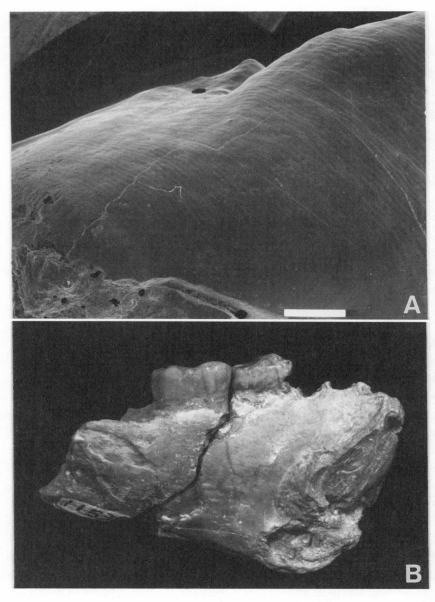

Fig. 11.2. Juvenile mandible of *Sivapithecus parvada*, GSP 11536. *A:* Perikymata on the upper central incisor. Note that perikymata are preserved all the way to the last formed enamel at the advancing cervical region (*lower left*). The incisive edge is to the *upper right*. *B:* Preservation of the two deciduous premolars. *C:* Radiograph of the posterior portion of the *Sivapithecus* juvenile mandible with the deciduous fourth premolar. Note the mesial root alveolus of the permanent first molar (*darkened area*) just posterior to the deciduous premolar root.

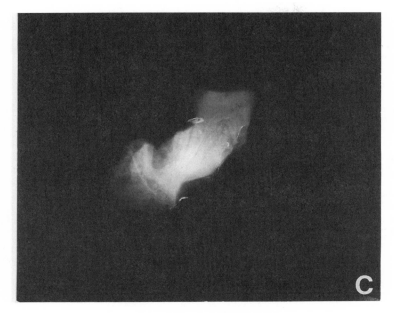

Fig. 11.2. (*Continued*)

that had elapsed between gingival emergence and death, which was then subtracted from the calculated age at death to arrive at an estimate of the age at first molar emergence. The amount of time since gingival emergence was estimated from the degree of root development evident in radiographs of the mandible (Fig. 11.2*C*). A comparative examination of root development in relation to tooth eruption in great apes and monkeys suggested that this individual probably died no more than a few months after first molar gingival emergence (Kelley 1997); from the comparative data, a value of four months was determined to be a reasonable estimate (Table 11.1).

The resulting modal estimate for age at first molar emergence, at nearly forty-one months, is slightly greater than the mean age for extant chimpanzees (Table 11.2). Importantly, the minimum estimate, which assumes that this individual expressed the minimum values for each of the included growth parameters, is still within the chimpanzee range but lies well above the ranges for all monkeys for which there are data. Based on these comparisons, it was concluded that the age of first molar emergence in 10 million-year-old *Siva-pithecus* was more apelike than monkeylike and may well have been essentially the same as in the extant great apes, which are all similar. This suggests a life-history profile approaching or equivalent to that of modern great apes.

More recent, as yet unpublished work on the *Sivapithecus* juvenile has

Table 11.1 Estimated Age at First Molar Emergence in GSP 11536, *Sivapithecus parvada*

			Months			
	Perikymata Count	Number of Cross-striations	Duration of Cuspal Enamel Growth	Length of I1 Postnatal Delay	Age at Death	Age at M1 Emergence
Minimum	133	6	4	0	30.1	26.1
Modal	133	8	6	4	44.7	40.7
Maximum	133	10	8	6	57.5	53.5

Source: For cross-striation number, duration of cuspal enamel formation, and length of I1 postnatal delay: Beynon and Reid (1987), Beynon et al. (1991), Bromage and Dean (1985), Boyde (1990), Dean (1987), Dean and Beynon (1991), Dean et al. (1986, 1993).

Note: Age at M1 emergence is set at four months before the death of the individual (see text).

Table 11.2 Age at First Molar Emergence in *Sivapithecus parvada*, *Afropithecus turkanensis*, *Pan troglodytes*, and Extant Monkeys

	Age (months)			
Species	Estimate	Mean	Minimum	Maximum
Fossils				
Sivapithecus parvada	40.7			
Afropithecus turkanensis	27.7			
Extant				
Pan troglodytes		38.9	25.7	48.0
Macaca mulatta		16.2	12.5	22.6
M. fascicularis		16.8	14	20
M. nemestrina		16.4	—	18.6+
M. fuscata		18.0	—	<24
Cercopithecus aethiops		9.9	7.9	12.0
Papio cynocephalus		20.0	—	—
P. anubis 1		20.0	>16	<25
P. anubis 2		16.7	15.7	<21

Source: Pan troglodytes (Kuykendall et al. 1992); *Macaca mulatta, Cercopithecus aethiops* (Hurme and van Wagenen 1961); *Macaca fascicularis* (Bowen and Koch 1970); *Macaca nemestrina* (Swindler 1985; B. H. Smith, pers. comm.—based on Swindler data); *Macaca fuscata* (Smith et al. 1994; B. H. Smith, pers. comm.—based on data in Iwamoto et al. 1987); *Papio cynocephalus* (Smith et al. 1994 —based on Reed in Phillips-Conroy and Jolly 1988); *Papio anubis* 1 (Smith et al. 1994; B. H. Smith, pers. comm.—based on data in Kahumbu and Eley 1991); *Papio anubis* 2 (J. Kelley, unpublished data).

Note: Reliability of range data varies considerably depending on method used and sample size.

strengthened the case for a modern great ape life-history profile. Using microcomputed tomography, a precise measurement of cuspal enamel thickness was obtained for the central incisor from this individual (Kelley et al., in preparation). The cuspal enamel is exceptionally thick, revealing that the duration of cuspal enamel formation was at least as great as the maximum value recorded in the literature for extant catarrhines. Substituting the maximum value for cuspal enamel formation in Table 11.1 for the modal and minimum values elevates the modal and minimum estimates of age at first molar emergence in the *Sivapithecus* juvenile by two and four months, respectively. This would bring even the minimum estimate well within the range for extant chimpanzees and strengthens the argument that life history in this individual was essentially modern ape–like.

Afropithecus. The mandible of an *Afropithecus* juvenile from Moruorot is shown in Figure 11.3. The mandible preserves a deciduous fourth premolar, an erupting first molar, and a developing lateral incisor still in its crypt. The incisor has clearly expressed perikymata from near the crown apex down to the most recently formed enamel at the advancing cervical margin (Fig. 11.4). In contrast to *Sivapithecus,* for which incisor growth parameters

Fig. 11.3. Juvenile mandible of *Afropithecus turkanensis,* MO 26, preserving the deciduous fourth premolar, the erupting first molar, and the permanent lateral incisor in its crypt.

had to be estimated from extant monkeys and apes, it was possible to make estimates for *Afropithecus* using recently published data on all of the relevant incisor growth parameters from several individuals of another, closely related Miocene genus, *Proconsul* (Beynon et al. 1998). These are compared to the ranges of values for extant great apes and humans in Table 11.3. Importantly, all of the growth parameter determinations for *Proconsul* but one—the number of daily increments between adjacent Retzius lines—were made on *Proconsul heseloni*, a species that was substantially smaller than *A. turkanensis*. This results in conservative estimates for the *Afropithecus* juvenile, since many aspects of tooth growth seem to scale with body size. Moreover, *Afropithecus* has much thicker enamel than even comparably sized *Proconsul* species (Leakey and Walker 1997), so the duration of cuspal enamel formation based on the *Proconsul* data is probably substantially underestimated. Last, the value for postnatal delay in incisor mineralization in *Proconsul* is much less than in most of the extant primate individuals examined (Table 11.3), including several individuals of the large Old World monkey *Theropithecus* (Swindler and Beynon 1993). Thus, the age at death of the *Afropithecus* juvenile based on the growth data retrieved from *Proconsul*, at just under twenty-five months (Table 11.4), is surely conservative and should be considered an absolute minimum estimate. It is likely that the actual age at death was anywhere from a few to several months greater.

The position of the first molar in relation to the alveolar margin in the *Afropithecus* mandible reveals that this tooth had not yet emerged from the gingiva, the event upon which age at tooth emergence is based in living animals. Therefore, it is necessary to estimate how much more time would have elapsed between death and gingival emergence, the opposite adjustment from that required for *Sivapithecus*. As with *Sivapithecus*, the amount of time involved was estimated from studies of tooth eruption in extant primates. Based on a longitudinal radiographic study of first molar development in anubis baboons that I have been conducting for the last few years, the amount of time remaining to gingival emergence from the stage of eruption represented by the first molar in the *Afropithecus* juvenile is consistently about three months (Kelley, unpublished data). Since both dental development and eruption seem to have been more prolonged in *Afropithecus* than in baboons, three months is probably again a slightly conservative estimate. Adding this estimate to the calculated age at death results in an age at first molar emergence of just under twenty-eight months. It is highly probable that this minimum value underestimates the actual age at first molar emergence by as much as several months, and a figure of at least thirty months is likely. Al-

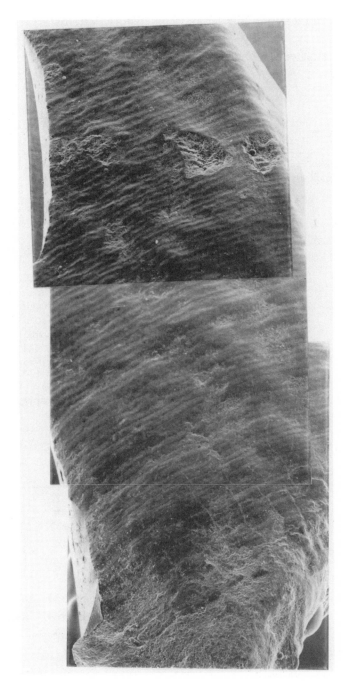

Fig. 11.4. Scanning electron micrograph photomontage of perikymata on the permanent lateral incisor of the *Afropithecus* juvenile mandible. Note that perikymata are preserved down to the last formed enamel at the advancing cervical region (*bottom*).

Table 11.3 Dental Growth Parameters in *Proconsul* and Extant Hominoids

Taxon	Months		Source
	Length of I2 Postnatal Delay	*Duration of Cuspal Enamel Growth*	
Proconsul heseloni	1.5	4.0	Beynon et al. 1998
Extant hominoids			
Hylobates	3.7	4.5	Dirks 1998
Pongo	13.0	—	Beynon et al. 1991
Gorilla	11.0	4.0	Beynon et al. 1991
Pan	2.4	6.4	Reid et al. 1998b
	7.9	6.4	Reid et al. 1998b
	2.3	5.8	Reid et al. 1998b
	8.4	6.8	Reid et al. 1998b
Homo	4.8	5.5	Reid et al. 1998a
	—	5.8	Reid et al. 1998a
	—	7.7	Reid et al. 1998a
	0	8.0	Dean and Beynon 1991
	8.3	—	Dean et al. 1993

Table 11.4 Estimated Age (in months) at First Molar Eruption in MO 26, *Afropithecus turkanensis*

Perikymata Count	Number of Cross-striations	Duration of Cuspal Enamel Growth	Length of I2 Postnatal Delay	Age at Death	Age at M1 Emergence
97	6	4	1.5	24.7	27.7

Source: Cross-striation number, duration of cuspal enamel formation, and length of I2 postnatal delay are based on data from *Proconsul* species (Beynon et al. 1998).

Note: Individual is estimated to have died three months before M1 gingival emergence (see text).

though this age at first molar emergence is substantially lower than that for *Sivapithecus* (as would be even a somewhat less conservative estimate), like that of *Sivapithecus* it is still within the range of chimpanzees and well outside the ranges of even large extant monkeys (Table 11.2).

Life-History Evolution in Miocene and Extant Apes

Interpreting life history in the two fossil ape species based on one individual of each certainly requires caution. However, as a simple matter of central tendency, it is more probable that each individual is closer to its respective species mean than to either extreme of the species range. Nevertheless, because we are dealing with single individuals, comparisons to extant catarrhine ranges of variation are more informative than are comparisons to means. Therefore, the fact that both individuals lie well outside the ranges of variation for age at first molar emergence of even the largest monkeys suggests that early Miocene *Afropithecus* was well along in a grade shift toward the more prolonged life histories of extant great apes and that *Sivapithecus* had achieved an essentially modern great ape life-history profile.

Broadly speaking, all extant great apes have similarly prolonged life histories. This may reflect inheritance from a last common ancestor, suggested from the degree of genetic divergence in the living apes to have existed some 12–13 million years ago (Pilbeam 1996). The prolonged life history inferred for the 10 million-year-old *Sivapithecus* individual, considered to be an early member of the orangutan lineage, is consistent with this interpretation. The establishment of a fundamental life-history pattern during the early history of the lineage, one that is subject to a limited amount of subsequent variation even with fairly large changes in body size and differences in habitus, would be consistent with more general mammalian patterns described by Martin and MacLarnon (1990) and Harvey and co-workers (e.g., Harvey et al. 1989; Read and Harvey 1989).

A plausible trajectory of life-history evolution in apes, therefore, would be for there to have been the beginning of a grade shift from the primitive catarrhine condition occurring fairly early in the history of the lineage. The continued transition to the very prolonged life histories characteristic of the extant great apes may have been relatively rapid and was perhaps established by the time of their last common ancestor. It is only with the appearance of the human lineage that there is another grade shift leading ultimately to the even more prolonged life history of modern humans. The major questions concerning this shift are (1) what was the tempo of further life-history change within the lineage, and (2) what were the selective factors that drove this change?

Life-History Evolution in Humans

For the last decade and a half, there has been a vigorous debate regarding the timing of life-history evolution in the human lineage. The debate has mostly been framed in terms of maturation rates rather than life history per se, although this has changed more recently (Smith and Tompkins 1995). According to one interpretation of dental development in early hominids, most of this change has occurred during the last 1.5 million years or so, which represents less than a third of the total duration of the human lineage (Anemone et al. 1996; Beynon and Dean 1988; Beynon and Wood 1987; Bromage and Dean 1985; Dean et al. 1993). The evidence includes analyses of the sequence and relative timing of individual tooth development, as well as direct histological measurement of tooth crown formation times as described above. According to the former evidence, certainly all australopithecines and perhaps the earliest species of *Homo* were still substantially apelike in their patterns and relative rates of dental development. This has been verified by direct measurement of crown formation times, which resulted in significant downward revisions of earlier estimates of age at death in several early hominid juveniles (Beynon and Dean 1988; Beynon and Wood 1987; Bromage and Dean 1985; Dean et al. 1993). These juveniles were found to have achieved any particular stage of dental development at roughly half the age of modern humans, an essentially chimpanzee rate of dental development. This suggests a chimpanzee-like rate of postnatal maturation more generally.

This interpretation has been aggressively challenged by those who argue that an equally plausible interpretation of the dental evidence is that maturation in late Pliocene/early Pleistocene hominids was more nearly like that of modern humans (Lampl et al. 1993; Mann 1975; Mann et al. 1991). In this view, the associated changes in life history came about much earlier in the history of the lineage. This challenge is mostly centered on issues relating to variation in dental development and eruption in both apes and humans, which, it is claimed, are sufficiently great to render doubtful any conclusions based on either the absolute or the relative timing of dental development. However, most of the objections raised by Mann and his coworkers have been successfully rebutted (Anemone et al. 1991, 1996; Dean and Beynon 1991; Smith 1994; Smith and Tompkins 1995).

In fact, the entire debate surrounding the validity of dental evidence for inferring maturation rates in early hominids has been more distracting than enlightening, since there is another, simpler means for making inferences about life history in fossil primates. As compelling as the correlations between dental development and life-history traits are those between these traits and brain size (Smith 1989). In a sense, brain size, like dental development, can be viewed as simply another life-history trait and, like dental development, one that also fossilizes. Although the relationship between brain size and individual life-history variables is complex and the causative role for the brain in life-history evolution suggested by Sacher (Sacher 1975; Sacher and Staffeldt 1974) is highly suspect (Harvey and Read 1988), brain size is still a robust measure of the overall pace of life among mammals generally (Harvey and Clutton-Brock 1985; Harvey and Read 1988).

Measurements of cranial capacity in numerous australopithecine skulls confirm the apelike life histories of these early hominids (Smith and Tompkins 1995; Smith et al. 1995). As noted by Smith et al., for species with cranial capacities as small as those of the australopithecines to have life histories even approaching that of modern humans would require a relationship between life history and brain size that does not exist in any other primate. Smith and colleagues caution that confidence intervals for the brain size/age-at-first-molar-emergence regression do not absolutely preclude modern humanlike life histories in late Pliocene/early Pleistocene hominids. However, this caution, while statistically accurate, seems unnecessary given the consistent corroborating evidence based on histologically determined ages at first molar emergence in the same species and is seemingly not given much weight even by Smith et al.

Smith et al. (1995) also note that the first evidence of more prolonged life history in the human lineage is in the latest Pliocene *Homo*. They describe maturation in *Homo habilis sensu lato* as being intermediate between that of australopithecines and extant great apes on the one hand and modern humans on the other (Smith et al. 1995, 165). However, brain size and inferred age at first molar eruption in *H. habilis* are still closer to australopithecine values, particularly contemporaneous robust australopithecines, than to those of modern humans or even to those of Pleistocene *Homo ergaster* or *Homo erectus*. Nevertheless, if the noted increase in brain size reflects an equivalent slowdown in life history, this marks the first significant change in life history in the ape and human fossil record since *Sivapithecus* more than 8 million years earlier.

Thus, for the vast majority of the roughly 6-million-year duration of the human lineage, there was essentially no change in life history from that which characterized its ape antecedents. Based on the evidence from *Sivapithecus*, the same life-history profile characterized apes for several million years before the appearance of the human lineage. This is a remarkable record of stasis in one of the most fundamental aspects of the biology of these species and says much about the biological identity, as opposed to the cladistic identity, of early humans. For life history, therefore, as in many other areas, the phyletic event of real biological significance is the emergence of *Homo* rather than the divergence of the human lineage from that of chimpanzees.

Causation in Human Life-History Evolution

Two attributes that are commonly cited as the focus of selection for life-history change in the human lineage are increased brain size and an extended adolescent learning period. In the first case, the driving mechanism is seen to be selection for cognitive abilities through increased brain size. Brain size is viewed as the pacemaker of life history (Sacher 1975), so that more prolonged life history is essentially a passive consequence of the selection for increased brain size. In the second case, selection is acting directly on life history, or at least on one particular life stage. Cognitive abilities are again implicated, however, since the perceived advantage of the extended adolescent period is to take advantage of the increased cognitive abilities bestowed by a larger brain to increase fitness through learned behavior. These explanations for more prolonged life history are certainly not mutually exclusive and have, in fact, been considered to be inextricably linked.

Several factors can be cited for the preoccupation with brain size and cognitive abilities. First among these is a tendency to view human life history as unique within the primate order. Unique attributes suggest unique causation, and the correlation between life history and brain size logically, seemingly almost inevitably, suggests to some a causal relationship with the unique cognitive abilities of humans. However, just as primate life histories (or those of bats, for that matter) are not unique with respect to other mammalian orders, in each case simply representing an extreme in the fast-slow life-history continuum, so, too, with modern human life history, which is at the extreme slow end of the primate life-history continuum. The correlation between brain size and the speed of life history is not unique to humans, and there is no reason to think that the arrow of causation in this general relationship in primates runs from brain size and cognitive ability to life history (Harvey and Read 1988; Read and Harvey 1989).

Also contributing to the focus on brain size and cognitive ability is the practice of framing discussions of human life history in terms of maturation—that is, how fast various early human species grew up. The emphasis on maturation has had a subtle but important influence by focusing attention on life stages themselves, especially adolescence, and away from the overwhelming importance of reproduction and demographics, which are the foundation of life-history theory. A final factor is the historically common tendency in human evolutionary studies to project modern human attributes back farther into the fossil record than the data warrant—thus, the impetus to focus on the emergent cognitive abilities of modern humans to explain the earliest phases of more prolonged life history in the human lineage.

In contrast to this *Homo*-centric perspective is the view that human evolutionary phenomena should be explainable by the same general theories and models that apply to other mammals (Cartmill 1990; Smith and Tompkins 1995). From this perspective, the evolution of human life history is about changes in reproductive scheduling in response to the selection pressures deriving from age-specific patterns of mortality. Within the framework of life-history theory, it is reasonable to presume that the extended life history inferred for *Homo* reflects further reduction in both juvenile and adult mortality from that typical of great apes. Any number of factors could be responsible for this. These could even include more intelligent behavior. It would not follow from this, however, that changes in life history were being driven by selection for enhanced cognitive ability. On the contrary, even though there might be positive feedback affecting brain size, selection would not be acting on brain size directly. Rather, it would be acting only indirectly through the contribution of cognitive ability toward reducing mortality. Moreover, since there is an expectation that the further slowing of life history would be accompanied by increased brain size (whatever the underlying basis for this relationship), it would be impossible to know which of these factors was, in fact, responsible for brain size increase. Likewise, reduction in mortality could result from any number of other factors having no necessary connection to intelligence, such as changes in sociality or foraging behavior and diet, and increased brain size would still result.

Therefore, regardless of causation, an increase in brain size would be expected to accompany the slowdown in life history in humans. Simply because early *Homo* has a larger brain than that of apes and further increase in brain size in the lineage ultimately leads to greatly enhanced cognitive abilities, there is no reason to presume that the initial change in life history in the human lineage resulted from selective pressures that were fundamentally different from those that produced the slowed life histories of Miocene and extant great apes.

Ackowledgments

I thank Nancy Minugh-Purvis and Ken McNamara for inviting me to contribute to this volume. Thanks also to Chris Dean, Paul Harvey, Don Reid, and Holly Smith for numerous helpful discussions about life history and dental development. I also acknowledge the Geological Survey of Pakistan for collaboration in the field work that led to the recovery of the *Sivapithecus* material and the National Museums of Kenya for access to the *Afropithecus* material. The research reported here was supported by National Science Foundation grants BNS-9196211 and SBR-9408664.

References

Andrews, P., and Cronin, J.E. 1982. The relationship of *Sivapithecus* and *Ramapithecus* and the evolution of the orangutan. *Nature* 297:541–546.

Anemone, R.L., Mooney, M.F., and Siegel, M.I. 1996. Longitudinal study of dental development in chimpanzees of known chronological age: implications for understanding the age at death of Plio-Pleistocene hominids. *American Journal of Physical Anthropology* 99:119–133.

Anemone, R.L., Watts, E.S., and Swindler, D.R. 1991. Dental development of known-age chimpanzees, *Pan troglodytes. American Journal of Physical Anthropology* 86:229–241.

Beynon, A.D., and Dean, M.C. 1988. Distinct dental development patterns in early fossil hominids. *Nature* 335:509–514.

Beynon, A.D., Dean, M.C., Leakey, M.G., Reid, D.J., and Walker, A. 1998. Comparative dental development and microstructure of *Proconsul* teeth from Rusinga Island, Kenya. *Journal of Human Evolution* 35:163–209.

Beynon, A.D., Dean, M.C., and Reid, D.J. 1991. Histological study on the chronology of the developing dentition in gorilla and orangutan. *American Journal of Physical Anthropology* 86:189–204.

Beynon, A.D., and Reid, D.J. 1987. Relationships between perikymata counts and crown formation times in the human permanent dentition. *Journal of Dental Research* 66:889–890.

Beynon, A.D., and Wood, B.A. 1987. Patterns and rates of enamel growth in the molar teeth of early hominids. *Nature* 326:493–496.

Boesch, C. 1991. The effects of leopard predation on grouping patterns in forest chimpanzees. *Behaviour* 117:220–242.

Bowen, W.H., and Koch, G. 1970. Determination of age in monkeys (*Macaca irus*) on the basis of dental development. *Laboratory Animals* 4:113–123.

Boyce, M.S. 1988. Evolution of life histories: theory and patterns from mammals. In M.S. Boyce (ed.), *Evolution of Life Histories of Mammals,* 3–30. New Haven: Yale University Press.

Boyde, A. 1990. Developmental interpretations of dental microstructure. In C.J. DeRousseau (ed.), *Primate Life History and Evolution,* 229–267. New York: Wiley-Liss.

Bromage, T.G. 1991. Enamel incremental periodicity in the pig-tailed macaque: a polychrome fluorescent labeling study of dental hard tissues. *American Journal of Physical Anthropology* 86:205–214.

Bromage, T.G., and Dean, M.C. 1985. Re-evaluation of the age at death of Plio-Pleistocene fossil hominids. *Nature* 317:525–528.

Cartmill, M. 1990. Human uniqueness and theoretical content in paleoanthropology. *International Journal of Primatology* 11:73–92.

Charlesworth, B. 1980. *Evolution in Age-Structured Populations.* Cambridge: Cambridge University Press.

Charnov, E.L.R. 1993. *Life-History Invariants.* Oxford: Oxford University Press.

Cheney, D.L., and Wrangham, R.W. 1987. Predation. In B.B. Smuts, D.L. Cheney, R.M. Seyfarth, R.W. Wrangham, and T.T. Struhsaker (eds.), *Primate Societies,* 227–239. Chicago: University of Chicago Press.

Dean, M.C. 1987. Growth layers and incremental markings in hard tissues: a review of the literature and some preliminary observations about enamel structure in *Paranthropus boisei. Journal of Human Evolution* 16:157–172.

———. 1989. The developing dentition and tooth structure in hominoids. *Folia Primatologica* 42:160–177.

Dean, M.C., and Beynon, A.D. 1991. Histological reconstruction of crown formation times and initial root formation times in a modern human child. *American Journal of Physical Anthropology* 86:215–228.

Dean, M.C., Beynon, A.D., Thackeray, J.F., and Macho, G.A. 1993. Histological reconstruction of dental development and age at death of a juvenile *Paranthropus robustus* specimen, SK 63, from Swartkrans, South Africa. *American Journal of Physical Anthropology* 91:401–419.

Dean, M.C., Stringer, C.B., and Bromage, T.G. 1986. A new age at death for the Neanderthal child from the Devil's Tower, Gibraltar and the implications for studies of general growth and development in Neanderthals. *American Journal of Physical Anthropology* 70:301–309.

Dirks, W. 1998. Histological reconstruction of dental development and age at death in a juvenile gibbon (*Hylobates lar*). *Journal of Human Evolution* 35:411–425.

Gage, T.B., and Dyke, B. 1988. Model life tables for the larger Old World monkeys. *American Journal of Physical Anthropology* 16:305–320.

Goodall, J. 1986. *The Chimpanzees of Gombe: Patterns of Behavior.* Cambridge: Harvard University Press.

Harvey, P.H., and Clutton-Brock, T.H. 1985. Life history variation in primates. *Evolution* 39:559–581.

Harvey, P.H., Martin, R.D., and Clutton-Brock, T.H. 1987. Life histories in comparative perspective. In B.B. Smuts, D.L. Cheney, R.M. Seyfarth, R.W. Wrangham, and T.T. Struhsaker (eds.), *Primate Societies,* 181–196. Chicago: Chicago University Press.

Harvey, P.H., Pagel, M.D., and Rees, J.A. 1991. Mammalian metabolism and life histories. *American Naturalist* 137:556–566.

Harvey, P.H., and Read, A.F. 1988. How and why do mammalian life histories vary? In M.S. Boyce (ed.), *Evolution of Life Histories of Mammals: Theory and Pattern*, 213–232. New Haven: Yale University Press.

Harvey, P.H., Read, A.F., and Promislow, D.E.L. 1989. Life history variation in placental mammals: unifying the data with the theory. *Oxford Surveys in Evolutionary Biology* 6:13–31.

Horn, H.S. 1978. Optimal tactics of reproduction and life history. In J.R. Krebs and N.B. Davies (eds.), *Behavioural Ecology: An Evolutionary Approach*, 272–294. Oxford: Blackwell.

Hurme, V.O., and van Wagenen, G. 1961. Basic data on the emergence of permanent teeth in the rhesus monkey (*Macaca mulatta*). *Proceedings of the American Philosophical Society* 105:105–140.

Isbell, L.A. 1994. Predation on primates: ecological patterns and evolutionary consequences. *Evolutionary Anthropology* 3:61–71.

Iwamoto, M., Watanabe, T., and Hamada, Y. 1987. Eruption of permanent teeth in Japanese monkeys (*Macaca fuscata*). *Primate Research* 3:18–28.

Kahumbu, P., and Eley, R.M. 1991. Teeth emergence in wild live baboons in Kenya and formulation of a dental schedule for aging wild baboon populations. *American Journal of Primatology* 23:1–9.

Kelley, J. 1997. Paleobiological and phylogenetic significance of life history in Miocene hominoids. In D.R. Begun, C.V. Ward, and M.D. Rose (eds.), *Function, Phylogeny, and Fossils: Miocene Hominoid Evolution and Adaptations*, 173–208. New York: Plenum Press.

Kuykendall, K.L., Mahoney, C.J., and Conroy, G.C. 1992. Probit and survival analysis of tooth emergence ages in a mixed-longitudinal sample of chimpanzees (*Pan troglodytes*). *American Journal of Physical Anthropology* 89:379–399.

Lampl, M., Monge, J.M., and Mann, A.E. 1993. Further observations on a method for estimating hominoid dental development patterns. *American Journal of Physical Anthropology* 90:113–127.

Leakey, M., and Walker, A. 1997. *Afropithecus* function and phylogeny. In D.R. Begun, C.V. Ward, and M.D. Rose (eds.), *Function, Phylogeny, and Fossils: Miocene Hominoid Evolution and Adaptations*, 225–239. New York: Plenum Press.

Mann, A.E. 1975. *Paleodemographic Aspects of the South African Australopithecines.* Philadelphia: University of Pennsylvania Publications in Anthropology, No. 1.

Mann, A.E., Monge, J., and Lampl, M. 1991. Investigation into the relationship between perikymata counts and crown formation times. *American Journal of Physical Anthropology* 86:175–188.

Martin, R.D. 1983. *Human Brain Evolution in an Ecological Context.* New York: American Museum of Natural History.

Martin, R.D., and MacLarnon, A.M. 1990. Reproductive patterns in primates and other mammals: the dichotomy between altricial and precocial offspring. In

C.J. DeRousseau (ed.), *Primate Life History and Evolution*, 47–79. New York: Wiley-Liss.

Nishida, T., Takasaki, H., and Takahata, Y. 1990. Demography and reproductive profiles. In T. Nishida (ed.), *The Chimpanzees of the Mahale Mountains: Sexual and Life History Strategies*, 63–97. Tokyo: University of Tokyo Press.

Partridge, L., and Harvey, P.H. 1988. The ecological context of life history evolution. *Science* 241:1449–1455.

Phillips-Conroy, J.E., and Jolly, C.J. 1988. Dental eruption schedules of wild and captive baboons. *American Journal of Primatology* 15:17–29.

Pilbeam, D. 1996. Genetic and morphological records of the Hominoidea and hominid origins: a synthesis. *Molecular and Phylogenetic Evolution* 5:155–168.

Promislow, D.E.L., and Harvey, P.H. 1990. Living fast and dying young: a comparative analysis of life-history variation among mammals. *Journal of Zoology, London* 220:417–437.

———. 1991. Mortality rates and the evolution of mammal life histories. *Acta Oecologica* 12:119–137.

Read, A.F., and Harvey, P.H. 1989. Life history differences among the eutherian radiations. *Journal of Zoology, London* 219:329–353.

Reid, D.J., Beynon, A.D., and Ramirez Rozzi, F.V. 1998a. Histological reconstruction of dental development in four individuals from a medieval site in Picardie, France. *Journal of Human Evolution* 35:463–477.

Reid, D.J., Schwartz, G.T., Dean, M.C., and Chandrasekera, M.S. 1998b. A histological reconstruction of dental development in the common chimpanzee, *Pan troglodytes. Journal of Human Evolution* 35:427–448.

Ross, C.R. 1998. Primate life histories. *Evolutionary Anthropology* 6:54–63.

Sacher, G.A. 1975. Maturation and longevity in relation to cranial capacity in human evolution. In R.H. Tutlle (ed.), *Primate Functional Morphology and Evolution*, 417–441. The Hague: Mouton.

Sacher, G.A., and Staffeldt, E.F. 1974. Relation of gestation time to brain weight for placental mammals: implications for the theory of vertebrate growth. *American Naturalist* 108:593–616.

Smith, A. 1988. Patterns of pika (genus *Ochotona*) life history variation. In M.S. Boyce (ed.), *Evolution of Life Histories of Mammals: Theory and Pattern*, 233–256. New Haven: Yale University Press.

Smith, B.H. 1989. Dental development as a measure of life history in primates. *Evolution* 43:683–688.

———. 1994. Patterns of dental development in *Homo, Australopithecus, Pan*, and Gorilla. *American Journal of Physical Anthropology* 94:307–325.

Smith, B.H., Crummet, T.L., and Brandt, K.L. 1994. Age of eruption of primate teeth: a compendium for aging individuals and comparing life histories. *Yearbook of Physical Anthropology* 37:177–231.

Smith, B.H., and Tompkins, R.L. 1995. Toward a life history of the Hominidae. *Annual Reviews of Anthropology* 24:257–279.

Smith, R.J., Gannon, P.J., and Smith, B.H. 1995. Ontogeny of australopithecines and early *Homo:* evidence from cranial capacity and dental eruption. *Journal of Human Evolution* 29:155–168.

Stanford, C.B., Wallis, J., Matama, H., and Goodall, J. 1994. Patterns of predation by chimpanzees on red colobus monkeys in Gombe National Park, 1982–1991. *American Journal of Physical Anthropology* 94:213–228.

Stearns, S. 1992. *The Evolution of Life Histories.* Oxford: Oxford University Press.

Swindler, D.R. 1985. Nonhuman primate dental development and its relationship to human dental development. In E.S. Watts (ed.), *Nonhuman Primate Models for Human Growth and Development,* 67–94. New York: Liss.

Swindler, D.R., and Beynon, D. 1993. The development and microstructure of the dentition of *Theropithecus.* In N.G. Jablonski (ed.), *Theropithecus: The Rise and Fall of a Primate Genus,* 351–381. Cambridge: Cambridge University Press.

Tsukahara, T. 1993. Lions eat chimpanzees: the first evidence of predation by lions on wild chimpanzees. *American Journal of Primatology* 29:1–11.

Ward, S.C., and Brown, B. 1986. Facial anatomy of Miocene hominoids. In D.R. Swindler and J. Erwin (eds.), *Comparative Primate Biology,* Vol. 1: *Systematics, Evolution, and Anatomy,* 413–452. New York: Liss.

Chapter 12

Dental Development and Life History in Hominid Evolution

ROBERT L. ANEMONE

The past decade has seen a resurgence of interest on the part of physical anthropologists and paleoanthropologists in the study of a series of related problems concerning the growth and development of living and fossil primates (Dean 1987b; Macho and Wood 1995). The major focus of this research has been the analysis of the developing hominoid dentition for clues to the origin of the unique ontogeny and life-history patterns that characterize *Homo sapiens*. Much of this research has been driven by the development and application of new analytical tools and techniques, including innovative approaches to imaging calcified tissues (e.g., confocal microscopy, scanning electron microscopy [SEM], and computed tomographic [CT] scanning), the discovery of immature fossil hominid specimens (e.g., WT-15000, the Nariokotome *Homo erectus* skeleton), and the application of heuristic theoretical models (e.g., heterochrony and life history) to these data. In addition, new data on hominoid development have been collected at a rapid pace using traditional tools and approaches (e.g., dental histology and radiography). These new data and approaches have led to significant improvements in our understanding of the evolution of human ontogeny and major shifts in our views concerning the ontogenetic and behavioral modernity of our Plio-Pleistocene hominid ancestors (Bromage 1987, 1990; Conroy and Kuykendall 1995; Dean 1987a; Mann et al. 1990; Smith 1989, 1992).

In this chapter I describe our knowledge of pattern and timing of dental development among modern hominoids and apply this knowledge to questions surrounding the use of modern apes or humans as referential models for reconstructing the ontogeny and life history of Plio-Pleistocene fossil hominids. In addition, I review the results from ontogenetic studies of the dentition of fossil hominids to determine whether their characteristic developmental patterns and rates more closely resemble the assumed primitive, apelike patterns of ontogeny seen among extant pongids or the derived, slow growth patterns of modern humans. The main goal of this chapter is to explore the evidence for the appearance of a major shift in ontogeny in human evolution from a rapid and short duration, apelike ontogeny to the slower and longer duration ontogeny characteristic of modern humans.

The Importance of Life History

An awareness of the interrelationships among variations in body size, rates of growth and development, and ecological, morphological, and behavioral variables across taxonomic groups has led many biological scientists to an appreciation of the importance of different life-history patterns (Charnov 1991; Charnov and Berrigan 1993; Harvey and Clutton-Brock 1985; Harvey et al. 1987). Primatologists and paleoanthropologists have been keen to apply ideas and methods of analysis from life-history theory to the study of extant (Alumna et al. 1981; Charnov and Berrigan 1993; Sigg et al. 1982) and extinct primates (Beynon et al. 1997) and, in particular, to the study of human evolution (Beynon and Dean 1987; Beynon and Wood 1987; Bromage 1987, 1990; Conroy and Kuykendall 1995; Conroy and Vannier 1988, 1991a, 1991b; Dean 1987a, 1989; Dean et al. 1993, 1986; Mann 1988; Mann et al. 1987; Ramirez Rozzi 1993).

The suite of life-history characteristics found among modern humans has long played a central role in evolutionary discussions of the origins of the distinctly human way of life, which is characterized to a large degree by culture, language, and learning (Smith 1992; Smith et al. 1995). The uniquely human pattern of life history includes the following traits, many of which are typically thought to be the result of *K*-selection (Harvey and Clutton-Brock 1985):

—long gestation
—large neonatal and adult brain size
—precocial neurological development at birth
—altricial motor development at birth

—large birth weight and adult body size
—slow development with long childhood
—adolescent growth spurt
—delayed reproduction
—enormous parental investment in few offspring
—long life span

The essential link between human life history and the human way of life has been the suggestion that a prolonged period of infant and childhood dependency upon caretaking adults allows humans a prolonged period of behavioral plasticity and leads to a reliance on learning and learned behavior as hallmarks of human existence. In addition, many human evolutionary studies have suggested that dependent infants requiring significant maternal care are a causal factor in the evolution of a home base, food sharing, male hunting, the sexual division of labor, the evolution of mating patterns, and family structure (Smith 1989, 65). Dobzhansky explicitly suggested that prolonged childhood dependence would typically be a disadvantage, since it would serve to prolong the period when individuals are most at risk of death by predators (Dobzhansky 1962). He further suggested that, among humans, this disadvantage is more than compensated for by the improved learning ability that results from prolonging this period of plasticity. Indeed, prolonged childhood dependence becomes one of the essential aspects of the human evolutionary success story in many secondary texts (Brace and Ashley-Montague 1977; Kelso and Trevathan 1984), as well as in the classic work of Alan Mann (Mann 1975) on the developing dentitions of robust australopithecines from Swartkrans. Mann's work used dental development to infer life history and suggested that evidence of humanlike prolonged development among australopithecines was prima facie evidence that australopithecines were culture-bearing creatures.

Ecological Theory, *r*- and *K*-Selection, and Life History

The concepts of *r*- and *K*-selection were first defined by Macarthur and Wilson (1967) in the context of their theory of island biogeography to distinguish between populations that maximize their intrinsic rate of increase (the ecologists' *r*) and those that maintain population size close to the carrying capacity (*K*) of their local environment. Pianka (1970) noted the presence of an ecological continuum of r- and K-selection, in which animals could be aligned in relative rank order. Pianka identified taxonomic correlates (e.g., most insects and terrestrial invertebrates were relatively r-selected, while most terrestrial vertebrates were primarily K-selected), but he also clearly saw that, within large taxonomic groups (e.g., mammals), lower level taxa could themselves be arrayed along a narrower continuum running from more r-selected forms to more K-selected forms.

Environmental correlates of r- and K-selection seem to be well understood also: r-selected species are more common in unstable, shifting, or highly seasonal environments, in which mortality is high and K (the carrying capacity) is rarely approached or attained. Primates are generally considered to be archetypal K-selected creatures, and Martin (1983) argued that this is a result of a long evolutionary history spent as rain-forest dwellers. K-selecting environments present stable environments with little seasonality, in which competition, both within and between species, is intense and the relatively few offspring produced are typically provided with enhanced parental investment of some kind (Martin 1983, 33). Pianka (1970, 593) suggested that, in maximally K-selecting environments, density effects are maximal and the environment is saturated with organisms. As a result, competition is keen and the optimal strategy is to channel all available matter and energy into maintenance and the production of a few extremely fit offspring. Replacement is the keynote here. K-selection leads to increasing efficiency of utilization of environmental resources (Pianka 1970, 593). As an environment fills up with organisms and taxa, the effects of natural selection can move a population from the r- toward the K-region of the ecological spectrum. Perhaps the most extremely K-selected extant primates are the great apes, especially chimpanzees, whose demographic dilemma of extremely low reproductive turnover was described by Lovejoy (1981) in the context of his speculations on hominid origins.

While sufficient data exist to establish the K-selected nature of the primates (Harvey and Clutton-Brock 1985; Martin and MacLarnon 1990), attempts to understand the nature and significance of this ecological strategy are less well developed. In his extensive attempts to develop an evolutionary-comparative biology of longevity, Sacher has demonstrated the existence of a strong linear relationship between maximum life span and both body and especially brain weight in a large sample of extant mammals (Sacher 1975, 1978, 1982; Sacher and Staffeldt 1974). Sacher (1975, 426) suggested that "the evolution of a larger brain imposes on the species an added metabolic and developmental burden, and a consequent decreased reproductive rate that can only be compensated by means of an extension of the reproductive span, and hence of the lifespan." He found support for this hypothesis in the relationships between gestation length, brain size, and litter size in a large sample of mammalian species, where larger-brained infants are typically born after longer gestations and in smaller litters. Thus, Sacher (1975, 427) argued that the brain is the "pacemaker" of mammalian growth, since "the characteristic mammalian rate of brain growth is the rate-limiting process for all other aspects of somatic growth."

In another analysis of primate life-history characteristics, Charnov and Berrigan (1993) suggested that primate longevity and low reproductive rates are best explained as a result of the fact that primates have slow individual growth rates, compared to other mammals of comparable body size. These slow growth rates lead directly to late reproduction and allow primates to attain large body size, the attainment of which leads to a shift of energy from growth to reproduction. Charnov's (1991) model of mammalian life history suggests that this slow growth rate can only be sustained with very low mortality rates and so, in a sense, mortality rates determine adult body size (Charnov and Berrigan 1993, 193).

Martin (1983) examined diet, brain size, metabolism, and life history in relation to human evolution. He suggested that maternal physiology sets limits on the size of the neonatal and, ultimately, the adult brain because of the high metabolic cost of the development of neural tissue. Maternal physiology is influenced importantly by the energetics of diet, and Martin explained relative brain size among primates in this context (e.g., frugivores tend to have larger brain sizes than folivores of a similar body size). Humans are unique in maintaining high, essentially fetal, growth rates for neural tissue through the first year of postnatal life, and this is linked to Portmann's (1941, cited in Martin 1983) suggestion that humans are secondarily altricial and that the real human gestation period could be considered to approach twenty-one months (i.e., nine months in utero and the first twelve months of postnatal life). The result is that humans are born at an extremely undeveloped stage in terms of motor and neural development, most likely because of biomechanical and obstetrical constraints in pelvic outlet size and shape resulting from bipedal adaptations. Martin used this logic to question the notion that an ecological shift to an r-selecting habitat like the African savanna was the cause for brain expansion in human evolution, since the combination of large brain size and slow reproductive turnover in *H. sapiens* is indicative of the operation of an extreme form of K-selection, not of exposure to r-selecting environments (Martin 1983, 36–37).

Although Martin's and Sacher's work suggests brain size as an important basis for determining correlated life-history characteristics, Smith (1988, 1989, 1991, 1992; Smith et al. 1995) demonstrated the extremely high correlation among primates between brain size and age at emergence of the first permanent molar (M1). This work suggests that brain size can be used as a predictor of age at emergence of the first molar and that, as a result of the high intercorrelations between most life-history traits, these data can be used to formulate and to test hypotheses concerning life-history traits of fossil hominids. As a result, the unique nature of human ontogeny, the associated suite of life-history traits that characterize humans, and the historical development and functional significance of the distinctive pattern of human growth can be approached from an analysis of the stories told by the hard tissues of the oral cavity among both living and fossil hominoids.

The Pattern and Timing of Dental Development

A clear distinction must be made between pattern and timing in dental development (Macho and Wood 1995). Studies of timing generally require some means of relating developmental events to an absolute chronology of development within individuals. Timing of dental development is best studied using material of known chronological age, preferably in longitudinal series, although cross-sectional data can also be useful. Thus, while cross-sectional studies can indicate species-typical aspects of the timing of growth, only longitudinal data can allow analysis of individual differences in growth (Tanner 1990). Patterns of dental development (e.g., eruption sequence polymorphisms [Kuykendall et al. 1992; Smith and Garn 1987]) can be studied cross-sectionally in material of unknown age, but studies of this sort cannot elucidate questions of developmental timing. The classic work of Adolph Schultz demonstrates the existence of similar patterns of dental and skeletal development among primates while also indicating significant interspecific timing differences in these developmental events (Schultz 1969).

Certainly, there are important pattern differences in various developing systems among primates, but the critical point is that similarity in developmental pattern cannot be taken as evidence of similar timing of development. This situation is well exemplified by a comparison of the immature dentition of a modern five- to six-year-old human, a modern three-year-old chimp, and the type specimen of *Australopithecus africanus* from the Pliocene site of Taung in South Africa. Each of these shares a similar pattern of dental development marked by the presence in the oral cavity of all deciduous teeth and the first permanent molars. However, we know that the human has taken approximately twice as long to reach this stage of dental development as the chimp. It would be equally unwarranted to assume that the australopithecine resembled either humans or chimps in dental development: this is what we hope to determine, not something that we can assume. As Macho and Wood (1995) suggested in their review, much of the disagreement between Alan Mann and colleagues on the one hand (Lampl et al. 1993; Mann et al. 1987, 1990) and Holly Smith on the other (Smith 1986, 1987, 1994) centers around their use of dental patterns to infer timing of dental events among australopithecines.

While radiography, histology, or CT scanning can be used to study the pattern and timing of dental development in extant hominids or pongids of known age, how can these questions be broached in the fossil record, where specimens do not come with age labels attached? The first part of this problem involves imaging techniques for studying fossil specimens, and in this respect, although each of these techniques has been applied to the fossil material, they all have drawbacks that limit their usefulness. The use of x-rays to image fossil material leaves much to be desired, since the mineralization of fossil specimens tends to make them opaque to x-rays (Mann 1975), while histological work often requires the sectioning of rare fossil material (but see work of Dean and colleagues that describes a new, nondestructive technique of microsectioning fossil teeth that may be more palatable to many museum curators [Beynon et al. 1997; Dean et al. 1993]). CT scanning, an expensive and complicated imaging technique, has the most potential to improve our ability to image dental developmental events deep within the jaws of even the most heavily mineralized fossils. The application of CT scanning to the visualization of anatomical details in fossil skulls was pioneered by Conroy and Vannier (1987, 1991a, 1991b) in their work on the Taung child and other South African australopithecines.

The second part of this problem involves age estimation in fossil specimens in order to study aspects of timing in dental development. Perhaps the most exciting approach to this problem involves the analysis of incremental lines in dental calcified tissues (especially enamel) whose periodicity has been hypothesized to be temporally based (for two reviews from very different perspectives, see Dean 1987b and Mann et al. 1990). The counting of apparently weekly growth intervals in enamel in the structures known as striae of Retzius and of daily increments in the so-called cross-striations found between striae of Retzius allows an independent estimation of the duration of growth intervals and the timing of developmental events during the formation of tooth crowns. Although most workers accept the temporal periodicity hypothesis for these incremental markers in dental tissues, the argument is still controversial in some quarters, partly because of the lack of understanding of the proximate, physiological mechanisms of temporal periodicity that may cause these lines. Different techniques and methods of specimen preparation, sectioning, and microscopy have also contributed to the confusion sometimes surrounding the interpretation of enamel incremental lines.

One of the central issues here is the question of whether modern humans or apes are a more appropriate referential model of dental development for comparison with fossil hominids. To answer this question, we must have some information concerning the pattern and timing of our two likely modern analogue taxa, humans and apes. There have been many studies of the dental development of modern humans, but until recently there have been few systematic studies of ape dentition. Much of my work (Anemone 1995; Anemone et al. 1991, 1996) and that of several other investigators (Kuykendall 1996; Kuykendall and Conroy 1996; Kuykendall et al. 1992) has been an attempt to fill in some of the gaps in our understanding of the pattern and timing of modern chimpanzee dental development, with the goal of improving our ability to model the dental development of extinct hominids. Only with a detailed understanding of the pattern and timing of dental development among modern humans and apes can we hope to determine empirically the goodness of fit between modern ape or human patterns of dental development and those of the fossils. Lacking this essential comparative database, we can only form a priori judgments as to the appropriateness of one or the other modern analogue for reconstructing the ontogeny and life history of extinct Plio-Pleistocene hominids.

Using a combination of the different methods and techniques outlined above, it is now possible to compare both pattern and timing of dental development in modern humans and chimpanzees with those of certain fossil hominids with the goals of determining when, where, and under what conditions ancestral hominids evolved the unique modern human pattern of ontogeny and life history. Assuming that (1) the ape pattern of ontogeny is primitive and the human pattern is derived and (2) the proposed connections between life history and the human way of life are real, the identification in the fossil record of the dental evidence for this life-history shift in the hominid lineage takes on real significance. Dental development can thus serve not only, as Mann (1975, 84) suggested, as direct morphological data for the presence of culture in australopithecines but also as a key predictor of a series of critical life-history parameters that can distinguish between a human and an apelike ontogenetic, behavioral, and ecological strategy (Smith 1992).

The Fossil Record of Hominid Dental Development

The history of the study of dental development as a tool to determine the age at death and other life-history parameters of fossil hominids begins with the first published account of an australopithecine, Dart's (1925) description of the type specimen of *A. africanus* from Taung. Dart seems to have simply assumed in an a priori fashion that the Taung child shared the human schedule of tooth eruption; thus, on the basis of its possession of a recently erupted first permanent molar, he estimated its age at death as roughly six years. One of the few early supporters of the hominid status of the australopithecines, Le Gros Clark (1947), concurred with Dart on the proposed human pattern and timing of australopithecine dental development, offering as evidence both the heavy differential wear on australopithecine permanent molars and the heavy wear on their deciduous molars. In his description of an immature gracile australopithecine mandible (MLD-2) from Makapansgat, Dart (1948) noted that its eruption pattern differed from the modern human pattern in that M2 was erupted before P4. Disregarding this difference, he then suggested that, due to the heavy wear on the milk dentition, the Makapansgat adolescent reinforces the evidence of a delayed infancy afforded by the Taungs (*sic*) infant by indicating that dental development in the australopithecine group was still further prolonged over an extended or retarded childhood, as is characteristic of humankind.

Broom and Robinson (1951, 443) suggested that tooth eruption in Swartkrans australopithecines agrees closely with that of modern humans, the only difference of note being that the second molar appears before the second premolar (in the fossils). Indeed, they suggested that the robust australopithecines were similar to humans of European ancestry in that their first incisors typically erupted before their first molars, while in *Sinanthropus*, Neanderthals, and Bushman, the sequence of eruption is almost as in all anthropoids and most monkeys. Interestingly, Broom and Robinson never made the leap from similar pattern to similar timing of development that so many other authors have made. Nor did they offer any speculations on ages at death or other life-history or ontogenetic parameters. Instead, they suggested that this developmental evidence supports two points of much more interest to these authors: First, the possession of a dental sequence that is unknown in any living anthropoid or monkey (Broom and Robinson 1951, 443) provides supporting evidence for the hominid status of the robust australopithecines. In an especially unfortunate phrase, they suggested that the dental eruption sequence of *Paranthropus* from Swartkrans is much nearer that of modern humans than is that of Pekin (*sic*) man, Neanderthals, or even the Bushman. Second, they emphasized the difference in dental eruption sequence between the robust australopithecines from Swartkrans and the gracile australopithecine from Makapansgat described by Dart (1948). This difference (essentially an earlier M2 and a later canine eruption in Dart's specimen) was taken by Broom and Robinson as support for their generic distinction between gracile (*Australopithecus*) and robust (*Paranthropus*) australopithecines.

The most influential modern analysis of australopithecine dental development and ontogeny was Alan Mann's (1975) monograph on the dentition and paleodemography of the South African australopithecines. Relying heavily on the sample of robust australopithecines from Swartkrans, Mann compared the patterns of dental calcification, eruption, and degree of occlusal wear of these fossils with what was known at the time about these parameters in humans and chimpanzees. He concluded that the pattern in australopithecines closely resembled the human pattern. Detailed pattern similarities in a whole series of events—crown calcification, root development and final closure, tooth eruption, and occlusal wear—suggested to Mann (1975, 77) a similar schedule or timing of developmental events. As a result, Mann calculated age at death for South African australopithecines according to a modern human schedule of dental development and eruption (e.g., specimens with M1 recently emerged like Taung were suggested to be approximately six years of age).

Most of the assumptions as well as many of the conclusions of these early attempts to reconstruct the life history of Plio-Pleistocene fossil hominids are poorly supported by the evidence that has accumulated in the interim. Dart's assertion that Taung followed a human schedule of dental development is seen today as just that: an unsupported assertion with very little to commend it and much evidence against it (Smith et al. 1995). Arguments linking humans and australopithecines based on eruption sequences have been seriously weakened by better data on species-typical eruption sequences (Kuykendall et al. 1992), by evidence of high levels of sequence polymorphisms within species (Smith and Garn 1987), and by the observation that gingival emergence and alveolar eruption sequences are often reversed within individuals (e.g., M2-P4 alveolar sequence reverses to P4-M2 gingival sequence in many humans [Garn et al. 1957]). Similarly, recent research indicates that, in the absence of population-specific rates of dental wear, dental attrition cannot be used to determine age at death (Walker et al. 1991).

It has long been suggested that humans and robust australopithecines share a synapomorphy involving the close approximation in time of emergence of the first permanent incisor and first permanent molar (Broom and Robinson 1951; Dart 1948; Dean 1985; Schultz 1935), but more recent evidence suggests that this trait is found only among humans. Grine (1987) and Conroy (1988) convincingly demonstrated that misidentification of deciduous incisors as permanent incisors in a robust jaw from Swartkrans (SK-61) is the basis for this confusion. Conroy's (1988) analysis of CT scans of the SK-61 jaw is particularly decisive in falsifying the notion of an I1/M1 eruption synapomorphy linking humans and robust australopithecines.

Dental Development among Apes

Although there is a voluminous literature on the calcification and development of the dentition among modern humans (citations in Smith 1992), until recently studies of ape dental development have been restricted to studies of emergence of the dentition into the oral cavity among chimpanzees (Nissen and Riesen 1945, 1964). The first reasonably successful attempt at developing standards of dental development among apes for comparison with the well-known human standards was the pioneering contribution of Dean and Wood (1981). These authors made a radiographic analysis of a cross-sectional sample of the skulls and jaws of 175 juvenile apes (chimpanzees, gorillas, and orangutans). Since the subjects of their study were wild-shot museum specimens, their chronological ages at the time of death were unknown. Dean and Wood's analytical approach purposely minimized within-group variation and sought to portray a modal or archetypal pattern of dental development by excluding radiographs with teeth whose developmental status was intermediate between the nine named stages of development they used. With the aid of published data on age at emergence into the oral cavity of the chimpanzee deciduous and permanent dentitions (Nissen and Riesen 1945, 1964), a series of extrapolations estimated the age at which dental developmental events occur among apes. Two critical assumptions were used to locate dental development among apes in "real time." The first assumption stated that the duration of crown calcification for all ape molars was approximately 2.5 years, and the second stated that root development time within similar tooth types (e.g., premolars or molars) was equal. Using these assumptions, Dean and Wood created a chart of ape dental development that has played an important role in the comparison of fossil hominid dentitions with those of modern pongids (e.g., Smith 1986, 1991).

More recently, chimpanzees of known chronological age have been used in mixed longitudinal radiographic studies of dental development (Anemone 1991, 1995; Anemone et al. 1991, 1996; Conroy and Kuykendall 1995; Conroy and Mahoney 1991; Kuykendall 1996; Kuykendall and Conroy 1996; Kuykendall et al. 1992). Using different populations of captive chimpanzees but similar calcification scoring techniques, researchers greatly increased the available database on ape dental development and made possible for the first time the comparison of pattern and timing of dental development among fossil hominids with these parameters in living pongids. Kuykendall and coworkers have made major advances in documenting patterns of intraspecific variation in the developing chimpanzee dentition, particularly with respect to sex differences and developmental relationships among different tooth types and among pairs of teeth. The work of Anemone and coworkers has been mostly concerned with determining the overall patterns and timing of chimpanzee dental development to facilitate comparisons with modern humans and Plio-Pleistocene hominids.

Dental Development in *Pan troglodytes:*
A Longitudinal Study

My work on chimpanzee dental development has relied upon the analysis of lateral head radiographs (cephalograms) to document, in longitudinal fashion, the calcification and growth of the mandibular dentition in two samples of captive-born common chimpanzees (*P. troglodytes*) of known chronological age. Although the limitations and biases of radiographic techniques for picking up early events in dental maturation are widely recognized (see Kuykendall, Chap. 13), radiography is still the preferred method for analysis of large samples of developing dentitions among both humans and nonhuman primates.

Materials and Methods

Sixteen Yerkes chimpanzees, born and reared at the Yerkes Laboratories of Primate Biology in Orange Park, Florida (now the Yerkes Regional Primate Research Center in Atlanta, Ga.), and raised specifically for studies of growth and development (Nissen 1942), were a portion of my sample. Data were also obtained from thirty-three chimpanzees born and reared at the Southwest Foundation for Biomedical Research in San Antonio, Texas. Lateral cephalograms of the San Antonio animals were taken as part of experimental studies on anterior craniofacial growth and the development of cleft palate (Mooney and Siegel 1991). An earlier study (Anemone et al. 1996) determined that the surgical interventions used on some of the San Antonio chimpanzees had no significant effects on molar dental maturation, and this conclusion is further strengthened by the closely congruent results obtained in studies of both samples.

The combined total of 398 radiographs from 49 individuals in the Yerkes and San Antonio samples provided dentitions scored by the author using the same methods, modified from the system for scoring human dental development of Demirjian (Demirjian and Goldstein 1976; Demirjian et al. 1973). Eight dental stages were used to describe the development of the mandibular molar teeth. The first four stages describe the development of the tooth crown from initial calcification (A) to completed crown formation (D). The next four stages reflect growth in root length (E-G) and culminate in apical root closure (H). The ages at onset and completion of the crowns and roots for antemolar teeth were also recorded, although these data are less complete because of difficulties in visualizing anterior teeth in lateral head radiographs.

What follows is an analysis of the combined datasets from the Yerkes and San Antonio chimpanzee samples (Fig. 12.1), designed to estimate more precisely the pattern and timing of chimpanzee dental development. More specifically, to facilitate comparisons with modern and fossil hominids, I use two different statistical techniques to estimate the ages at which chimpanzees attain various stages of dental development. The age at which a dental stage was attained was recorded as the averaged age of the individual at first radiographic appearance of the stage and the age of the ani-

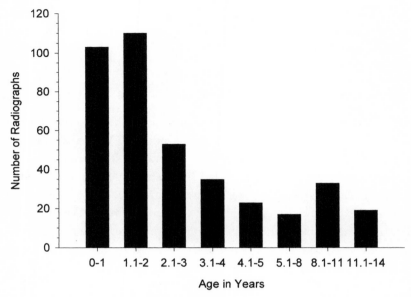

Fig. 12.1. Age distribution of the combined Yerkes and San Antonio sample of 398 lateral head radiographs from 49 common chimpanzees (*Pan troglodytes*).

mal in its previous radiograph. Means and standard deviations of these corrected age-at-stage data are presented (Table 12.1; Fig. 12.2) for the combined sample (n = 49 chimpanzees and n = 398 radiographs).

In addition, probit analysis was used to obtain cumulative distribution functions as an alternative approach to the problem of determining age at attainment of dental stages. Probit analysis has been widely used for many years in studies of age at stages of dental formation among humans and is generally recognized as the preferred method for analysis of a dosage-response relationship between one or more independent variables and a dichotomous dependent variable (Dahlberg and Menegaz-Bock 1958; Friedlaender and Bailit 1969; Kuykendall et al. 1992). For my purposes, the independent variable (dose) is the age of the individual for each radiograph, and the dichotomous, dependent variable (response) is the presence or absence of a given stage of dental development (e.g., crown completion or stage D of the second molar).

Results from the probit analyses are presented as the median, 25th percentile, and 75th percentile of the age for each attained stage of molar development (Table 12.2; Fig. 12.3). The median, or 50th percentile, represents the age at which 50 percent of the population will have attained a

Table 12.1 Descriptive Statistics for Corrected
Ages at Dental Development Stages for Mandibular
Molars, Combined Yerkes and San Antonio Samples

| Tooth Stage | Age (years) | | n |
	Mean	Standard Deviation	
M1A	0.33[a]	0.09	28
B	0.75	0.18	33
C	1.21	0.17	29
D	1.70	0.25	26
E	2.28	0.43	23
F	2.92	0.44	20
G	3.67	0.43	14
H	5.31	1.15	17
M2A	1.31	0.27	28
B	1.84	0.34	25
C	2.60	0.39	23
D	3.43	0.38	16
E	4.48	0.14	5
F	6.51	NA	1
G	5.97	1.28	10
H	8.25	2.29	13
M3A	3.58	0.38	13
B	4.45	0.14	4
C	3.50	NA	1
D	5.59	1.10	7
E	7.30	1.99	6
F	8.04	1.78	11
G	10.11	0.84	10
H	11.44	1.02	7

[a]M1 actually appears at birth in chimpanzees, as it does in humans. The value of 0.33 year is an artifact resulting from the lack of prenatal radiographs in these samples.

Note: n = 398 radiographs from 49 chimpanzees.

given stage of dental development. The current analysis goes beyond that presented in several earlier papers (Anemone 1991, 1995; Anemone et al. 1991, 1996) by presenting complete data for central tendency and variation for all eight stages of development of all three mandibular molars in the combined Yerkes–San Antonio samples and by comparing results derived from two different statistical techniques (i.e., corrected ages and probit analysis). Together, these results provide the most complete picture of chimpanzee dental development to date.

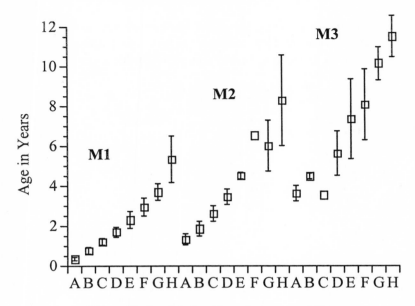

Fig. 12.2. Means ± 1 standard deviation for corrected age at dental stages for mandibular molars from the combined Yerkes and San Antonio chimpanzee sample of 398 lateral head radiographs from 49 common chimpanzees (*P. troglodytes*). Corrected age is calculated as the mean of the age of an individual at first radiographic appearance of a dental stage and the age of the animal in its previous radiograph. Eight molar stages mark the development of the crown (A-D) and root (E-H).

Results and Discussion

The first molar in chimpanzees is radiographically present at birth (Anemone et al. 1991), and its crown has completed its calcification on average before the chimp reaches two years of age (mean of 1.70 years for corrected ages, median of 1.73 years in probit analysis). The second molar is present before 1.5 years of age (mean = 1.31 years, median = 1.40 years) and crown complete two years later (mean = 3.43 years, median = 3.73 years), while M3 appears between three and a half and four years of age (mean = 3.58 years, median = 3.93 years) and likewise takes roughly two years to calcify completely (mean = 5.59 years, median = 6.01 years).

Chimpanzee molar roots take progressively longer to develop further distally in the jaw so that the M1 root develops on average in 3.05 years, the M2 in 3.77 years, and the M3 in 4.13 years (all ages are corrected mean ages; see Table 12.1 and Fig. 12.2). A comparison with human data (Moor-

Table 12.2 Descriptive Statistics Based on Probit
Analysis of Dental Development Data for Mandibular
Molars, Combined Yerkes and San Antonio Samples

	Age (years)		
Tooth Stage	*25th percentile*	*Median*	*75th percentile*
M1A	NA	NA	NA
B	0.46	0.66	0.86
C	1.05	1.25	1.75
D	1.48	1.73	1.98
E	2.21	2.43	2.65
F	2.80	3.07	3.34
G	3.48	3.79	4.09
H	5.57	6.34	7.10
M2A	1.11	1.40	1.69
B	1.77	2.05	2.34
C	2.52	2.77	3.01
D	3.29	3.73	4.17
E	5.12	5.58	6.04
F	5.86	6.36	6.85
G	6.37	6.91	7.45
H	7.92	8.75	9.58
M3A	3.58	3.93	4.27
B	4.93	5.40	5.87
C	5.54	6.01	6.48
D	6.02	6.53	7.04
E	7.71	8.37	9.03
F	8.31	9.02	9.74
G	9.55	10.39	11.22
H	11.32	12.31	13.31

Note: n = 398 radiographs from 49 chimpanzees.

rees et al. 1963) suggests that, while human molar crowns take longer to
calcify than do those of chimpanzees (mean duration of 7.7 years for humans and 5.8 years for chimps), much of the difference between these
species is the result of longer periods of root growth among humans (Table
12.3; Fig. 12.4). In the human (i.e., middle-class white American) data presented by Moorrees et al. (1963), root growth takes 3.9 years for M1, 5.1
years for M2, and 4.8 years for M3 (total root growth time = 13.8 years
for humans and 10.95 years for chimps).

The data also indicate another significant difference between chimpanzees and humans in the temporal patterning of adjacent molar crown development. Adjacent molar crowns in chimpanzees are typically marked by

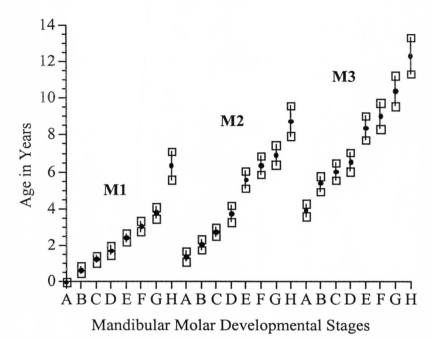

Fig. 12.3. Twenty-fifth, 50th, and 75th percentiles of age at dental stages for mandibular molars from the combined Yerkes and San Antonio chimpanzee sample of 398 lateral head radiographs from 49 common chimpanzees (*P. troglodytes*) based on probit analysis.

Table 12.3 Age at Crown Initiation (Stage A) and Crown Completion (Stage D) in Human and Chimpanzee Mandibular Molars

		Age (years)			
		Human		*Chimpanzee*	
		Mean	*Standard Deviation*	*Mean*	*Standard Deviation*
M1	crown initiation	Birth	0.1	Birth	NA
	crown completion	2.2	0.3	1.7	0.2
	root completion	7.1		5.31	1.15
M2	crown initiation	3.7	0.4	1.3	0.3
	crown completion	6.5	0.7	3.4	0.4
	root completion	11.6		8.25	2.29
M3	crown initiation	9.3	1.0	3.6	0.4
	crown completion	12.0	1.3	5.6	1.1
	root completion	16.8		11.44	1.02

Note: Human data are based on radiographic studies of 246 middle-class white children from Ohio (136 boys and 110 girls) from Moorrees et al. (1963). Chimpanzee data (n = 49 chimpanzees) are based on corrected ages from this study.

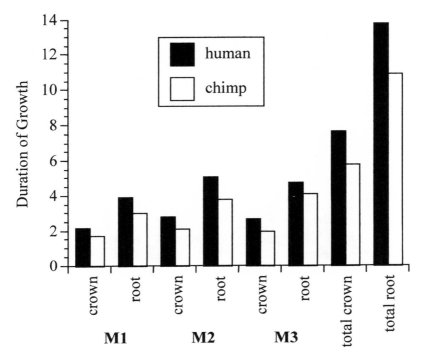

Fig. 12.4. Comparison of duration of root and crown growth in years among chimpanzees and humans. Human data are based on radiographic studies of 246 middle-class white children from Ohio (Moorrees et al. 1963). Chimpanzee data are from corrected ages presented in this study. Duration of crown growth among chimpanzees is calculated as elapsed time from attainment of stage A to stage D; duration of root growth is calculated as elapsed time from stage D to stage H (apical closure).

extensive temporal overlap in their development, while there is a substantial temporal gap between the completion of crown calcification of the mesial molar and the initiation of crown calcification of its distal neighbor among humans (Anemone 1995; Anemone et al. 1991, 1996). Thus, the chimpanzee M2 crown is typically well developed (stage A or B) before M1 is crown complete (stage D), and the M3 crown is similarly developed at the time of initiation of the crown of M2. Conversely, there is a variable but significant temporal gap between crown completion of mesial and crown initiation of distal molar pairs in humans. The chimpanzee pattern can be clearly seen in both the corrected age data and the probit data presented in Tables 12.1 and 12.2 and plotted in Figures 12.2 and 12.3, respectively. The contrasting human pattern can be discerned in the data from Moorrees et al. (1963) presented in Table 12.3 and plotted in Figure

12.4. These patterns are also clearly seen in the comparative dental development chart in Figure 12.5, as well as in individual chimp or human dental radiographs of the proper ages. This difference is one of the clear and consistent pattern differences between modern human and chimp developmental patterns.

Comparison of the stage of development of incisors and canines relative to that of the first molar illustrates another difference between human and chimpanzee dental development that is relevant to the question of australopithecine affinities in dental developmental pattern (Fig. 12.5). The typical pattern seen in chimpanzees is that incisors and canines are less than crown complete with no root visible when the first molar emerges into the oral cavity (typically at 3–3.5 years of age) (Nissen and Riesen 1964). This stands in stark contrast to the typical human pattern, in which incisors and canines are crown complete with substantial root development at the time of M1 emergence. Since human and chimpanzee incisors are not particu-

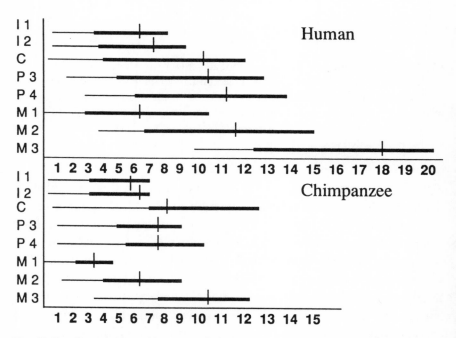

Fig. 12.5.   Comparison of human and chimpanzee patterns and timing of dental development. Human data are based on Dean and Wood (1981); chimpanzee data are based on the Yerkes sample (Anemone et al. 1991). *Vertical lines* indicate age at dental emergence; *thin horizontal lines* represent crown calcification; *thick horizontal lines* represent root development.

larly different in the duration of their development (both taking roughly eight years *in toto*), the standard interpretation of this difference has been that human molar development is prolonged or delayed vis-á-vis the pongid pattern (Anemone 1991, 1995; Anemone et al. 1991, 1996; Dean and Wood 1981). In particular, the prolonged period of crown and especially root development of human molars has led to this change in developmental pattern relative to faster root extension rate among apes (Dean and Beynon 1991; Dean et al. 1993). A result of this change is that, while human first molars emerge close in time to both incisors, it is the second molar of the chimpanzee that emerges at around the same time as the incisors. Conroy and Vannier (1991a) showed that gracile australopithecines resemble chimpanzees in this respect, while robusts more closely resemble the human pattern. Figure 12.5 presents an overview of human and chimpanzee dental development that invites comparison with immature fossil hominid dentitions to determine which of these two modern analogue taxa offers closer approximations to ontogenetic patterns in the fossils.

Back to the Fossils: Studies of Timing and Pattern in Dental Development

Beginning in the middle 1980s, a series of new approaches to the study of dental development among fossil hominids led to a reevaluation of earlier ideas concerning the life history of Plio-Pleistocene hominids. As mentioned above, the earlier work of Alan Mann had convinced most anthropologists that australopithecines resembled modern humans in having a prolonged period of development as a result of selective pressures involved in the acquisition of a tool-using, culture-bearing way of life (Mann 1975). By the early 1990s this consensus had been replaced by disagreement between Mann and colleagues and several other research groups concerning the origins of the human life-history pattern and the nature of australopithecine ontogeny.

The first challenge to Mann's position was raised by Bromage and Dean (1985). Using histological and microscopic approaches pioneered by Alan Boyde (Boyde 1990; Boyde and Martin 1984), Bromage and Dean counted incremental growth markers known as striae of Retzius in the dental enamel of a series of Plio-Pleistocene hominid incisors to establish a time scale of development in fossil hominids believed to be independent of the rate of dental development in modern humans. The key observation concerning the two major incremental markers in dental enamel (i.e., striae of Retzius and cross-striations) is that they have a consistent temporal periodicity (Beynon and Dean 1988; Dean 1987b).

Cross-striations are undulations along the length of the enamel prisms that are seen under the light or scanning electron microscope and are thought to result from circadian changes in the rates of enamel formation by the advancing wave of ameloblasts (Boyde 1990). Striae of Retzius and their surface manifestations, known as perikymata, are essentially exaggerated cross-striations separated by seven to ten regular cross-striations. These are thought to represent circa-septan, or roughly weekly, increments of ameloblast activity. Although the temporal nature of these enamel markers is still questioned by a few critics (Mann et al. 1990; Warshawsky and Bai 1983; Warshawsky et al. 1984) and their physiological basis is still unknown, many investigators have been persuaded of their regular temporal periodicity by a series of careful labeling (Bromage 1991) and other microscopic studies (Beynon and Dean 1987, 1988; Beynon et al. 1991; Beynon and Reid 1987; Beynon and Wood 1987; Boyde 1990).

The results of Bromage and Dean (1985) suggested that the incisor teeth of gracile and robust australopithecines, as well as those of *Homo habilis*, formed in a much shorter time interval than those of modern humans. Their estimates of age at death for these early hominids suggested that neither australopithecines nor the earliest members of the genus *Homo* had yet undergone the ontogenetic shift toward the longer period of development characteristic of modern humans. Rather, this aspect of the life history of all Plio-Pleistocene hominids more closely resembled modern apes than modern humans. Early hominids like the Taung child, whose first adult molar had recently emerged into its oral cavity, were thus considered to have died not at five or six years of age (like a modern human at this developmental stage), but at roughly three or four years of age, like modern apes with a comparable immature dentition. Subsequent papers from these authors and their colleagues presented perikymata counts from a variety of taxa ranging from Miocene apes (Beynon et al. 1997) to Neanderthals (Dean et al. 1986; Stringer et al. 1990) and historical human populations (Beynon and Reid 1987; Dean and Beynon 1991; Stringer et al. 1990). In addition to incisor teeth, these authors have analyzed incremental lines in fossil hominid canines (Dean et al. 1993), premolars (Beynon and Dean 1987), and molars (Beynon and Wood 1987). These results consistently suggest that fossil hominids did not begin to develop in a characteristically human fashion until very recently in human evolution, while australopithecines and early members of genus *Homo* were much more similar in life history to living apes than to living humans.

Corroboration of the results of Dean and colleagues has come from the work of Fernando Ramirez Rozzi on the Plio-Pleistocene Omo Group hominids from southern Ethiopia (see Chap. 15). By counting perikymata exposed on the lateral surfaces of naturally fractured teeth, Ramirez Rozzi demonstrated that robust australopithecines formed their thick-enameled premolar crowns in a much shorter period than did modern humans or apes, while their molar crowns developed over a period similar to that seen in humans and chimps. Ramirez Rozzi also suggested a regression approach to estimating the total number of striae of Retzius based on counts of the imbricational striae (i.e., those that reach the enamel surface as perikymata). If this relationship can be consistently determined for different fossil taxa, it will allow estimation of the total period of crown development of many other unbroken fossil teeth.

Smith (1986) compared patterns of dental development among fossil hominids and living apes and humans using a method that she calls *central tendency discrimination* (CTD) (Smith 1994). She compared the stage of development of the entire preserved dentition of immature fossil hominids to standards of dental development for modern apes and humans to determine whether the fossils more closely approximated the patterns seen in modern apes or humans. By plotting the developmental status of each tooth in a fossil jaw onto human and chimpanzee standards, Smith calculated the predicted individual age based on each individual tooth according to the two different sets of dental standards. Calculation of the standard deviation of age estimates for each dental standard allows one to determine quantitatively whether the chimpanzee or the human model provides a better fit to the particular fossil.

Although Smith has been criticized for attempting to infer timing from pattern of development (Anemone et al. 1991; Macho and Wood 1995), her work has played a central role in the debate concerning the ape or human resemblances of fossil hominid dental development. Her results indicate that all australopithecines do not share the same patterns of dental development, but that most resemble apes more than humans. The major exception seems to be the robust australopithecines, who may superficially resemble modern humans more than apes. The differences she demonstrated between dental development patterns in most australopithecines and modern humans invalidate attempts to determine age at death of the fossils based solely on human analogies and led Smith to question Mann's

(1975) published ages at death for australopithecines. Smith's work further supports the suggestion that the shift to a prolonged pattern of extended maturation had not yet occurred in the hominid lineage in the early Pleistocene. Her analysis of the developmental status of the Nariokotome *H. erectus* (*ergaster?*) juvenile from West Turkana (WT-15000) shows a better fit with modern human than ape standards, but also indicates that hominids were not yet developing in exactly the same prolonged fashion as modern humans (Smith 1993).

More recently, Smith explored the complex statistical relationships between a range of life-history variables (e.g., brain and body size, gestational length, neonatal brain and body size) and dental development among higher primates (Smith 1992; Smith et al. 1995). In a fascinating series of papers, she used regression analysis to predict age at death and other life-history variables of fossil hominids based on the empirical relationships between those variables among extant primates. Building upon the classic work of George Sacher, which suggested that the brain is the pacemaker of life span and growth rates in modern mammals (Sacher 1975, 1978, 1982; Sacher and Staffeldt 1974), Smith persuasively argued that, if the relationship between age at M1 emergence and brain size among modern anthropoids holds for Plio-Pleistocene hominids, small-brained creatures like the australopithecines would almost certainly have first molars emerging at three to four years of age, like modern chimpanzees, rather than at five or six years, as in modern humans.

One of the difficulties in determining the developmental status of immature fossil hominid dentitions has been the inability of standard radiography to successfully resolve details deep within the alveolar crypts of heavily mineralized fossil skulls and jaws (Mann 1975). Conroy and Vannier (1984, 1987, 1991a, 1991b) solved this problem by using CT scans to visualize the developing teeth in the Taung skull. Their results demonstrate that the Taung child resembled modern apes in having incisors that were not yet crown complete and that had no root development, while the typical human pattern would be to have more than half of the root already developed by the time the first molar emerged. Further work by Conroy and Vannier (1991a) also confirmed the presence of pattern differences between gracile and robust australopithecines. These findings, in light of the enamel incremental line data suggesting that all australopithecines had an apelike period of dental development and life span (Beynon and Dean 1988; Bromage 1987, 1990; Bromage and Dean 1985; Dean 1987a), clearly stress the importance of distinguishing between pattern and timing of dental development. Overall, these results lend further support to the increasingly popular position that the australopithecines had not yet undergone the ontogenetic shift to a humanlike, prolonged period of development.

Mann and colleagues have criticized much of the evidence for an apelike ontogeny for Plio-Pleistocene hominids and continue to argue that australopithecines and early members of *Homo* were developing according to a typically human schedule (Lampl et al. 1993; Mann 1988; Mann et al. 1987, 1990, 1991). Citing the work of dental histologists (Warshawsky and Bai 1983; Warshawsky et al. 1984), they argued that cross-striations and striae of Retzius may be the result of artifacts of the geometry of enamel prisms viewed microscopically rather than temporal periodicities in enamel formation. In particular, they suggested two alternative explanations for the presence of cross-striations: (1) they may result from enamel rods or prisms cut perpendicular to their length, or (2) they may be the result of knife chatter, an artifact of the process of sectioning the teeth during preparation for microscopy. Similarly, striae of Retzius have been interpreted as structural rather than temporal features in dental enamel resulting from the interposition of planes of interrod enamel between layers of enamel rods. Many dental histologists, microscopists, and anatomists dismiss these critiques as poorly supported or already falsified. Moreover, evidence from studies of labeled and of naturally fractured teeth provides strong support for the temporal periodicity hypothesis (Beynon and Dean 1988; Beynon et al. 1991; Beynon and Reid 1987; Boyde 1990; Bromage 1991; Dean 1987b; Ramirez Rozzi 1993).

In spite of their critical assessment of the temporal periodicity hypothesis for enamel incremental markers, Mann and colleagues have spent considerable effort counting perikymata on SEM montages of Neanderthal and modern human incisor crowns, using these numbers to estimate crown formation times and ages at death (Mann et al. 1991). Their findings led them to argue that Plio-Pleistocene hominids and archaic and modern humans broadly overlap in the number of perikymata found on their incisor teeth and thus in their species-typical crown formation times. While these criticisms have been partially met (Dean and Beynon 1991), disagreement over the significance of variability in perikymata counts persists. Mann and colleagues have similarly argued that the true ranges of species level variability in patterns of dental development make distinctions between humans and chimpanzees difficult or impossible to apply consistently. (For a different view, see Anemone et al. 1991, 1996, and Anemone and Watts 1992.)

Summary

A new consensus seems to be emerging concerning the life-history characteristics of Plio-Pleistocene hominids based on recent research on dental development among both living and fossil apes and humans. Australopithecines and early members of the genus *Homo* are increasingly seen as more closely resembling the modern great apes in many aspects of life history, rather than as early versions of *H. sapiens*. An improved picture of the developmental differences between modern human and chimpanzee dentitions contributed to this new paradigm by allowing clear comparisons between the living organisms and the fossils. In addition, new histological and imaging approaches have greatly improved our ability to read the story of life history that is laid down in the calcified dentitions of primates and hominids. While controversies and unresolved issues remain to be settled, life-history work has enriched our understanding of hominid evolution, and the continuing rapid pace of work in this field in many labs around the world promises to yield further insights in the future.

Acknowledgments

I thank all the colleagues with whom I have worked on dental development: Daris Swindler of the University of Washington, Michael Siegel and Mark Mooney of the University of Pittsburgh, and the late Elizabeth Watts of Tulane University. For interesting and helpful discussions on matters ontogenetic and dental, I thank Chris Dean, Alan Mann, Michelle Lampl, Janet Monge, Fernando Ramirez Rozzi, Linda Winkler, and Kevin Kuykendall. A special thanks to Jeff Koch and Jim Bearden of SUNY at Geneseo for help with the probit analysis. Finally, I thank the editors of this volume for inviting my contribution and for their perseverance.

References

Alumna, J., Altmann, S.A., and Hausfater, G. 1981. Physical maturation and age estimates of yellow baboons, *Papio cynocephalus*, in Amboseli National Park, Kenya. *American Journal of Primatology* 1:389–399.

Anemone, R.L. 1991. Can we determine the age at death of Plio-Pleistocene fossil hominids? *Human Mosaic* 25:10–24.

———. 1995. Dental development in chimpanzees of known chronological age: implications for understanding the age at death of Plio-Pleistocene hominids. In J. Moggi-Cecchi (ed.), *Aspects of Dental Biology: Paleontology, Anthropology, and Evolution,* 201–215. Florence: International Institute for the Study of Man.

Anemone, R.L., Mooney, M.F., and Siegel, M.I. 1996. Longitudinal study of dental development in chimpanzees of known chronological age: implications for understanding the age at death of Plio-Pleistocene hominids. *American Journal of Physical Anthropology* 99:119–133.

Anemone, R.L., and Watts, E.S. 1992. Dental development in apes and humans: a comment on Simpson, Lovejoy and Meindl (1990). *Journal of Human Evolution* 22:149–153.

Anemone, R.L., Watts, E.S., and Swindler, D.R. 1991. Dental development of known-age chimpanzees, *Pan troglodytes* (Primates, Pongidae). *American Journal of Physical Anthropology* 86:229–241.

Beynon, A.D., and Dean, M.C. 1987. Crown-formation time of a fossil hominid premolar tooth. *Archives of Oral Biology* 32:773–780.

———. 1988. Distinct dental development patterns in early fossil hominids. *Nature* 335:509–514.

Beynon, A.D., Dean, M.C., Leakey, M.G., Reid, D.J., and Walker, A. 1997. Histological study of *Proconsul* teeth from Rusinga Island, Kenya. *American Journal of Physical Anthropology* 24 (suppl): 77.

Beynon, A.D., Dean, M.C., and Reid, D.J. 1991. Histological study on the chronology of the developing dentition in gorilla and orangutan. *American Journal of Physical Anthropology* 86:189–203.

Beynon, A.D., and Reid, D.J. 1987. Relationships between perikymata counts and crown formation times in the human permanent dentition. *Journal of Dental Research* 66:889–890.

Beynon, A.D., and Wood, B.A. 1987. Patterns and rates of enamel growth in the molar teeth of early hominids. *Nature* 326:493–496.

Boyde, A. 1990. Developmental interpretations of dental microstructure. In C.J. DeRousseau (ed.), *Primate Life History and Evolution,* 229–267. New York: Wiley-Liss.

Boyde, A., and Martin, L.B. 1984. The microstructure of primate dental enamel. In D. Chivers, B. Wood, and A. Bilsborough (eds.), *Food Processing in Primates,* 341–367. New York: Plenum Press.

Brace, C.L., and Ashley-Montague, M.F. 1977. *Human Evolution: An Introduction to Biological Anthropology.* New York: Macmillan Publishing Co.

Bromage, T.G. 1987. The biological and chronological maturation of early hominids. *Journal of Human Evolution* 16:257–272.

———. 1990. Early hominid development and life history. In C.J. DeRousseau (ed.), *Primate Life History and Evolution,* 105–113. New York: Wiley-Liss.

———. 1991. Enamel incremental periodicity in the pig-tailed macaque: a polychrome fluorescent labeling study of dental hard tissues. *American Journal of Physical Anthropology* 86:205–214.

Bromage, T.G, and Dean, M.C. 1985. Re-evaluation of the age at death of immature fossil hominids. *Nature* 317:525–527.

Broom, R., and Robinson, J.T. 1951. Eruption of the permanent teeth in the South African fossil ape-man. *Nature* 167:443.

Charnov, E.L. 1991. Evolution of life history variation among female mammals. *Proceedings of the National Academy of Sciences, USA* 88:1134–1137.

Charnov, E.L., and Berrigan, D. 1993. Why do female primates have such long lifespans and so few babies? or Life in the slow lane. *Evolutionary Anthropology* 1:191–194.

Conroy, G.C. 1988. Alleged synapomorphy of the M1/I1 eruption pattern in australopithecines and *Homo:* evidence from high-resolution computed tomography. *American Journal of Physical Anthropology* 75:487–492.

Conroy, G.C., and Kuykendall, K. 1995. Paleopediatrics: Or when did human infants really become human? *American Journal of Physical Anthropology* 98:121–131.

Conroy, G.C., and Mahoney, C.J. 1991. Mixed longitudinal study of dental emergence in the chimpanzee, *Pan troglodytes* (Primates, Pongidae). *American Journal of Physical Anthropology* 86:243–254.

Conroy, G.C., and Vannier, M.W. 1984. Noninvasive three-dimensional computer imaging of matrix-filled fossil skulls by high-resolution computed tomography. *Science* 226:456–458.

———. 1987. Dental development of the Taung skull from computerized tomography. *Nature* 329:625–627.

———. 1988. The nature of Taung dental maturation continued. *Nature* 333:808.

———. 1991a. Dental development in South African Australopithecines. Part 1: Problems of pattern and ontogeny. *American Journal of Physical Anthropology* 86:121–136.

———. 1991b. Dental development in South African Australopithecines. Part 2: Dental stage assessment. *American Journal of Physical Anthropology* 86:137–156.

Dahlberg, A.A., and Menegaz-Bock, R.M. 1958. Emergence of the permanent teeth in Pima Indian children. *Journal of Dental Research* 37:1123–1140.

Dart, R.A. 1925. *Australopithecus africanus:* the ape-man of South Africa. *Nature* 115:195–197.

———. 1948. The adolescent mandible of *Australopithecus prometheus. American Journal of Physical Anthropology* 6:391–411.

Dean, M.C. 1985. The eruption pattern of the permanent incisors and first permanent molars in *Australopithecus (Paranthropus) robustus. American Journal of Physical Anthropology* 67:251–257.

———. 1987a. The dental developmental status of six East African juvenile fossil hominids. *Journal of Human Evolution* 16:197–213.

———. 1987b. Growth layers and incremental markers in hard tissues: a review of the literature and some preliminary observations about enamel structure in *Paranthropus boisei. Journal of Human Evolution* 16:157–172.

————. 1989. The developing dentition and tooth structure in hominoids. *Folia Primatologica* 53:160–176.

Dean, M.C., and Beynon, A.D. 1991. Histological reconstruction of crown formation times and initial root formation times in a modern child. *American Journal of Physical Anthropology* 86:215–228.

Dean, M.C., Beynon, A.D., Thackeray, J.F., and Macho, G.A. 1993. Histological reconstruction of dental development and age at death of a juvenile *Paranthropus robustus* specimen, SK 63, from Swartkrans, South Africa. *American Journal of Physical Anthropology* 91:401–419.

Dean, M.C., Stringer, C.B., and Bromage, T.G. 1986. Age at death of the Neanderthal child from Devil's Tower, Gibraltar, and the implications for studies of general growth and development in Neanderthals. *American Journal of Physical Anthropology* 70:301–309.

Dean, M.C., and Wood, B.A. 1981. Developing pongid dentition and its use for ageing individual crania in comparative cross-sectional growth studies. *Folia Primatologica* 36:111–127.

Demirjian, A., and Goldstein, H. 1976. New systems for dental maturity based on seven and four teeth. *Annals of Human Biology* 3:411–421.

Demirjian, A., Goldstein, H., and Tanner, J.M. 1973. A new system of dental age assessment. *Human Biology* 42:211–217.

Dobzhansky, T. 1962. *Mankind Evolving.* New Haven: Yale University Press.

Friedlaender, J.S., and Bailit, H.L. 1969. Eruption times of the deciduous and permanent teeth of natives on Bougainville Island, Territory of New Guinea: a study of racial variation. *Human Biology* 41:51–65.

Garn, S.M., Koski, K., and Lewis, A.B. 1957. Problems in determining the tooth eruption sequence in fossil and modern man. *American Journal of Physical Anthropology* 15:313–331.

Grine, F.E. 1987. On the eruption pattern of the permanent incisors and first permanent molars in *Paranthropus*. *American Journal of Physical Anthropology* 72:353–359.

Harvey, P.H., and Clutton-Brock, T.H. 1985. Life history variation in primates. *Evolution* 39:559–581.

Harvey, P.H., Martin, P.D., and Clutton-Brock, T.H. 1987. Life histories in comparative perspective. In B.B. Smuts, D.L. Cheney, R.M. Seyfarth, R.W. Wrangham, and T.T. Struhsaker (eds.), *Primate Societies*, 181–196. Chicago: University of Chicago Press.

Kelso, A.J., and Trevathan, W.R. 1984. *Physical Anthropology.* Englewood Cliffs, N.J.: Prentice Hall.

Kuykendall, K.L. 1996. Dental development in chimpanzees (*Pan troglodytes*): the timing of tooth calcification stages. *American Journal of Physical Anthropology* 99:135–157.

Kuykendall, K.L., and Conroy, G.C. 1996. Permanent tooth calcification in chimpanzees (*Pan troglodytes*): patterns and polymorphisms. *American Journal of Physical Anthropology* 99:159–174.

Kuykendall, K.L., Mahoney, C.J., and Conroy, G.C. 1992. Probit and survival analysis of tooth emergence ages in a mixed-longitudinal sample of chimpanzees (*Pan troglodytes*). *American Journal of Physical Anthropology* 89:379–399.

Lampl, M., Monge, J.M., and Mann, A.E. 1993. Further observations on a method for estimating hominoid dental developmental patterns. *American Journal of Physical Anthropology* 90:113–127.

LeGros Clark, W.E. 1947. Observations on the anatomy of the fossil Australopithecinae. *Journal of Anatomy* 81:300–333.

Lovejoy, C.O. 1981. The origin of man. *Science* 211:341–350.

Macarthur, R.H., and Wilson, E.O. 1967. *The Theory of Island Biogeography.* Princeton: Princeton University Press.

Macho, G.A., and Wood, B.A. 1995. The role of time and timing in hominid dental development. *Evolutionary Anthropology* 4:17–31.

Mann, A.E. 1975. *Paleodemographic Aspects of the South African Australopithecines.* Philadelphia: University of Pennsylvania Publications in Anthropology, No. 1.

———. 1988. The nature of Taung dental maturation. *Nature* 333:123.

Mann, A.E., Lampl, M., and Monge, J. 1987. Maturational patterns in early hominids. *Nature* 328:673–674.

———. 1990. Decomptes de Perikymaties chez les enfants Néandertaliens de Krapina. *Bulletin et Mémoires du Societé Anthropologie, Paris* 2:213–220.

Mann, A.E., Monge, J., and Lampl, M. 1991. Investigation into the relationship between perikymata counts and crown formation times. *American Journal of Physical Anthropology* 86:175–188.

Martin, R.D. 1983. *Human Brain Evolution in an Ecological Context.* New York: American Museum of Natural History.

Martin, R.D., and MacLarnon, A.M. 1990. Reproductive patterns in primates and other mammals: the dichotomy between altricial and precocial offspring. In C.J. DeRousseau (ed.), *Primate Life History and Evolution*, 47–79. New York: Wiley-Liss.

Mooney, M.P., and Siegel, M.I. 1991. Premaxillary-maxillary suture fusion and anterior nasal tubercle morphology in the chimpanzee. *American Journal of Physical Anthropology* 85:451–456.

Moorrees, C.F.A., Fanning, E.A., and Hunt, E.E. 1963. Age variation of formation stages for ten permanent teeth. *Journal of Dental Research* 42:1490–1502.

Nissen, H.W. 1942. Studies of infant chimpanzees. *Science* 95:159–161.

Nissen, H.W., and Riesen, A.H. 1945. The deciduous dentition of chimpanzees. *Growth* 9:265–274.

———. 1964. The eruption of the permanent dentition of chimpanzees. *American Journal of Physical Anthropology* 22:285–294.

Pianka, E.R. 1970. On *r*- and *K*-selection. *American Naturalist* 104:592–597.

Ramirez Rozzi, F. 1993. Tooth development in East African *Paranthropus. Journal of Human Evolution* 24:429–454.

Sacher, G.A. 1975. Maturation and longevity in relation to cranial capacity in hominid evolution. In R.H. Tuttle (ed.), *Primate Functional Morphology and Evolution*, 417–441. The Hague: Mouton Publishers.

———. 1978. Longevity, aging, and death: an evolutionary perspective. *Gerontologist* 18:112–119.

———. 1982. The role of brain evolution in the maturation of primates. In D. Falk (ed.), *Primate Brain Evolution*, 97–112. New York: Plenum Press.

Sacher, G.A., and Staffeldt, E.F. 1974. The relation of gestation time to brain weight for placental mammals: implications for the theory of vertebrate growth. *American Naturalist* 108:593–616.

Schultz, A.H. 1935. The eruption and decay of the permanent teeth in primates. *American Journal of Physical Anthropology* 19:489–581.

———. 1969. *The Life of Primates*. New York: Universe Books.

Sigg, H., Stolba, A., Abegglen, J., and Dasser, V. 1982. Life history of hamadryas baboons: physical development, infant mortality, reproductive parameters and family relationships. *Primates* 23:473–487.

Smith, B.H. 1986. Dental development in *Australopithecus* and early *Homo*. *Nature* 323:327–330.

———. 1987. Maturational patterns in early hominids. *Nature* 328:674–675.

———. 1988. Dental development as a measure of life history in primates. *Evolution* 43:683–688.

———. 1989. Growth and development and its significance for early hominid behaviour. *Ossa* 14:63–96.

———. 1991. Dental development and the evolution of life history in Hominidae. *American Journal of Physical Anthropology* 86:157–174.

———. 1992. Life history and the evolution of human maturation. *Evolutionary Anthropology* 1:134–142.

———. 1993. The physiological age of KNM-WT 15000. In A. Walker and R. Leakey (eds.), *The Nariokotome* Homo erectus *skeleton*, 195–220. Cambridge: Harvard University Press.

———. 1994. Patterns of dental development in *Homo, Australopithecus, Pan* and *Gorilla. American Journal of Physical Anthropology* 94:307–325.

Smith, B.H., and Garn, S.M. 1987. Polymorphisms in eruption sequence of permanent teeth in American children. *American Journal of Physical Anthropology* 74:289–303.

Smith, R.J., Gannon, P.J., and Smith, B.H. 1995. Ontogeny of australopithecines and early *Homo:* evidence from cranial capacity and dental eruption. *Journal of Human Evolution* 29:155–168.

Stringer, C.B., Dean, M.C., and Martin, R.D. 1990. A comparative study of cranial and dental development within a recent British sample and among Neandertals. In C.J. DeRousseau (ed.), *Primate Life History and Evolution*, 115–152. New York: Wiley-Liss.

Tanner, J.M. 1990. *Fetus into Man*. Cambridge: Harvard University Press.

Walker, P.L., Dean, G., and Shapiro, P. 1991. Estimating age from tooth wear in archaeological populations. In M.A. Kelley and C.S. Larsen (eds.), *Advances in Dental Anthropology*, 169–178. New York: Alan R. Liss.

Warshawsky, H., and Bai, P. 1983. Chatter during thin sectioning of rat incisor enamel can cause periodicities resembling cross striations. *Anatomical Records* 207:533–538.

Warshawsky, H., Bai, P., and Nanci, A. 1984. Lack of evidence for rhythmicity in enamel development. In A.B. Belcourt and J.V. Ruch (eds.), *Tooth Morphogenesis and Differentiation*, 241–255. Paris: INSERM.

Chapter 13

An Assessment of Radiographic and Histological Standards of Dental Development in Chimpanzees

KEVIN L. KUYKENDALL

The dentition provides the means for reliable assessment of developmental maturity in both living and skeletal populations, and because of preservational conditions it may be the only clue to an individual's maturity status in archeological and paleontological samples (Miles 1963). Thus, fossilized infant and juvenile dentitions serve as a unique comparative interface between long-dead and living representatives of ancestral populations or lineages. Because of the close taxonomic and phylogenetic relationship among modern hominoids, humans, and fossil hominids, the developing dentition forms the basis for comparative and evolutionary studies of hominoid and hominid life history.

Life history has been defined as "the allocation of an organism's energy toward growth, maintenance, reproduction, raising offspring to independence, and avoiding death" (Smith and Tompkins 1995, 257). Phrased rather differently, life history is the normal course of events affecting an individual's existence from the moment of conception until death. Thus, knowledge of the complexities involved in a particular species' life history is critical to understanding its ecological role and adaptive strategy. In studying fossil or extinct species, it is the reconstruction of such adaptive strategies that we use as a framework for evolutionary models to bring an extinct species "back to life."

Life-history reconstructions for extinct taxa require comparative data relating ontogenetic events in extant species to some kind of time scale (e.g., the length of the gestation period, the age at weaning, the age at M1 tooth emergence), particularly data allowing the use of dental development schedules as a surrogate for other events in life history and ontogeny (Smith 1989a, 1989b; see also Harvey and Clutton-Brock 1985). The fact that dental structures are so well represented in the fossil record (since enamel is the hardest tissue in the body) only increases their value for those studying evolutionary aspects of life history and ontogeny. Finally, dental structures and their development seem to be fairly conservative because of their critical role in food processing, in that "fueling" the organism is prerequisite to any subsequent strategies for survival (see Smith 1989a). Nevertheless, the application of such data to the hominid fossil record is far from straightforward, and it is likely that finer scale differences in developmental timing among closely related species are not detectable even if general trends among anthropoid primates are (Smith et al. 1995).

The prolonged period of human growth and maturation is a period of intellectual, social, and physical development and has long been cited as a key feature in ancestral hominid life history as their brains and cultures evolved greater complexity. In fact, this developmental feature has been viewed with such significance that it has been invoked in many evolutionary scenarios for the origin of typically "human" traits, such as our enlarged brains, complex social organization, home bases, tool manufacture and utilization, and extensive learning and socialization (e.g., Washburn 1960; Lancaster 1975; Mann 1975), even at very early stages of hominid evolutionary history.

In this context, dental developmental status has been used to interpret early hominid life history since the discovery of the Taung fossil, subsequently assigned to *Australpithecus africanus* (e.g., Dart 1925, 1948; Broom and Robinson 1952; Le Gros Clark 1967; Wallace 1977). Mann's (1975) paleodemographic reconstruction of South African (robust) australopithecines included a comparison of patterns of dental development in modern hominoids, humans, and fossil hominids. His conclusions supported the then-prevailing view that even early hominids could be characterized by a prolonged "humanlike" period of childhood dependency (see also Mann et al. 1990a; Mann 1988; Wolpoff et al. 1988). Subsequent reevaluations of the Plio-Pleistocene archeological record (e.g., Binford 1985; Potts 1984; Shipman 1986), evidence for hominid brain evolution (e.g., Falk 1987; Holloway 1972; Sacher 1975), and evidence relating to dental development and other indicators of life history (e.g., Bromage 1987; Bromage and Dean

1985; Conroy and Vannier 1991a, 1991b; Conroy and Kuykendall 1995; Dean 1987; Smith 1991a, 1992, 1993; Smith and Tompkins 1995) suggested that this "essential core of humanness" (Mann et al. 1991) did not evolve until considerably later in hominid evolutionary history.

Thus, in roughly the past two decades, research into the life history and ontogeny of early hominids has seen an intensification of interest (and stimulated an often heated debate), resulting in what has been termed a "paradigm shift" (Anemone 1995) in our understanding of Plio-Pleistocene hominid ontogeny. Much of this research has focused on aspects of dental development in extant (usually African) hominoids and fossil hominids (see reviews by Dean 1987, 1989; Macho and Wood 1995; Beynon et al. 1998a; other references cited below).

Most researchers would now concede that early hominid growth and development was more similar in duration to that of modern apes than to that of modern humans, but that the pattern of events during "early hominid ontogeny" was unique to each taxon. Researchers have largely moved away from the simplistic dichotomy describing "humanlike" versus "apelike" patterns of ontogeny and dental development in hominid evolutionary models (Aiello and Dean 1990; Beynon and Dean 1988; Bromage 1987; Conroy and Vannier 1991a, 1991b; Conroy and Kuykendall 1995; Macho and Wood 1995; Moggi-Cecchi 2001; Moggi-Cecchi et al. 1998; Smith 1991a), recognizing subtle but complex differences among closely related extant and extinct species. Although the final implications of these new interpretations are yet to be fully assimilated, their significance has begun to emerge in both popular and scientific writings about early hominid evolution (Smith and Tompkins 1995; Conroy 1997; Johanson and Edgar 1996; Walker and Shipman 1996).

As a result of such studies, it now seems that the fundamental questions concerning the general duration of early hominid growth (was it humanlike or apelike?) and the evolutionary appearance of a humanlike developmental period (did it evolve early or late in hominid phylogeny?) have been resolved. But since current interpretations of early hominid ontogeny are in one way or another reliant on radiographic assessments of dental development in modern apes and humans, it is important to understand the nature and significance of differences among such studies. Thus, this chapter provides a brief overview and appraisal of radiographic studies documenting chimpanzee dental development. I then review recent studies using histological techniques that analyze incremental growth markers in tooth enamel that have produced developmental schedules that differ from those derived radiographically. It is necessary to examine these findings to understand how they might affect our interpretations of radiographic studies.

Dental Development Standards for Hominoids

The dental development studies of relevance here generally take one of two forms: (1) assessment of gross aspects of tooth eruption and calcification using visual inspection, radiography, or computed tomography (see Smith 1991b for a review of such studies in humans and references below for hominoids) or (2) microscopic or histological studies documenting incremental growth features of hard tissue microstructure (usually enamel) that are formed during tooth mineralization (see reviews by Dean 1987; Mann et al. 1990a; Beynon et al. 1998a). The advantages of the former type of study lie in the feasibility of obtaining large samples (potentially numbering in the hundreds or thousands), the widespread availability of comparative data from a variety of taxa, and its versatility for noninvasive application to living, skeletal, or fossil samples. Disadvantages include geometric and anatomical difficulties in obtaining useful radiographs of all teeth (many radiographic studies are limited to the mandibular dentition) (Kuykendall 1996; Beynon et al. 1998a), dangers of radiation exposure for living subjects, and the lack of consistent radiographic, assessment, and statistical methods among studies, which hinders comparison of results (see Smith 1991b; Beynon et al. 1998a; Simpson and Kunos 1998).

Histological studies, on the other hand, are more accurate in that calcification is detectable at earlier stages (Beynon et al. 1998a; Dirks 1998; Winkler 1995; Winkler et al. 1996) and the methods are well established and consistent (Reid et al. 1998a). Age estimates attain higher levels of precision, though their accuracy has been questioned (Mann et al. 1990a, 1990b; but see Fitzgerald 1998) and the nature of variation in the underlying processes may not be fully understood (Schwartz and Kuykendall 1996). Furthermore, information about tooth microstructure seems to correspond with morphological and functional features such as enamel thickness and cusp morphology that have an adaptive and evolutionary basis (Reid et al. 1998a). Thus, histological studies may inherently place information about the timing of dental development into a useful life-history context. Perhaps the primary disadvantages of histological studies are that the procedures are invasive (requiring destruction of specimens; note that this does not refer to study of perikymata [see Aiello and Dean 1990; Dean

1987; and Ramirez Rozzi, Chap. 15, this volume] though this noninvasive procedure is conceptually similar to histological techniques and thus subject to some of the same limitations), as well as time-consuming, meaning that it has been impossible thus far to produce large comparative samples for use as reference standards.

As discussed below, the methodological differences between radiographic and histological studies have created certain problems for comparison of results derived from each. On the other hand, the availability of two quite different methods of documenting dental developmental processes fortuitously provides a framework for assessing correspondence between results and for independent confirmation of hypotheses and results derived from either method.

Studies documenting various aspects of permanent dental development in extant populations of chimpanzees (and some studies including other primate species) provide valuable data on tooth emergence (Clements and Zuckerman 1953; Dean and Wood 1981; Conroy and Mahoney 1991; Kraemer et al. 1982; Kuykendall et al. 1992; Schultz 1935; Smith 1994; Smith et al. 1994), timing and pattern of tooth calcification (Anemone 1995; Anemone and Watts 1992; Anemone et al. 1991, 1996; Kuykendall 1996; Kuykendall and Conroy 1996), and relative developmental relationships among teeth (Simpson et al. 1992). Winkler and colleagues (Winkler 1995; Winkler et al. 1996) documented orangutan dental development using both radiographic and histological techniques, and other studies provided histological analyses of dental development in the chimpanzee (Chandrasekera et al. 1993; Reid et al. 1998a) and the gibbon (Dirks 1998). In addition, Beynon et al. (1998b) and Dean and Shellis (1998), using histological techniques, recently documented aspects of dental development in fossil samples of Miocene hominoids. In comparison to the state of research even a decade ago, there is now a relative wealth of data documenting different aspects of dental development in hominoids, especially in chimpanzees.

Although Dean and Wood (1981) published the first detailed chronology of dental development in great apes, their data consisted of cross-sectional tooth emergence estimates derived from a skeletal sample of *Pan, Gorilla,* and *Pongo.* Perhaps, since chimpanzees are both a more appropriate model for representation of early hominids and more accessible for study, much of the subsequent research focused on this genus. Anemone and colleagues used data from a longitudinal radiographic study of living chimpanzees (e.g., Anemone et al. 1991, 1996; Anemone and Watts 1992) to modify Dean and Wood's dental development chronology for *Pan*. Documentation of tooth emergence in a mixed-longitudinal sample (Conroy and Mahoney 1991; Kuykendall et al. 1992) and tooth calcification in a cross-sectional radiographic study (Kuykendall 1992, 1996; Kuykendall and Conroy 1996) resulted in further minor modifications to the chimpanzee developmental schedule. Figure 13.1 presents the general schedule for chimpanzee tooth calcification in comparison to the human schedule, as currently documented from radiographic studies.

Radiological Studies of Chimpanzee Tooth Calcification

The studies by Anemone and Kuykendall with their respective colleagues are the most appropriate available sources for data about the timing and pattern of chimpanzee dental development. However, these studies differ, respectively, in the type of sample (longitudinal vs. cross-sectional), sample size (n = 33 vs. 118), type of radiographs used (lateral cephalograms vs. periapical intraoral films), and statistical methods. As noted by Kuykendall (1996), these differences may largely explain the discrepancies between results for the timing of some stages of crown and root formation.

Both researchers' studies document dental development on the basis of eight arbitrary stages of crown and root formation (four stages describing crown calcification and four describing root elongation) originally used by Demirjian et al. (1973; also see studies cited in Table 13.1). Table 13.1 compares mean and median age estimates for these eight tooth calcification stages from Kuykendall (1996) and Anemone (Anemone et al. 1991, 1996; Anemone and Watts 1992). Most differences in age estimates between these studies are less than one year. If a "significant difference" (whether biological or methodological) is arbitrarily defined as one of one year or greater, most differences noted are within acceptable limits of variation. However, discrepancies of greater than one year are found for stages representing crown completion (stage 4) in M2 means and premolar medians. Age estimates for root completion (stage 8) differed by more than one year for M1 means and incisor, M1, and M2 medians. Additional differences of this magnitude are seen for M2 stage 6 and M3 stages 2, 3, and 5 (data for M3 development typically being inadequate in all studies).

Many of the differences exceeding one year involve crown or root completion—major events during tooth calcification. Kuykendall (1996) sug-

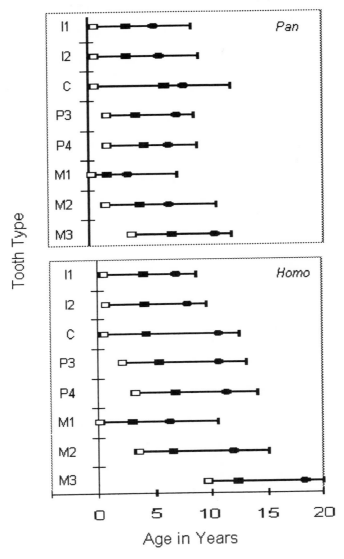

Fig. 13.1. Summary charts of dental development in chimpanzees (*top*) and humans (*bottom*), after Anemone et al. (1991), Anemone and Watts (1992), and Kuykendall (1996). The *horizontal scale* shows age in years; the *vertical scale* in each chart represents permanent tooth types for the mandibular dentition. Symbols on each bar represent the average ages documented for the initiation of crown formation (*open rectangles*), the completion of crown formation (*closed rectangles*), tooth emergence (*solid ovals*), and root completion (*vertical bars*). Thus, the period of crown formation is represented between the *open* and *closed rectangles*, and that for root formation between the *closed rectangle* and *vertical bar*.

Table 13.1 Summary of Mean and Median Age Estimates from Kuykendall (1996) and Anemone (Anemone et al. 1991, 1996; Anemone and Watts 1992) and Differences between Such Estimates in Those Studies

| Tooth | Devel. Stage[a] | Age (years) | | | | | |
| | | Median | | | Mean | | |
		Kuykendall	Anemone	Diff.	Kuykendall	Anemone	Diff.
I1	1	0.42	0.66	−0.24			
	4	3.13	3.1	0.03			
	8	8.64	7	1.64[b]			
I2	1	0.42	0.75	−0.33			
	4	3.13	3.1	0.03			
	8	9.24	7	2.24			
C	1	0.42	1.25	−0.83			
	4	6.32	7	−0.68			
	8	—	12	—			
P3	1	1.34	1.75	−0.41			
	4	3.87	6	−2.13			
	8	8.77	9.5	−0.73			
P4	1	1.37	1.75	−0.38			
	4	4.58	6	−1.42			
	8	9.06	9.5	−0.44			
M1	1	0.13	—	—	0.13	0.33	−0.2
	2	0.26	0.83	−0.57	0.46	0.75	−0.29
	3	0.6	1.5	−0.9	1.21	1.22	−0.01
	4	1.4	2	−0.6	1.69	1.71	−0.02
	5	2.17	2.75	−0.58	2.72	2.36	0.36
	6	3.02	3.25	−0.23	3.63	3	0.63
	7	4.58	4	0.58	5.75	3.67	2.08
	8	7.29	5	2.29	8.3	4.44	3.86
M2	1	1.29	1.5	−0.21	1.34	1.31	0.03
	2	1.53	2	−0.47	1.81	1.94	−0.13
	3	2.34	3	−0.66	3.21	2.69	0.52
	4	4.2	4	0.2	4.59	3.47	1.12
	5	4.99	6	−1.01	5.2	4.48	0.72
	6	5.31	7	−1.69	6.78	—	—
	7	7.67	8	−0.33	8.74	—	—
	8	10.74	9	1.74	—	—	—
M3	1	3.41	3.5	−0.09	3.5	3.64	−0.14
	2	4.2	6	−1.8	4.45	4.52	−0.07
	3	4.85	7	−2.15	6.08		
	4	6.88	7	−0.12	7.28		
	5	7.65	9	−1.35	8.35		

(continued)

Table 13.1 (*Continued*)

Tooth	Devel. Stage[a]	Age (years)					
		Median			Mean		
		Kuykendall	Anemone	Diff.	Kuykendall	Anemone	Diff.
	6	9.33	10	−0.67	10.02		
	7	—	11	—	—		
	8	—	12	—	—		

[a]Developmental stages refer to radiographic stage assessments based on the system developed by Demirjian et al. (1973).
[b]Differences of more than one per year between means or medians are italic.

gested that these might result from difficulties in assessing developmental stages, especially since different kinds of radiographs were used in each study. In addition, Beynon et al. (1998a) demonstrated that the morphological complexity of the enamel crown at the cervix results in difficult or inaccurate assessment of crown completion on radiographs.

Perhaps more important than such minor differences in the timing of tooth calcification stages, assessments that distinguish patterns of relative dental development (including emergence as well as calcification) between chimpanzee and human dentitions have also stimulated debate. Such patterns have proven useful for assessing aspects of early hominid dental development in attempts to reconstruct evolutionary patterns of life history.

The three most-cited pattern features used in making distinctions between "ape" and "human" dental development (and thus for assessing patterns from hominid fossils) are (1) relative advancement of human anterior teeth compared to molars, typically expressed in terms of the advanced state of I1 development at M1 emergence in comparison with I1-M1 in chimpanzees (e.g., Dean and Wood 1981); (2) delayed chimpanzee canine development relative to M2 and other teeth (e.g., Clements and Zuckerman 1953; Schultz 1935; Swindler 1985); and (3) relative delays in molar development in humans compared with chimpanzee molar teeth (e.g., Anemone et al. 1991; Mann et al. 1987; Swindler 1985) (see Fig. 13.1).

Following these ideas, Anemone (1995; Anemone et al. 1991, 1996) concluded that modern chimpanzee and human patterns of dental development can be distinguished by two clear-cut features (see Fig. 13.1). First, chimpanzee molar crown calcification is marked by temporal overlap between adjacent molar crowns (e.g., between M1 and M2), while adjacent

human molar crowns demonstrate significant periods of delay between calcification periods (e.g., Moorrees 1959; Demirjian 1986).

However, Fanning and Moorrees (1969) documented temporal overlap
in development between adjacent molar crowns in both Australian aborigines and Australian whites, recording overlap between M1-M2 and between M2-M3 at respective frequencies of 29 percent and 14 percent in
aborigines and 54 percent and 4 percent in whites. More recently, Reid et
al. (1998b) documented overlapping periods of M1-M2 crown calcification (but not for M2-M3) in a histological study of dental development in
medieval human skeletal material. In a histological examination of tooth
calcification in a single African individual, Dean et al. (1993) documented
only minimal separation (not a significant delay) between M2-M3 crown
formation periods. Taken together, analyses of molar tooth calcification
and tooth emergence patterns in chimpanzees and humans suggest that relative molar development is ambiguous for distinguishing between these
species (Simpson et al. 1992; Kuykendall 1992, 1996; Kuykendall and Conroy 1996). The conflict between Anemone's view and others may stem from
the fact that he reported an overlap in the temporal ranges for *average ages*
of crown formation, while others were examining tabulations of *individual
sequences* of crown calcification stages. In summary, while temporal overlap
in crown formation periods between successive molars has been documented in all chimpanzees studied thus far, human individuals and populations may demonstrate either delay or overlap.

The second feature described by Anemone et al. (1996) to distinguish
between chimpanzee and human dental development is that, "at emergence
of the first molar, human anterior teeth are developmentally advanced compared to chimpanzees, or conversely . . . human molar development is delayed compared to that of the chimpanzee" (Anemone et al. 1996, 129).
Typically, human first molars and central incisors demonstrate similar or
even equivalent degrees of dental development, and in some cases the I1 is
advanced compared to the M1 (e.g., the I1-M1 emergence sequence is
commonly reported [Smith 1994]). In chimpanzees, incisor development
is always considerably delayed compared with emergence of the first molar
(Kuykendall and Conroy 1996). These contrasting developmental relationships of the I1 *at M1 emergence* are often cited to describe ape, human,
and early hominid dental development (Schultz 1935; Broom and Robinson 1952; Clements and Zuckerman 1953; Mann 1975, 1988; Wallace
1977; Dean and Wood 1981; Bromage 1987; Conroy and Vannier 1991a,
1991b; Mann et al. 1990a; Wolpoff et al. 1988). However, Kuykendall and

Conroy (1996) demonstrated that this developmental relationship is not as distinctive throughout all stages of M1 development, as the chimpanzee M1 and I1 were documented at the same developmental stage in 28 percent of recorded cases in their sample.

Thus, despite overall differences in the pattern and timing of dental development between chimpanzees and humans, the observed variation in relative developmental relationships among teeth makes it difficult to define a specific trait that distinguishes reliably between chimpanzee and human patterns of dental development (also see Winkler et al. 1996).

Radiological versus Histological Studies of Chimpanzee Dental Development

The recent study by Beynon et al. (1998a) addressing known discrepancies among estimates of dental developmental timing in studies using radiographic, dissection, and histological techniques indicates that radiographic assessment lacks the capability for accurate detection of either initial crown calcification or crown completion. Thus, Reid et al. (1998a) reported that crown formation periods derived from histological analysis of a small sample of chimpanzee teeth were greater than published radiological estimates by as much as 3.1 years, and similar differences were detected in a separate histological study of dental development in a medieval human sample (Reid et al. 1998b). If confirmed, these findings may affect our interpretations of relative developmental relationships among teeth, both intra- and interspecifically, and certainly have implications for assessing age at death on the basis of dental development.

It should also be noted that properly calculated population estimates for periods of crown calcification (i.e., the period from crown initiation to crown completion) in chimpanzees are not available from either radiographic or histological studies. Normally, summary statistics for attainment of arbitrary (radiographic) developmental stages are included, and developmental periods of crown and root formation are estimated from these. Such estimates should be calculated from individual longitudinal records, averaged for population estimates. Kuykendall (1996) used a cross-sectional sample and thus could not document actual periods of development in individuals, whereas Anemone's longitudinal analyses (Anemone et al. 1991, 1996; Anemone and Watts 1992; Anemone, Chap. 12, this volume) focused largely on the molar teeth and provided only means and standard deviations for their developmental stages. In addition, all available studies have very small samples for certain teeth, such as the canine and third molar, and lack data for some developmental stages.

The differences in crown formation periods reported by Reid et al. (1998a, 440) were calculated as the difference in mean estimates between their histological sample and radiographic studies (G. Schwartz, pers. comm.). Table 13.2 and Figure 13.2 compare published estimates for the ranges (rather than means) of crown initiation ages, crown completion ages, and crown calcification periods in chimpanzees from radiographic (Table 3 in Kuykendall 1996) and histological sources (Table 6 in Reid et al. 1998a). Figure 13.2 also includes mean ages and midpoint ages of attainment for crown initiation (stage 1) and crown completion (stage 4) as provided by Kuykendall (1996). The estimates calculated in this way do not represent an actual period of crown calcification for any individual but do provide a comparison of the timing estimates for developmental stages involved.

According to Table 13.2 and Figure 13.2, the ranges of histologically documented estimates for crown initiation encompass younger ages, and those for crown completion are older, supporting the conclusion that histological data document longer periods for crown formation than do radiographically obtained data (Reid et al. 1998a). However, the radiological ages for crown initiation in M2 and M3 are younger and those for crown completion in the canine and the M3 are older than their respective histological ages. This may be a sampling effect, in that histological data may comprise a single individual in some cases, whereas radiological data include four to ten chimpanzees, depending on the tooth and stage.

Figure 13.2 also demonstrates some degree of overlap in the histological and radiographic age ranges for both crown initiation and crown completion in most teeth. Insofar as this comparison is valid, Figure 13.2 suggests that the basic timing relationships for tooth crown development are similar whether described by histological or radiographic methods, in contrast to Reid et al. (1998a). However, the mean and midpoint attainment (MA) ages for crown formation stages are well below the histological age ranges from Reid et al. (1998a), suggesting that the magnitude of differences they reported is in part due to the method of calculation used. Overall, basing estimates for periods of crown formation (stage 1 to stage 4) on mean or MA ages will result in much shorter calculated periods of crown formation. Table 13.2C shows that crown formation periods calculated from minimum-maximum ranges in either radiographic or histological studies are much more similar than those reported by Reid et al. (1998a).

Table 13.2 Comparison of Crown Formation Estimates in Chimpanzees from Radiographic (Kuykendall 1996) and Histological (Reid et al. 1998a) Studies

A. Estimates for Crown Initiation Ages from Histology and Radiography

Tooth Type	Mean CI	Att age CI	CI min R	CI max R	CI min H	CI max H
I1	0.48	0.42	0.45	0.51	0.15	0.46
I2	0.48	0.42	0.45	0.51	0.19	0.7
C	0.48	0.42	0.45	0.51	0.38	0.57
P3	1.52	1.34	1.31	1.73	1.11	1.4
P4	1.5	1.37	1.31	1.69	1.19	1.95
M1	0.13	0.13	0.13	0.13	-0.15	-0.5
M2	1.34	1.29	1.15	1.48	1.67	1.95
M3	3.08	2.68	3	4.61	3.6	3.63

B. Estimates for Crown Completion Ages from Histology and Radiography

Tooth Type	Mean CC	Att age CC	CC min R	CC max R	CC min H	CC max H
I1	3.8	3.13	2.89	5.05	4.6	5.81
I2	3.8	3.13	2.89	5.05	5.2	5.81
C	6.62	6.32	5.6	8.2	5.85	6.93
P3	4.72	3.87	3.93	5.36	5.45	5.7
P4	4.6	4.58	3.81	5.78	5.55	5.65
M1	1.69	1.4	1.21	2.19	2.4	3.05
M2	4.59	4.2	3.81	5.36	4.52	5.45
M3	7.28	6.88	6.06	8.15	7	7

(continued)

Table 13.2 (Continued)

C. Estimates for Crown Formation Periods from Histology and Radiography

Tooth Type	CI min R	CC max R	R cr fm	CI min H	CC max H	H cr fm	H-R
I1	0.45	5.05	4.6	0.15	5.81	5.66	1.06
I2	0.45	5.05	4.6	0.19	5.81	5.62	1.02
C	0.45	8.2	7.75	0.38	6.93	6.55	−1.2
P3	1.31	5.36	4.04	1.11	5.7	4.59	0.55
P4	1.31	5.78	4.47	1.19	5.65	4.46	−0.01
M1	0.13	2.19	2.06	−0.15	3.05	3.2	1.14
M2	1.15	5.36	4.20	1.67	5.45	3.78	−0.42
M3	3	8.15	5.15	3.6	7	3.4	−1.75

Note: Section A is a comparison of crown initiation age estimates (in years), including mean age (Mean CI), attainment age (Att age CI), minimum age from radiography (CI min R) and histology (CI min H), and maximum age from radiography (CI max R) and histology (CI max H). Section B is a comparison of crown completion estimates (abbreviations similar), and Section C is a comparison of crown formation periods estimated from these minimum and maximum age estimates in radiographic (R cr fm) and histological (H cr fm) studies and the difference between them (H-R).

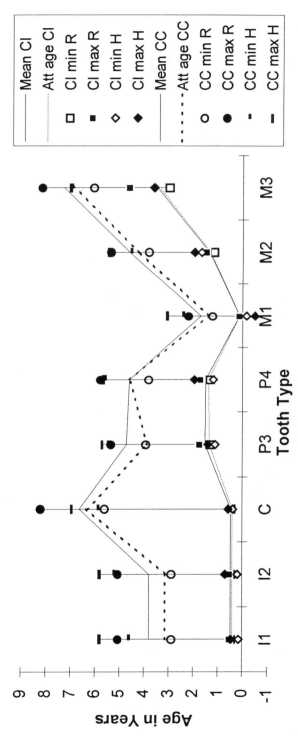

Fig. 13.2. Published ranges for crown initiation (*CI*) and crown completion (*CC*) from radiographic (*R*) and histological (*H*) studies, using data from Kuykendall (1996) for radiography and Reid et al. (1998a) for histology. The *horizontal axis* represents tooth types for the permanent mandibular dentition, and the *vertical axis* shows age in years. The graph shows the minimum ages from the radiographic and histological studies (e.g., *open squares* represent the minimum documented age for crown initiation from the radiographic data, and *longer horizontal hash marks* represent the maximum documented age for crown completion from the histological data). The area delineated by *solid lines* represents the period of crown formation defined by the mean ages for crown initiation and crown completion from the radiographic study (Kuykendall 1996), and the area defined by the *dashed lines* represents the same for midpoint ages of attainment in that study.

As mentioned, truly valid estimates for such periods can be derived only from longitudinal data.

This comparison suggests that, although there are certain technical advantages to assessment of dental development stages by histological methods, issues relating to sample size and statistical analysis must be addressed before it is possible to determine the effect of recent findings on our interpretations from reference standards of dental development.

Discussion

This chapter provides an overview and assessment of developmental standards for chimpanzee tooth calcification, primarily the more recent studies using radiographic methods. Available standards would benefit from improved documentation of both crown initiation and crown completion for most teeth, calcification of the third molar overall, and clarification of relative relationships in crown calcification between adjacent molars. Improved data will help to clarify variation in features and processes used to make interspecific distinctions in patterns of tooth calcification, including those for fossil hominids. Also, as noted by Reid et al. (1998a), maxillary dentition developmental standards do not exist, since radiographic studies are generally restricted to the mandible because of geometric complications in obtaining and analyzing radiographs of maxillary teeth. Finally, studies incorporating deciduous or mixed dentitions, or both, are sorely needed, as previous studies involving only the permanent dentition indicate that interspecific differences between apes and humans (Smith 1991a) and between different fossil hominid species (Conroy and Kuykendall 1995) are more clearly discriminated at later stages of development.

The observed differences notwithstanding, one of the most striking conclusions from comparisons between radiographic studies documenting permanent tooth calcification in chimpanzees is the overall similarity of results. Despite differences in sample size and structure, method of scoring radiographs, and statistical analysis, the published population estimates from different samples correspond broadly, and Figure 13.1 remains an accurate representation of the schedule of chimpanzee dental development for radiographic assessment. The problems remaining do not apply as much to issues of timing as to detection of reliable patterns for discrimination between closely related taxa. We are fortunate that dental development is robust in defending against developmental as well as methodological insults!

Beynon et al. (1998a) and Reid et al. (1998a) discussed the differences between histological and radiographic standards for the timing of chimpanzee tooth calcification, resulting from more accurate detection of calcified tissue by histological techniques. Beynon et al. (1998a) also clearly identified several important factors in radiographic assessment that hinder the precision of such studies, particularly the complex relationship between tooth geometry, mineral composition of the different tissues, and differences in subject contrast on radiographs due to the resultant variations in the intensity of radiation transmission. If such technical aspects of radiographic assessment for dental development could become more standardized, comparisons between studies would be more reliable (also see Garn et al. 1957; Dahlberg and Menegaz-Bock 1958; Bailit 1976; Demirjian 1986; Smith 1991b).

Because of these problems, Beynon et al. (1998a) and Reid et al. (1998a, 1998b) gave straightforward warnings against direct comparison of histological versus radiological data, echoing those of previous researchers (Beynon et al. 1991; Chandrasekera et al. 1993; Winkler 1995; Winkler et al. 1996). In addition, Beynon et al. (1998a) noted the distinction between radiographic studies that document dental developmental patterns and sequences to infer chronological information and histological studies analyzing incremental markers in dental tissues to establish developmental chronologies directly. As they stated (Beynon et al. 1998a, 352), "the two methods are diametrically opposed, in one case with individuals being referred to known populations, and in the other case inferring population standards from individuals"; histological and radiographic methods produce "fundamentally dissimilar results" (Beynon et al. 1998a, 66). In effect, comparisons between radiological and histological age estimates for tooth calcification events are subject to conceptual limitations similar to those governing comparisons of any developmental phenomenon (e.g., skeletal ossification, sexual maturity) with chronological age. Radiological studies of modern chimp or human tooth calcification are reference standards against which developmental ages of individuals are to be compared, and histological methods produce data thought to approach a true chronology of tooth calcification events in the individuals assessed.

Finally, histological assessment of calcified tissues produces information about successive events that are long past—for initiation of crown formation in some teeth, these events occurred prenatally. In this sense, histological studies produce longitudinal data (Dean et al. 1993), while radiographic assessment can only document the developmental status of an individual at one specific point in time (obviously, longitudinal data are produced by repeated observations over long periods; see, e.g., Anemone, Chap. 12). Since most radiographic studies available do not include prenatal or neonatal individuals, the age estimates produced are biased toward later periods of development.

The complexity of this situation is enhanced by the fact that studies of dental development in fossil hominids, which are ultimately the reason for these studies in modern species, have of necessity relied on an opportunistic strategy examining either gross or microstructural features of developing dentition via some combination of direct observation, radiography, various forms of microscopy, computed tomography, and histology (e.g., Beynon and Dean 1988; Bromage and Dean 1985; Conroy and Vannier 1991a, 1991b; Dean 1987; Koski and Garn 1957; Mann 1975; Mann et al. 1990a, 1990b; Wallace 1977; Zuckerman 1928). At present, such studies are ultimately reliant upon radiographically derived reference standards of dental development to interpret ages at death, the timing or pattern of tooth calcification stages, and other characteristics of extinct hominid dental development.

The general implication of studies like Smith (1991b), Beynon et al. (1998a), and Reid et al. (1998a) is that the presumed compatibility of data obtained by different means, such as histology versus conventional radiography and computed tomography, or even those using differing samples consisting of living versus skeletal versus fossilized specimens must be demonstrated experimentally before comparative results are accepted wholesale. It is necessary to document the magnitude and patterns of error resulting from methodological considerations, in addition to naturally occurring variation among individuals and populations.

Despite the problems discussed, basic differences in the timing and pattern of tooth development do characterize ape, human, and fossil hominid species. To proceed to new questions about ontogenetic, functional, and ultimately adaptive considerations of early hominid life history, one must first standardize the comparative methods for data collection so that existing dental developmental standards for extant hominoids and extinct hominids can be verified and refined.

Acknowledgments

I express my great appreciation to Nancy Minugh-Purvis and Ken McNamara for their invitation to contribute to this volume and especially for their good humor and patience above and beyond the call of duty while waiting for my manuscript to arrive. In addition, I thank Jacopo Moggi-Cecchi and Prof. J. C. Allan for their comments on the early draft and Gary Schwartz for e-mail discussions about this chapter. Finally, some of the ideas in this chapter were planted at the Paris International Workshop, "Enamel Structure and Development: Its Application in Hominid Evolution and Taxonomy," convened by Fernando Ramirez Rozzi, and I thank all the participants for their stimulating discussions.

References

Aiello, L., and Dean, M.C. 1990. *Introduction to Human Evolutionary Anatomy.* New York: Academic Press.

Anemone, R.L. 1995. Dental development in chimpanzees of known chronological age: implications for understanding the age at death of Plio-Pleistocene hominids. In J. Moggi-Cecchi (ed.), *Aspects of Dental Biology: Palaeontology, Anthropology, and Evolution,* 201–215. Florence: International Institute for the Study of Man.

Anemone, R.L., Mooney, M.P., and Siegel, M.I. 1996. Longitudinal study of dental development in chimpanzees of known chronological age: implications for understanding the age at death of Plio-Pleistocene hominids. *American Journal of Physical Anthropology* 99:119–133.

Anemone, R.L., and Watts, E.S. 1992. Dental development in apes and humans: a comment on Simpson, Lovejoy and Meindl (1990). *Journal of Human Evolution* 22:149–153.

Anemone, R.L., Watts, E.S., and Swindler, D.S. 1991. Dental development of known-age chimpanzees, *Pan troglodytes* (Primates, Pongidae). *American Journal of Physical Anthropology* 86:229–241.

Bailit, H.L. 1976. Variation in tooth eruption: a field guide. In E. Jiles and J.S. Friedlander (eds.), *The Measures of Man: Methodologies in Biological Anthropology,* 321–336. New Haven: Peabody Museum Press.

Beynon, A.D., Clayton, C.B., Ramirez Rozzi, F.V., and Reid, D.J. 1998a. Radiographic and histological methodologies in estimating the chronology of crown development in modern humans and great apes: a review, with some applications for studies on juvenile hominids. *Journal of Human Evolution* 35: 351–370.

Beynon, A.D., and Dean, M.C. 1988. Distinct dental development patterns in early fossil hominids. *Nature* 335:509–514.

Beynon, A.D., Dean, M.C., Leakey, M.G., Reid, D.J., and Walker, A. 1998b. Comparative dental development and microstructure of *Proconsul* teeth from Rusinga Island, Kenya. *Journal of Human Evolution* 35:163–209.

Beynon, A.D., Dean, M.C., and Reid, D.J. 1991. Histological study on the chronology of the developing dentition in gorilla and orang-utan. *American Journal of Physical Anthropology* 86:189–203.

Binford, L.R. 1985. Human ancestors: changing views of their behaviour. *Journal of Anthropological Archaeology* 4:292–327.

Bromage, T.G. 1987. The biological and chronological maturation of early hominids. *Journal of Human Evolution* 16:257–272.

Bromage, T.G., and Dean, M.C. 1985. Re-evaluation of the age at death of immature fossil hominids. *Nature* 317:525–527.

Broom, R., and Robinson, J.T. 1952. The Swartkrans ape-men *Paranthropus crassidens. Transvaal Museum Memoirs* 6.

Chandrasekera, M.S., Reid, D.J., and Beynon, A.D. 1993. Dental development in chimpanzee (*Pan trogolodytes*). *Journal of Dental Research* 72:729.

Clements, E.M.B., and Zuckerman, S. 1953. The order of eruption of the permanent teeth in the Hominoidea. *American Journal of Physical Anthropology* 11: 313–337.

Conroy, G.C. 1997. *Reconstructing Human Origins.* New York: W. W. Norton.

Conroy, G.C., and Kuykendall, K.L. 1995. Paleopediatrics: or when did human infants really become human? *American Journal of Physical Anthropology* 98: 121–131.

Conroy, G.C., and Mahoney, C.J. 1991. Mixed longitudinal study of dental emergence in the chimpanzee, *Pan trogolodytes* (Primates, Pongidae). *American Journal of Physical Anthropology* 86:243–254.

Conroy, G.C., and Vannier, M.W. 1991a. Dental development in South African Australopithecines: I. Problems of pattern and chronology. *American Journal of Physical Anthropology* 86:121–136.

———. 1991b. Dental development in South African Australopithecines: II. Dental stage assessment. *American Journal of Physical Anthropology* 86:121–156.

Dahlberg, A.A., and Menegaz-Bock, R. 1958. Emergence of the permanent teeth in Pima Indian children. *Journal of Dental Research* 37:1123–1140.

Dart, R.A. 1925. *Australopithecus africanus:* the man-ape of South Africa. *Nature* 115:195–199.

———. 1948. The adolescent mandible of *Australopithecus prometheus. American Journal of Physical Anthropology* 6:391–411.

Dean, M.C. 1987. Growth layers and incremental markings in hard tissues: a review of the literature and some preliminary observations about enamel structure in *Paranthropus boisei. Journal of Human Evolution* 16:157–172.

———. 1989. The developing dentition and tooth structure in hominoids. *Folia Primatologia* 53:160–176.

Dean, M.C., Beynon, A.D., Reid, D.J., and Whittaker, D.K. 1993. A longitudinal study of tooth growth in a single individual based on long- and short-period incremental markings in dentine and enamel. *International Journal of Osteoarchaeology* 3:249–264.

Dean, M.C., and Shellis, R.P. 1998. Observations on stria morphology in the lateral enamel of *Pongo, Hylobates* and *Proconsul* teeth. *Journal of Human Evolution* 35:401–410.

Dean, M.C., and Wood, B.A. 1981. Developing pongid dentition and its use for ageing individual crania in comparative cross-sectional growth studies. *Folia Primatologia* 36:111–127.

Demirjian, A. 1986. Dentition. In F. Falkner and J.M. Tanner (eds.), *Human Growth: A Comprehensive Treatise,* 269–295. New York: Plenum Press.

Demirjian, A., Goldstein, H., and Tanner, J.M. 1973. A new system of dental age assessment. *Human Biology* 45:211–227.

Dirks, W. 1998. Histological reconstruction of dental development and age at death in a juvenile gibbon (*Hylobates lar*). *Journal of Human Evolution* 35: 411–425.

Falk, D. 1987. Hominid paleoneurology. *Annual Review of Anthropology* 16:13–30.

Fanning, E.A., and Moorrees, C.F.A. 1969. A comparison of permanent mandibular molar formation in Australian Aborigines and Caucasoids. *Archives of Oral Biology* 14:999–1006.

Fitzgerald, C.M. 1998. Do enamel microstructures have regular time dependency? Conclusions from the literature and a large-scale study. *Journal of Human Evolution* 35:371–386.

Garn, S.M., Koski, K., and Lewis, A.B. 1957. Problems in determining the tooth eruption sequence in fossil and modern man. *American Journal of Physical Anthropology* 15:313–331.

Harvey, P.H., and Clutton-Brock, T.H. 1985. Life-history variation in primates. *Evolution* 39:559–581.

Holloway, R.L. 1972. Australopithecine endocasts, brain evolution in the Hominoidea, and a model of hominid evolution. In R.H. Tuttle (ed.), *The Functional and Evolutionary Biology of Primates,* 185–203. Chicago: Aldine.

Johanson, D., and Edgar, B. 1996. *From Lucy to Language.* London: Weidenfeld and Nicolson.

Koski, K., and Garn, S.M. 1957. Tooth eruption sequences in fossil and modern man. *American Journal of Physical Anthropology* 15:469–488.

Kraemer, H.C., Horvat, J.R., Doering, C., and McGinnis, P.R. 1982. Male chimpanzee development focusing on adolescence: integration of behavioral with physiological changes. *Primates* 23:393–405.

Kuykendall, K.L. 1992. Dental development in chimpanzees (*Pan troglodytes*) and implications for dental development patterns in fossil hominids. Ph.D. diss., Washington University.

———. 1996. Dental development in chimpanzees (*Pan troglodytes*): the timing of tooth calcification stages. *American Journal of Physical Anthropology* 99: 135–157.

Kuykendall, K.L., and Conroy, G.C. 1996. Permanent tooth calcification in chimpanzees (*Pan troglodytes*): patterns and polymorphisms. *American Journal of Physical Anthropology* 99:159–174.

Kuykendall, K.L., Mahoney, C.J., and Conroy, G.C. 1992. Probit and survival analysis of tooth emergence ages in a mixed-longitudinal sample of chimpanzees (*Pan troglodytes*). *American Journal of Physical Anthropology* 89:379–399.

Lancaster, J.B. 1975. *Primate Behaviour and the Emergence of Human Culture.* New York: Holt, Rinehart and Winston.

Le Gros Clarke, W.E. 1967. *Man-apes or Ape-men?* New York: Holt, Rinehart and Winston.

Macho, G.A., and Wood, B.A. 1995. The role of time and timing in hominid dental evolution. *Evolutionary Anthropology* 4:17–31.

Mann, A.E. 1975. *Paleodemographic Aspects of the South African Australopithecines.* Philadelphia: University of Pennsylvania Publications in Anthropology, No. 1.

———. 1988. The nature of Taung dental maturation. *Nature* 333:123.

Mann, A.E., Lampl, M., and Monge, J.M. 1987. Maturational patterns in early hominids. *Nature* 328:673–674.

———. 1990a. Patterns of ontogeny in human evolution: evidence from dental development. *Yearbook of Physical Anthropology* 33:111–150.

Mann, A.E., Monge, J.M., and Lampl, M. 1990b. Dental caution. *Nature* 348:202.

———. 1991. Investigation into the relationship between perikymata counts and crown formation times. *American Journal of Physical Anthropology* 86:175–188.

Miles, A.E.W. 1963. The dentition in the assessment of individual age in skeletal material. In D.R. Brothewell (ed.), *Dental Anthropology*, 191–209. New York: Pergamon Press.

Moggi-Cecchi, J. 2001. Patterns of dental development of *Australopithecus africanus*, with some inferences on their evolution with the origin of the genus *Homo*. In P.V. Tobias, M.A. Raath, J. Moggi-Cecchi, and G.A. Doyle (eds.), *Humanity from African Naissance to Coming Millennia*, Coloquia in Human Biology and Paleoanthropology, 125–133. Firenze, Italy: Firenze University Press; Johannesburg: Witwatersrand University Press.

Moggi-Cecchi, J., Tobias, P.V., and Beynon, A.D. 1998. The mixed dentition and associated skull fragments of a juvenile fossil hominid from Sterkfontein, South Africa. *American Journal of Physical Anthropology* 106:425–465.

Moorrees, C.F.A. 1959. *The Dentition of the Growing Child: A Longitudinal Study of Dental Development between 3 and 18 Years of Age.* Cambridge: Harvard University Press.

Potts, R. 1984. Hominid ecology? Problems of identifying the earliest hunter/ gatherers. In R. Foley (ed.), *Hominid Evolution and Community Ecology*, 129–166. New York: Academic Press.

Reid, D.J., Beynon, A.D., and Ramirez Rozzi, F.V. 1998b. Histological reconstruction of dental development in four individuals from a medieval site in Picardie, France. *Journal of Human Evolution* 35:463–477.

Reid, D.J., Schwartz, G.T., Dean, C., and Chandrasekera, M.S. 1998a. A histological reconstruction of dental development in the common chimpanzee, *Pan troglodytes. Journal of Human Evolution* 35:427–448.

Sacher, G.A. 1975. Maturation and longevity in relation to cranial capacity in hominid evolution. In R.H. Tuttle (ed.), *Primate Functional Morphology and Evolution*, 417–441. The Hague: Mouton.

Schultz, A. 1935. Eruption and decay of the permanent teeth in primates. *American Journal of Physical Anthropology* 19:489–581.

Schwartz, G.T., and Kuykendall, K.L. 1996. Enamel structure and development. *Evolutionary Anthropology* 5:150–151.

Shipman, P. 1986. Scavenging or hunting in early hominids: theoretical framework and tests. *American Anthropologist* 88:27–43.

Simpson, S.W., and Kunos, C.A. 1998. A radiographic study of the development of the human mandibular dentition. *Journal of Human Evolution* 35:479–505.

Simpson, S.W., Lovejoy, C.O., and Meindl, R.S. 1992. Further evidence on relative dental maturation and somatic developmental rate in hominoids. *American Journal of Physical Anthropology* 87:29–38.

Smith, B.H. 1989a. Dental development as a measure of life history in primates. *Evolution* 43:683–688.

———. 1989b. Growth and development and its significance for early hominid behavior. *Ossa* 14:63–96.

———. 1991a. Dental development and the evolution of life history in Hominidae. *American Journal of Physical Anthropology* 86:157–174.

———. 1991b. Standards of human tooth formation and dental age assessment. In M.A. Kelley and C.S. Larsen (eds.), *Advances in Dental Anthropology*, 143–168. New York: Wiley-Liss.

———. 1992. Life history and the evolution of human maturation. *Evolutionary Anthropology* 1:134–142.

———. 1993. The physiological age of KNM-WT 15000. In A. Walker and R. Leakey (eds.), *The Nariokotome* Homo erectus *Skeleton*, 195–220. Cambridge: Harvard University Press.

———. 1994. Sequence of emergence of the permanent teeth in *Macaca, Pan, Homo*, and *Australopithecus:* its evolutionary significance. *American Journal of Human Biology* 6:61–76.

Smith, B.H., Crummett, T.L., and Brandt, K.L. 1994. Ages of eruption of primate teeth: a compendium for aging individuals and comparing life histories. *Yearbook of Physical Anthropology* 37:177–231.

Smith, B.H., and Tompkins, R.L. 1995. Toward a life history of the hominidae. *Annual Review of Anthropology* 24:257–279.

Smith, R.J., Gannon, P.J., and Smith, B.H. 1995. Ontogeny of australopithecines and early *Homo:* evidence from cranial capacity and dental eruption. *Journal of Human Evolution* 29:155–168.

Swindler, D. 1985. Nonhuman primate dental development and its relationship to human dental development. In E.S. Watts (ed.), *Nonhuman Primate Models for Human Growth and Development*, 67–94. New York: Alan R. Liss.

Walker, A., and Shipman, P. 1996. *The Wisdom of the Bones: In Search of Human Origins.* London: Weidenfeld and Nicolson.

Wallace, J.A. 1977. Gingival eruption sequences of permanent teeth in early hominids. *American Journal of Physical Anthropology* 46:483–494.

Washburn, S.L. 1960. Tools and human evolution. *Scientific American* 203:63–75.

Winkler, L.A. 1995. A comparison of radiographic and anatomical evidence of tooth development in infant apes. *Folia Primatologia* 65:1–13.

Winkler, L.A., Schwartz, J.H., and Swindler, D.R. 1996. Development of the orangutan permanent dentition: assessing patterns and variation in tooth development. *American Journal of Physical Anthropology* 99:205–220.

Wolpoff, M., Monge, J., and Lampl, M. 1988. Was Taung human or an ape? *Nature* 335:501.

Zuckerman, S. 1928. Age-changes in the chimpanzee, with special reference to growth of brain, eruption of teeth, and estimation of age; with a note on the Taungs ape. *Proceedings of the Zoological Society of London* 1928:1–42.

Chapter 14

Evolutionary Relationships between Molar Eruption and Cognitive Development in Anthropoid Primates

SUE TAYLOR PARKER

Following in the footsteps of evolutionary biologists, biological anthropologists have shown an increasing interest in the evolution of primate life histories. Life-history theory is particularly relevant to understanding the evolution of primate intelligence because large brains and higher mental processes correlate with the prolonged gestation, immaturity, and life span characteristic of anthropoid and especially hominoid primates. Therefore, if we are to understand the evolution of hominid intelligence, we must place comparative studies of intellectual development in monkeys, apes, and hominids in the context of life-history data.

Comparative data on the life history and intellectual abilities of living monkeys and apes are prerequisite for reconstructing the evolution of hominid intelligence. Equally important are comparative data on the life history and intellectual abilities of fossil hominid species. Although behavior does not fossilize, it can be inferred from morphological and archeological remains of some hominid species. Tool kits of species of *Homo*, for example, provide some indications of the mental abilities of their makers. Life histories can be inferred from comparative data on molar eruption sequences and enamel growth patterns of fossil species that include immature forms.

In this chapter, I briefly review comparative data on life history as measured by molar eruption and on cognitive development as measured by Piagetian stages. The data show a pattern of sequential addition of new stages at the end of the cognitive development in a series of hominoid ancestors. This pattern of sequential addition in the cognitive realm is associated with a pattern of sequential delay and extension of the period of molar development in these species.

Evolutionary reconstruction of this kind requires a variety of techniques. First, it requires an internally consistent model for identifying relevant character states, in this case cognitive abilities and life-history patterns. Second, it requires a systematic method for discovering the ancestry of new characteristic states. Third, it requires a framework for interpreting the changes in developmental patterns that have occurred during evolution.

Piagetian theory provides the scale for distinguishing various levels of cognitive development. Molar eruption sequence provides the framework marking such life-history variables as age at weaning and age at sexual maturity in living and fossil species. Cladistic theory provides the method for reconstructing the evolution of new developmental patterns. Finally, evolutionary developmental biology provides the model for classifying the processes that generate new developmental patterns.

Cladistics provides a systematic procedure for phylogenetic reconstruction of the common ancestry of a given pattern from comparative data on closely related species. The procedure involves mapping these comparative data onto independently derived phylogenies of the focal species (known as the *ingroup*) and their close relatives (known as the *outgroup*) (Brooks and McLennan 1991). This procedure reveals which character states are unique to the ingroup (shared derived characters) and which are shared with the larger outgroup. Only shared derived character states indicate new adaptations (Coddington 1988).

Fortunately, an independently derived phylogeny is available for anthropoid primates. Phylogenetic relationships among families and superfamilies in this suborder are well established (Maddison and Maddison 1992). Considerable controversy, however, surrounds the question of the branching relations among African apes and the hominids. The phylogeny in Figure 14.1 assumes that chimpanzees and humans shared the last common ancestor (Ruvolo 1994).

Developmental evolutionary biology provides a system for classifying processes and outcomes of evolutionary changes in the timing of development. These heterochronic changes in rates and timing are (1) acceleration or deceleration of developmental rates, (2) earlier or later onset of

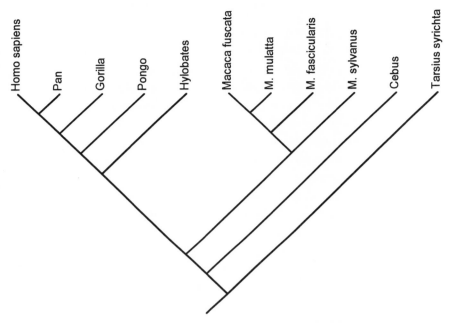

Fig. 14.1. Primate phylogeny. *Source:* After Maddison and Maddison 1992.

development, and (3) earlier or later offset of development. Later onset, earlier offset, and decelerated rates relative to the ancestor produce juvenilization, or paedomorphosis, of the descendant. Earlier onset, later offset, and accelerated developmental rates produce adultification, or peramorphosis (see Fig. 4.1). See McKinney and McNamara (1991) and Alba (Chap. 2) for a detailed discussion of these processes and their outcomes.

Together, the cladistic and evolutionary developmental frameworks provide the analytic tools for reconstructing the evolution of human life history and cognitive development from comparative data on patterns of cognitive and molar development. Developmental patterns used in such an analysis are known as *ontogenetic characters*. Reconstructing the evolution of ontogenetic characters as opposed to nondevelopmental, or *simultaneous, characters* demands unitary treatment of the relevant developmental sequence of transformations in each taxonomic group (de Queiroz 1985; Mabee and Humphries 1993). Ontogenetic sequences are specified below for molar development and cognitive development.

A Review of Comparative Data on the Evolution of Life History in Anthropoids

Dental development is a prime measure of somatic maturation. Data on molar tooth development are particularly useful for comparative studies of life history because it is highly correlated with brain size. Brain size is a pacemaker of life history in mammals, which is often difficult to measure. Molars are doubly useful because they are accessible for both living and fossil species. Since molars fossilize more frequently than brains do, they offer valuable surrogate measures of brain size in early hominids. For these reasons and because comprehensive data on molar tooth development are available for many primate species, this is the measure of life history I use in this chapter.

Smith and her colleagues (e.g., Smith 1993; Smith et al. 1994; Smith et al. 1995; Anemone, Chap. 12, this volume) provide comprehensive data on dental development in monkeys, great apes, humans, and some early hominids. The eruption of the three molar teeth mark, respectively, the end of infancy, the end of childhood, and the achievement of full maturity (Smith 1993). Table 14.1 shows a comparison of the (mandibular) molar eruption sequences for macaque monkeys, chimpanzees (*Pan*), humans, and *Homo erectus*. (Because the data are similar for great apes, I present only the data on chimpanzees.) These developmental patterns constitute the ontogenetic characters for anthropoid molar development. In other words, each developmental pattern is characteristic of a particular taxonomic group and originated in its ancestor. See Table 14.2 for a summary of these ontogenetic characters.

The developmental sequence begins about two years earlier in macaques than in chimpanzees and occurs at approximately two-year intervals. The sequence in chimpanzees begins about two years later than that in macaques and occurs at approximately three-year intervals. The sequence in

Table 14.1 Molar Eruption Sequences (in years) in Selected Primates

Species	Last Deciduous	Molar 1	Molar 2	Molar 3
Macaca mulatta	0.44	1.36	3.2	5.5
Pan troglodytes	1.2	3.15	6.5	10.5
Homo erectus	—	4.5	9.5	14.5
Homo sapiens	2.3	5.4	12.5	18

Source: Data from Smith et al. 1994; Smith 1992, 1994.

Table 14.2 Ontogenetic Characters for Molar Development

Taxon	Ontogeny (Age in Years)
H. sapiens	M1 (5.4)-----M2 (12.5)-----M3 (18)
H. erectus	M1 (4.5)----M2 (9.5)----M3 (14.5)
Pan	M1 (3.15)---M2 (6.5)---M3 (10.5)
Macaca	M1 (1.36)-M2 (3.2)-M3 (5.5)

humans begins about two years later than that in chimpanzees and occurs in approximately six-year intervals.

Molar development in *H. erectus* apparently began about one year later than that in chimpanzees and one year earlier than that in humans and occurred at approximately five-year intervals (Smith 1993). In other words, in a sequence from macaque to chimpanzee to human, the timing of molar eruption in each of these species is relatively delayed (Fig. 14.2). As I men-

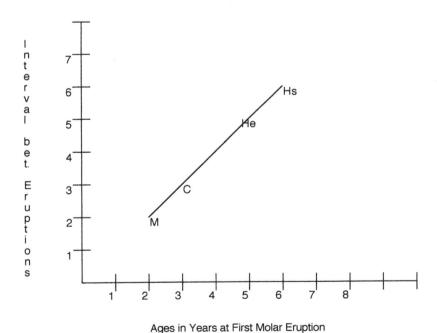

Fig. 14.2. Deceleration in rate of molar development from macaque (*M*) to chimpanzee (*C*) to human (*He, Homo erectus; Hs, Homo sapiens*); postdisplacement in later onset of molar development; sequential hypermorphosis in later offset of molar development.

tioned previously, the pace of molar development is systematically related to brain size. Increasing age of molar eruption correlates strongly with increasing brain size (0.98) (Smith et al. 1995). Such a pattern of heterochrony has been called *sequential hypermorphosis* (McKinney and McNamara 1991; McNamara, Chap. 5, this volume), reflecting its iteratively delayed offset times. It could also be called *sequential postdisplacement,* reflecting its later onset times. Either way it is a sequential delay in the time of expression in the descendant of an ancestral developmental event.

A Review of Comparative Data on the Evolution of Cognitive Development in Anthropoid Primates

Stages of cognitive development in human infants and children provide useful yardsticks for comparing the highest levels of intellectual ability in monkeys, great apes, and humans. Piagetian and neopiagetian stages are particularly useful because their sequential epigenetic nature provides the rationale for ranking abilities. Table 14.3 shows a summary of Piaget's major developmental periods (Piaget and Inhelder 1969). Piagetian and neopiagetian stages provide a systematic framework for comparing cognitive development in social and nonsocial domains in anthropoid species (Parker 1990; Parker and McKinney 1999). Although such studies are limited in number, they provide some clear data. Specifically, they reveal species differences in sequences of cognitive development in macaques, great apes, and humans. These patterns are summarized in Table 14.4 and Figure 14.3.

Human children traverse four periods and seven subperiods of cognitive development, in several domains: early and late sensorimotor period development; early and late preoperational period development; early and late operational period development; and formal operations period development. Great apes traverse only three of these subperiods of cognitive development: early sensorimotor period development, from birth to almost 12 months; late sensorimotor period development, from 12 through about 40 months; and early preoperational period (symbolic) development, from 40 to about 96 months. Macaque monkeys traverse just the first subperiod, the early sensorimotor period, from birth to about 5 months, and partially traverse the late sensorimotor period, from about 5 months to some unknown time.

These patterns constitute the ontogenetic characters for cognitive development in these anthropoid primates. In other words, each pattern is

Table 14.3 Piaget's Periods and Subperiods of Cognitive Development

Periods and Subperiods	Physical Knowledge	Logicomathematic Knowledge	Interpersonal Knowledge
Sensorimotor Period			
Early subperiod (stages 1–4): birth to ⅔ yr	discovery of practical properties of objects, space, time, & causality		discovery of personal efficacy through circular reactions
Late subperiod (stages 5 & 6): ⅔ to 2 yr	trial & error groping toward goals; insight	construction of first order relations among objects	imitation of novel schemes; deferred imitation of same
Preoperations Period			
Symbolic subperiod: 2–4 yr	symbolic representation of objects & actions	formation of second-order relations among objects	construction of new routines through symbolic play
Intuitive subperiod: 4–6 yr	symbolic representation of events	construction of nonreversible classes	construction of social roles through symbolic play
Concrete Operations Period			
Early subperiod: 6–9 yr	anticipation and reconstruction of reversible physical phenomena	construction of reversible classes and relations	construction of locally applicable rules
Late subperiod: 9–12 yr	understanding of mediated causes	construction of multiplicative classes	construction of mutually agreed rules
Formal Operations Period			
Early subperiod: 12–15 yr	measurement and testing of variables	construction of hypothetical classes & relations	understanding of universality of rules

characteristic of a particular taxonomic group and originated in its ancestor. Summaries of these ontogenetic characters is given in Table 14.5 and Figure 14.4.

Examination of these data reveal the following patterns. First, all these anthropoid species traverse the same sequence of stages and subperiods up to the terminal level for their group. Macaques reach their highest level in the fourth and fifth stages of the sensorimotor period. Chimpanzees complete all six sensorimotor stages and reach their highest level in the symbolic subperiod of preoperations. Humans reach their highest level in for-

Table 14.4 Subperiods (in Years) of Cognitive Development in Selected Primates

Subperiod[a]	Macaca	Pan	Homo erectus	Homo sapiens
ESM	0.5	1	1	1
LSM	2.5? (incomplete)	3.5	3	2
EPO		8.5	6	4
LPO			9	6
ECO			14	9
LCO				12
FO				16

[a]ESM, Early Sensorimotor Period; LSM, Late Sensorimotor Period; EPO, Early Preoperations; LPO, Late Preoperations; ECO, Early Concrete Operations; LCO, Late Concrete Operations; FO, Formal Operations.

mal operations. Second, macaques complete the sensorimotor period earlier than do great apes or humans. In contrast, great apes complete the late sensorimotor period later than do human infants. They also begin and end the symbolic subperiod of preoperations later than do human children. Although not apparent from these data, macaques and great apes show asyn-

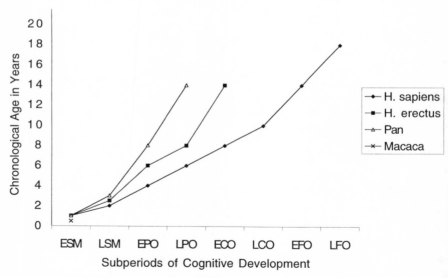

Fig. 14.3. Chart of primate comparative cognitive development. *ESM*, Early Sensorimotor; *LSM*, Late Sensorimotor; *EPO*, Early Preoperations; *LPO*, Late Preoperations; *ECO*, Early Concrete Operations; *LCO*, Late Concrete Operations; *EFO*, Early Formal Operations; *LFO*, Late Formal Operations.

Table 14.5 Ontogentic Characters for Cognitive Development in Anthropoid Primates

Taxon	Ontogeny
Humans	ESM-LSM-EPO-LPO-ECO-LCO-FO
Great apes	ESM-LSM-EPO
Macaques	ESM-LSM*

Note: ESM, Early Sensorimotor Subperiod; LSM, Late Sensorimotor Subperiod; EPO, Symbolic Subperiod of Preoperations; LPO, Intuitive Subperiod of Preoperations; ECO, Early Concrete Operations; LCO, Late Concrete Operations; FO, Formal Operations.
*Incomplete.

chronies or temporal displacements among various cognitive domains that develop synchronously in humans. Specifically, they develop the object concept and space earlier than other sensorimotor series.

These ontogenetic patterns reveal that, since the divergence of chim-

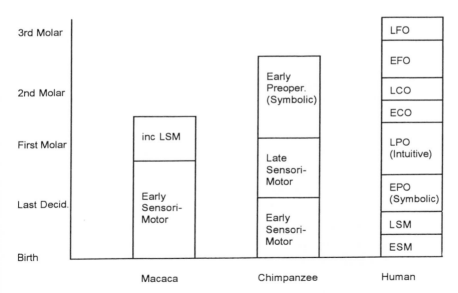

Fig. 14.4. Cognitive development in relation to molar maturation in selected primates. *Decid.*, deciduous; *inc*, incomplete; *ESM*, Early Sensorimotor; *LSM*, Late Sensorimotor; *EPO*, Early Preoperations; *LPO*, Late Preoperations; *ECO*, Early Concrete Operations; *LCO*, Late Concrete Operations; *EFO*, Early Formal Operations; *LFO*, Late Formal Operations.

panzees and humans, human ancestors have added four new subperiods of cognitive development, namely, late preoperations, early and late concrete operations, and formal operations. Analysis of stone tools associated with late *H. erectus* suggests that this species may have displayed early concrete operational cognition at least in the domain of spatial reasoning (Wynn 1989). In terms of heterochrony, this pattern is one of peramorphosis through terminal addition. Such a pattern of heterochrony results in developmental recapitulation.

These patterns also reveal that humans have earlier onset and offset of preoperations than do great apes. Consequently, these periods of development are of shorter duration and occur more rapidly than those of the great apes. In terms of developmental evolutionary biology, humans show classic peramorphic patterns of predisplacement and acceleration in the arena of cognitive development.

Summary and Conclusions

Examination of cognitive development in relation to molar development in monkeys, great apes, and humans reveals that the longer the period of molar development in a species, the greater the number of cognitive subperiods that are traversed. Humans, who begin the molar eruption sequence at about 5 years and end at about 18 years, traverse seven subperiods of cognitive development. Great apes, who begin molar eruption at 3 years and end at 10 years, traverse three cognitive subperiods. Macaques, who begin molar development at less then 1.5 years and end at 5.5 years, traverse only one cognitive subperiod. Preliminary analysis suggests that *H. erectus*, who began molar eruption at about 4.5 years and ended at 14.5 years, traversed five subperiods of cognitive development.

Relating these patterns back to life-history patterns revealed by molar eruption, we see the following. Macaques have come close to completing their cognitive development by the end of infancy, great apes come close to completing their cognitive development by the end of childhood, and humans complete their cognitive development only at the onset of full maturity (Smith 1993). In terms of evolutionary developmental biology, hominid cognitive development has evolved through terminal addition leading to developmental recapitulation of evolutionary stages. Given that, through the sequence macaque-great ape-hominid, the onset of each subperiod was earlier and therefore the period spent in each subperiod was shorter, this is an example of sequential progenesis, as each of the subperiods is terminated earlier. In combination with terminal hypermorphosis, this allowed more subperiods to be accommodated. Hominid somatic development, on the other hand, has evolved through sequential hypermorphosis leading to longer developmental stages.

References

Brooks, D., and McLennan, D. 1991. *Phylogeny, Ecology, and Behavior.* Chicago: University of Chicago Press.

Coddington, J.A. 1988. Cladistic tests of adaptational hypotheses. *Cladistics* 4:3–22.

de Queiroz, K. 1985. The ontogenetic method for determining character polarity and its relevance to phylogenetic systematics. *Systematic Zoology* 34:280–299.

Mabee, P., and Humphries, J. 1993. Coding polymorphic data: examples from allozymes and ontogeny. *Systematic Zoology* 42:166–181.

Maddison, D., and Maddison, W. 1992. *MacClade.* Sunderland, Mass.: Sinaur Associates.

McKinney, M.L., and McNamara, K.J. 1991. *Heterochrony: The Evolution of Ontogeny.* New York: Plenum.

Parker, S.T. 1990. The origins of comparative developmental evolutionary studies of primate mental abilities. In S.T. Parker and K.R. Gibson (eds.), *"Language" and Intelligence in Monkeys and Apes,* 3–74. Cambridge: Cambridge University Press.

Parker, S.T., and McKinney, M.L. 1999. *Origins of Intelligence: The Evolution of Cognitive Development in Monkeys, Apes, and Humans.* Baltimore: Johns Hopkins University Press.

Piaget, J., and Inhelder, B. 1969. *The Psychology of the Child.* New York: Basic Books.

Ruvolo, M. 1994. Molecular evolutionary processes and conflicting gene trees: the hominoid case. *American Journal of Physical Anthropology* 94:89–113.

Smith, B.H. 1992. Life history and the evolution of human maturation. *Evolutionary Anthropology* 1:134–142.

———. 1993. The physiological age of KNM-WT 15000. In A. Walker and R. Leakey (eds.), *The Nariokotome* Homo erectus *Skeleton,* 195–220. Cambridge: Harvard University Press.

———. 1994. Sequence of emergence of the permanent teeth in *Macaca, Pan, Homo,* and *Australopithecus:* its evolutionary significance. *American Journal of Human Biology* 6:61–76.

Smith, B.H., Crummett, T.L., and Brandt, K.L. 1994. Ages of eruption of primate teeth: a compendium for aging individuals and comparing life histories. *Yearbook of Physical Anthropology* 37:177–231.

Smith, R.J., Gannon, P.J., and Smith, B.H. 1995. Ontogeny of australopithecines and early *Homo* evidence from cranial capacity and dental eruption. *Journal of Human Evolution* 29:155–168.

Wynn, T. 1989. *The Evolution of Spatial Competence.* Urbana: University of Illinois Press.

Part III

The Evolution of Hominid Development

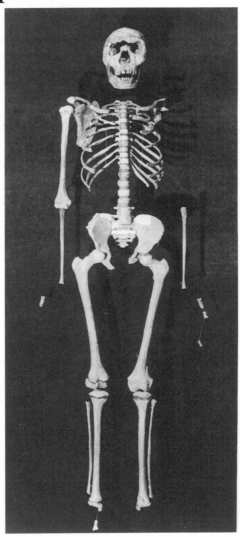

KNM-WT 15000, the skeleton of the eleven-year-old youth from Nariokotome, Kenya. This *Homo ergaster/erectus* specimen was discovered in 1984 on the west side of Lake Turkana. *Source:* Photograph courtesy of Dr. Alan Walker.

Chapter 15

Enamel Microstructure in Hominids

New Characteristics for a New Paradigm

FERNANDO RAMIREZ ROZZI

> A paradigm is not a particular scientific theory, but rather a scientific frame or "worldview." It is through a paradigm that an archaeologist perceives what sorts of things may be learned (and are worth learning) from the archaeological record and what sorts of evidence and analytical methods are relevant to learning them.
>
> Harrold 1991, 164

To change a paradigm implies the adoption of a new point of view, new methods, and new data and/or a new way to interpret evidence available to resolve new problems. Bromage et al. (1995, 105) called for a new paradigm in paleoanthropology. They defined it as an "ecocentric" paradigm: hominid fossils have to be studied in their ecological context, that is, from an ecological and biogeographical point of view. From this perspective, morphology (raw data from the fossil record) is not limited to the analysis of a final form but perceived as a final stage of a process of growth and development (i.e., ontogeny), which develops in an environmental context within a specific habitat (i.e., adaptation). An ecological and biogeographical context enables the necessary formulation of a paradigm before the emergence of modern evolutionary theories (Bowler 1989). From an ecological perspective, it is imperative to integrate hominid evolution within environmental evolution as a fundamental step toward understanding the history of hominids in a holistic evolutionary way.

Such an ecocentric paradigm needs data. It is advantageous if the characteristics analyzed show habitat specificity and give some information on histological/physiological processes and life-history patterns. Such characteristics are certainly almost exclusively genetic and would have, therefore, a taxonomic value. The dentition is part of the body capable of providing such data. Indeed, not only are teeth very abundant in the fossil record but the microstructure of dentine and enamel is the result of growth and development patterns minimally influenced by exogenic factors (Smith 1991a, 1991b). Because teeth participate in the provision of nourishment, they are under direct environmental pressures and therefore are prime targets of natural selection.

This chapter explores enamel microstructure traits that provide data needed by an ecocentric paradigm in hominid evolution. Three analyses of enamel microstructural characteristics of Plio-Pleistocene hominid teeth—dental histogenesis, life histories, and dental development patterns—are presented on the basis of three subsamples from the Omo Group in Ethiopia.

Dental Histogenesis

In paleoanthropology, dental development is usually analyzed using radiographic images. However, this method is often less than ideal, as it enables only a few aspects of dental development to be studied and with limited accuracy (Beynon et al. 1998; Macho and Wood 1995; Kuykendall, Chap. 13, this volume). Enamel (and dentine) microstructure analysis (e.g., histological methods), even if they also present some constraints, have become more widely used in dental development studies. They do not present the problems of radiographic studies and do lend themselves to the identification of several patterns of tooth development in both extant and extinct hominoids.[1]

Enamel is secreted in both lateral and longitudinal directions (Fig. 15.1). Three enamel microfeatures (Hunter Schreger bands, cross-striations, striae of Retzius) enable estimations of the timing of enamel formation (Fig. 15.2, Table 15.1) (Ramirez Rozzi 1997a). Hunter Schreger bands are zones in the enamel where prisms have a common orientation in contrast to those in adjacent zones (Boyde and Fortelius 1986). Their disposition is oblique or perpendicular to the enamel-dentine junction (EDJ), and their course is more or less curved in the vertical plane. The curvature of Hunter Schreger bands is related to the decussation of prisms, and their width gives an estimation of the size of a prism cluster (Table 15.1) (Beynon and Wood

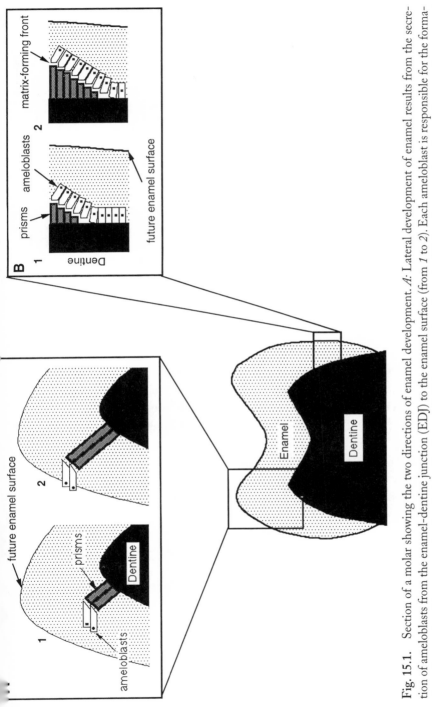

Fig. 15.1. Section of a molar showing the two directions of enamel development. *A*: Lateral development of enamel results from the secretion of ameloblasts from the enamel–dentine junction (EDJ) to the enamel surface (from *1* to *2*). Each ameloblast is responsible for the formation of one prism (the structural unit of mineralized enamel). Prisms run, thus, from the EDJ to the enamel surface, indicating the course of the ameloblast responsible for its production during enamel formation. *B*: Longitudinal development of enamel occurs from the dentine horn to the cervix and results in ameloblasts becoming active cervically. Newly active ameloblasts become part of the matrix-forming front of the enamel (from *1* to *2*).

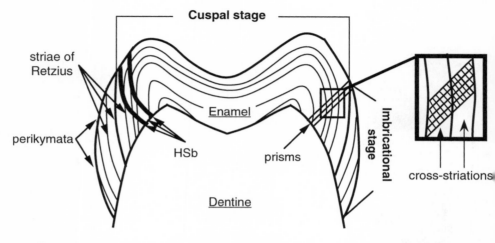

Fig. 15.2. Schematic section of a tooth showing incremental lines and possible Hunter Schreger bands (*HSb*) in the enamel. The crown can be divided into two stages according to the arrangement of striae of Retzius. The cuspal (appositional) stage comprises striae that have not yet reached the enamel surface and involves successive layers of appositional enamel. The imbricational stage shows the enamel layers in a tile-like fashion, with the striae having reached the enamel surface to produce a slight depression: the perikymata.

1986; Osborn 1968; Risnes 1986). Cross-striations, slight striations transverse to the axis of the enamel prism, result from a circadian variation in enamel secretion by ameloblasts (Risnes 1986; Shellis 1984; Dean 1987; Ramirez Rozzi 1992). The distance between successive cross-striations corresponds to the amount of enamel formed by an ameloblast in a day (the incremental or secretion rate of the enamel) (Table 15.1) (Beynon and Dean 1987). Striae of Retzius give a two-dimensional representation of successive and bent planes in the enamel (Risnes 1990, 1998). They are caused by repetitive, systematic, and regular modifications in the metabolic activity of ameloblasts or changes in ameloblast course, or both (Rose 1979). Striae of Retzius reflect approximately a circaseptan rhythm ranging from six to eleven days, depending on the individual (Beynon and Reid 1987; Dean 1987; Fitzgerald 1998; Shellis and Poole 1977). Therefore, cross-striations and striae of Retzius are incremental lines in the enamel with a circadian and approximately circaseptan periodicity. The total number of cross-striations and/or striae of Retzius thus enable estimation of the duration of crown formation.

Striae of Retzius correspond to successive advances of the enamel matrix forming-front (Fig. 15.1) (Baker 1972; Boyde 1964; Bromage 1991; Risnes

Table 15.1   Relations between Histological/Physiological Processes and Enamel Microstructure Traits

Histological/ Physiological Aspect	Enamel Microstructure	Method
Apposition of the enamel		
Incremental rate	cross-striations	distance between adjacent cross-striations
	striae of Retzius	distance between adjacent striae divided by 7 (6–11)
Pattern	Hunter Schreger bands	curvature angle with the EDJ
Extension rate of the enamel	striae of Retzius	angle with the EDJ distance between adjacent striae of Retzius along the EDJ
Time of crown formation	striae of Reztius	number of striae multiplied by 7 (6–11) and divided by 365
Active life of ameloblasts	cross-striations	number of cross-striations in a prism
	striae of Retzius	number of striae in a prism multiplied by 7 (6–11)
Number of active ameloblasts	striae of Retzius	stria's length
Activity of enamel organ		
Time	striae of Retzius	proportion of the cuspal stage in crown formation time
Type	striae of Retzius	proportion of the cuspal stage in dentine height
Shape of the enamel matrix forming-front	striae of Retzius	stria's course

Note: EDJ, enamel-dentine junction.

1990; Shellis 1984). The enamel matrix forming-front is produced and advances by the activation of newly matured ameloblasts. Since striae of Retzius present an approximately circaseptan periodicity, the distance between adjacent striae along the EDJ approximates the advancement of the enamel matrix forming-front in a week and is a measure of the extension rate of the enamel. Shellis (1984) suggested that the slope of striae reaching the EDJ gives an estimation of the velocity of the advancing matrix forming-front; the more open the slope of striae, the slower the advancement velocity and, hence, the lower the extension rate of the enamel (Table 15.1). Striae are more easily observed in inner enamel than where they contact the EDJ. Thus, measuring striae-EDJ angles is an easier way of estimating the extension rate. Although the slope of striae with the EDJ does not enable the value of the extension rate to be assessed, its variation from the cusp tip to the cervix indicates how the extension rate changes during crown formation. The study of striae-EDJ angles in extant hominoids has revealed that the extension rate, high at the first stage of crown formation, becomes lower and lower toward the cervix (Beynon and Reid 1987).

In this part of the investigation, the slope of striae with the EDJ has been measured to estimate the enamel extension rate in Plio-Pleistocene hominids, following Beynon and Wood (1986), Grine and Martin (1988), and Ramirez Rozzi (1993a). These authors assumed that the relationship between striae slope and the extension rate suggested by Shellis (1984; see Risnes 1998) from a study of modern human teeth was similar in extinct hominids. In fact, it has not been proven that this relationship is the same in Plio-Pleistocene hominid teeth. Nevertheless, since it is very difficult to measure enamel extension rates directly, it is very important to know whether they can be accurately estimated using the slope of striae. Furthermore, as any given enamel extension rate is related to crown completion, modifications of the extension rate during crown formation must be documented to understand patterns of tooth development in Plio-Pleistocene hominids.

Materials and Methods

The sedimentary sequence of the Omo Group in Ethiopia is continuous from 3.5 to 1.05 million years. Deltaic, lacustrine, and fluvial sediments are intercalated with layers of volcanic ash that have allowed workers to establish a stratigraphic sequence of Members and Submembers named from A (oldest) to L (youngest). An accurate chronology, well-known lithostratigraphy and biostratigraphy, well-recorded evolution of fauna and flora, and an abundance of hominid fossils make the Omo Group a reference site for the Plio-Pleistocene (e.g., Brown et al. 1965; Brown and Feibel 1988; Coppens 1972; Feibel et al. 1989; Harris et al. 1988).

A number of workers have documented changing faunal communities in East Africa during the Pliocene (Turner and Wood 1993). The replacement of species through time permits the identification of four faunistic associations in the Omo Group (Coppens 1972, 1975). These successive faunistic associations show increasing adaptations to open environments with low precipitation, indicating a shift in climatic conditions through time.

This is confirmed by paleoclimatic data (see Denton 1999). Members C and E correspond to periods of change in faunal communities (Coppens 1972, 1975).

In this portion of the study, the striae-EDJ angle and the distance between adjacent striae along the EDJ of twelve naturally broken Plio-Pleistocene hominid molars from the Omo Group (Table 15.2) (see Ramirez Rozzi 1997a), dating from between 3.36 and 2.1 million years, were analyzed. The angle of the striae with the EDJ was measured between the tangent to the EDJ and the tangent to the striae at their point of intersection (Fig. 15.3) (Beynon and Wood 1986). Angles were taken in the occlusal, central, and cervical thirds of the lateral tooth faces. For each area, five angles were measured on three separate occasions and the average of all measurements was obtained for each third. The three striae-EDJ angle averages for each lateral face permit documentation of how extension rate changed as the crown formation process proceeded. However, the broken surface is often neither perpendicular nor vertical to the EDJ and, hence, prevents an accurate measurement of the slope of striae with the EDJ. Nevertheless, the broken surface keeps the same inclination to the EDJ and to a vertical line along the lateral face. Thus, attention was given principally not to the real values of the angles but to the changes of the angles along the EDJ (Ramirez Rozzi 1992, 1994, 1997a, 1997b).

To observe whether changes of angles along the EDJ result from changes in the extension rate, the distance along the EDJ between adjacent striae

Table 15.2   Molars from Omo Included in the Study of Striae-EDJ Angles

Specimen	Geol. Mem.	Tooth	Cusp	Face
W7-559i	B	M3/2 ur	protocone	P
L2-79	B	M2/3 ll	entoconid	L
L849-1	C	M2/1 lr	metaconid	L
L9-11	D	M3 lr	entoconid	L
L9-92	D	M3 ul	protocone	M
L296-1	D	M3 lr	protoconid	L
L209-17	E–F	M2/1 lr	entoconid	D
L398-14	F	M lr	entoconid	L
L398-266	F	M3 lr	entoconid	L
L398-847	F	M3 ll	protoconid	M
L465-112	F	M2 ul	entoconid	B
Omo 76-37	G	M3 ll	hypoconulid	M

Note: Geo. Mem., geological Member of Omo Group; P, palatal; L, lingual; B, buccal; M, mesial; D, distal; u, upper; l, lower; r, right; l, left.

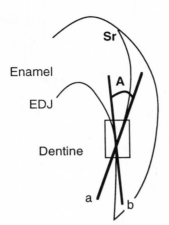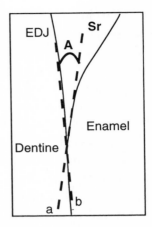

Fig. 15.3. Angle striae-EDJ (*A*) was measured between the tangent (*a*) to the striae (*Sr*) and the tangent (*b*) to the enamel-dentine junction (*EDJ*) at their point of intersection.

was obtained. Five adjacent striae of Retzius were measured along the EDJ, and an average distance between them was calculated for each one-third of the lateral tooth face. Changes in average distance between striae from cusp tip to cervix were compared with those of the striae-EDJ angles for the same broken surface. A similar pattern of change in both characteristics would confirm that the striae-EDJ angle allows an estimation of the enamel extension rate.

Results

Two changes of striae slope along the EDJ were observed (Table 15.3). In one, the striae-EDJ angle increases from cusp to cervix, while another pat-

Table 15.3 Striae-EDJ Angle and Average Distance between Adjacent

	W7-559i		L2-79		L849-1		L9-11		L9-92		L296-1	
	a	*x*	*a*	*x*	*a*	*x*	*a*	*x*	*a*	*x*	*a*	*x*
Cuspal	12		11	87.2	13		18	55.8	10	68.6	20	
Central	28	43	23	57.4	17	40.4	36	37.2	28	35.4	32	35.6
Cervical	20	60.6	31	45	25	34	24	58.6	16	51.6	36	41.2

Note: a, stria-EDJ angle; x, average distance between adjacent striae along the EDJ (μm). In L296-1, the angle increases, as does the distance between striae. It is possible that distances were measured between six and not five striae.

tern shows an initial increase from the cusp centrally but then a decrease cervically (Fig. 15.4). Two patterns of change in the distance between adjacent striae were found: The distance between striae along the EDJ diminishes from cusp to cervix or diminishes from the cusp centrally, then increases cervically.

Comparison of results shows that the first kind of striae-EDJ angle is found in teeth where the distance between adjacent striae diminishes toward the cervix. This indicates that, as the extension rate slows, it is accompained by a higher slope of the striae with the EDJ. In teeth with the second pattern, the distance between adjacent striae increases cervically, suggesting that the extension rate increases near the cervix. In these teeth, striae-EDJ angles are more acute in the cervical than in the central one-third of the crown. Therefore, a higher extension rate is accompained by more acute striae-EDJ angles, and a lower extension rate by more open slopes of striae.

In the teeth examined here, those with a high extension rate in the cervical one-third are more abundant in strata 2.4 million years old than from 3 million years. The number of striae of Retzius probably increased through time, indicating a longer crown formation time (see below). In contrast, a high extension rate in the cervical one-third of the crown is probably an adaptation to decreased crown formation time.

In Omo specimen 76-37, the striae-EDJ angles as well as the distance between adjacent striae were measured in many places on the mesial and distal faces. Results are similar for both faces. Even if the interstriae distance increases in the cervical one-third, the last striae are separated by the shortest distance (30 μm), suggesting that the lowest extension rate occurred during the last periods of crown formation. Macho and Wood (1995) suggested that hominids are characterized by a slow extension rate in the cervical re-

Striae Measured along the EDJ

L209-17		L398-14		L398-266		L398-847		L465-112		Omo 76-37	
a	*x*	*a*	*x*	*a*	*x*	*a*	*x*	*a*	*x*	*a*	*x*
9	47.8	7	51.4	10		13	66.2	11	62	25	70.4
19	34.8	20	36.4	20	44.6	32	40.2	28	37.6	38	35.8
29		15	44	25	27.4	27	56.2	33		31	45.4

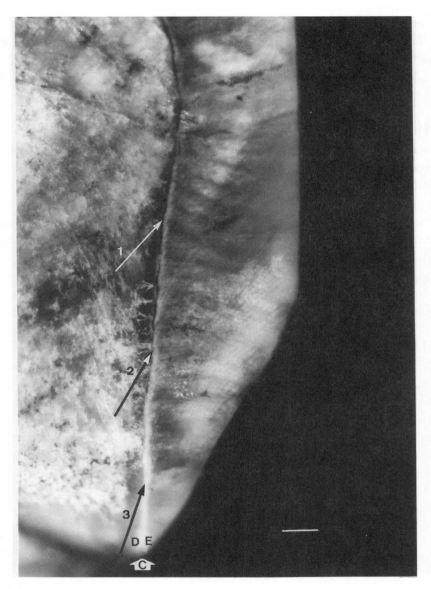

Fig. 15.4. Mesial face fracture of L9-92. The white line in the tooth cervix (*C*) corresponds to the junction between the enamel (*E*) and the dentine (*D*). The slope of striae with the EDJ becomes more acute from the central one-third (*arrow 1*) to the cervical one-third (*arrow 3*). The *bar* corresponds to 0.25 mm.

gion. Thus, although a high extension rate was measured in the cervical one-third of some specimens, it is possible, as seen in Omo 76-37, that the last periods of crown formation show the lowest extension rate.

Conclusions

This portion of the study confirms that, in Plio-Pleistocene hominid teeth, as in modern human teeth, the striae-EDJ angle gives an estimation of the distance between adjacent striae. In other words, the striae-EDJ angle is appropriate for estimating the enamel extension rate. However, in modern human teeth, the extension rate decreases toward the cervix (Beynon and Reid 1987; Shellis 1984), whereas in Plio-Pleistocene hominid teeth, the extension rate decreases toward the cervix or decreases initially and then increases in the cervical one-third of the crown.

Life History

Heterochronic Processes

The characteristics of enamel microstructure permit assessment of the rate and time of crown development (Table 15.1). The incremental rate operates in the lateral development of the enamel, with the final product being enamel thickness (Fig. 15.1). The extension rate participates in the longitudinal development of the enamel that produces the height of the apparent dentine cap.[2] Incremental lines in enamel allow the total time of crown formation, the time of imbricational stage formation, and the time of cuspal stage formation to be calculated. The time of cuspal stage formation corresponds to the active life of ameloblasts during the first phase of amelogenesis and operates in the lateral development of the enamel (Table 15.1) (Ramirez Rozzi 1997a).

The analysis of enamel modification through the Omo Group hominids shows evolutionary changes in some characteristics of lateral and longitudinal enamel development in the molar teeth. These are thought to relate to adaptation to a more abrasive, harder, or more resistant diet as a result of dryer environmental conditions (Ramirez Rozzi 1993b, 1997a; Ramirez Rozzi et al. 1999). Here, I assess how heterochronic processes can be used to explain adaptive changes observed in Plio-Pleistocene hominid molar formation.

Materials and Methods

The twenty-four molars used in this portion of the study are from the Usno and Shungura Formations of the Omo Group (Feibel et al. 1989) and are a subsample of those analyzed in Ramirez Rozzi (1993c, 1995, 1997a, 1998) and Ramirez Rozzi et al. (1999). They are naturally broken and date between 3.36 million years (the base of Member B) and 2.1 million years (the base of Submember G9) old. Molars were grouped by time spans (Periods) of approximately 0.4 million years that correspond to faunal associations. Period 1 (four teeth) corresponds to Member B (3.36–2.85 million years), Period 2 (four teeth) to Members C and D (2.85–2.4 million years), and Period 3 (16 teeth) to Members E, F, and G (2.4–2.1 million years) (Table 15.4).

Taxonomic attributions of hominids from the Omo deposits are very

Table 15.4 Molars Included in the Analysis of Heterochronies

Specimen	Geol. Mem.	Period	Tooth	Cusp	Face
B7-39c	B	1	M1 ul	hypocone	P
W7-559i	B	1	M3/2 ur	protocone	P
W7-578	B	1	M2/1 ll	metaconid	L
L2-79	B	1	M2/3 ll	entoconid	L
L183-11	C	2	M1 ul	paracone	B
L849-1	C	2	M2/1 lr	metaconid	L
L9-11	D	2	M3 lr	entoconid	L
L296-1	D	2	M3 lr	protoconid	L
L10-21	E	3	M3 lr	entoconid	L
Omo 57.5-320	E	3	M2/1 lr	hypoconid	B
L209-18	E-F	3	M ll	hypoconid	B
F22-1a	F	3	M2 lr	entoconid	L
F22-1b	F	3	M3 lr	entoconid	L
Omo 33-65	F	3	M3 ul	protocone	P
Omo 33-3325	F	3	M3 ul	metacone	B
Omo 33-3721	F	3	M3 ur	metacone	B
Omo 33-6172	F	3	M3 lr	metaconid	L
Omo 57.6-244	F	3	M2/1 ur	hypocone	B
L398-266	F	3	M3 lr	entoconid	L
L465-112	F	3	M2 ul	entoconid	B
Omo 35-4024	G	3	M2/1 ur	metacone	B
Omo 136-2	G	3	M2 ll	metaconid	B
Omo 136-3	G	3	M2 lr	hypoconid	B
Omo 141-1	G	3	M3 ur	protocone	B

Note: Geol. Mem., geological Member of Omo Group; P, palatal; L, lingual; B, buccal; u, upper; l, lower; r, right; l, left.

confused, especially for those represented only by teeth. Although several works on the Omo material have given taxonomic attributions to hominid teeth, only Suwa et al. (1996) explain which characteristics were used to assign teeth to Plio-Pleistocene hominid species.

Suwa (1990) originally assigned teeth from Member B (Period 1) to *Australopithecus afarensis*. However, more recently these teeth have been assigned to *Australopithecus/Homo* sp. indet. (Suwa et al. 1996). All other authors refer teeth from Member B to *A. afarensis* (i.e., Howell et al. 1987; White 1988). Teeth from Members C through G (Periods 2 and 3) show great variability in morphology. A very few of these have been attributed to *Homo*, and several to robust australopithecines (Coppens 1980; Howell et al. 1987; Ramirez Rozzi 1993a; Suwa 1990; Suwa et al. 1996; White 1988). While the presence of *Australopithecus boisei* or *Australopithecus aethiopicus*, or both, would be expected in the Omo deposits, no consensus has been reached on specific attributions of "robust" teeth.

In this study, teeth from Period 1 are assumed to be *A. afarensis*. Teeth from Periods 2 and 3 are considered to be from robust australopithecines. Teeth from Period 1 were assumed to be ancestral to those from Period 2 and those, in turn, to be the ancestors of those from Period 3. Despite the small sample sizes from Periods 1 and 2, it is expected that any patterns evident with this material would still be present, given more specimens.

Two criteria were used to include teeth in this exercise. Tooth faces analyzed had to present equivalent enamel development, and teeth from different periods had to be attributed to species with a (probable) ancestor/descendant relationship. Enamel development differs at various locations on the tooth. On proximal faces (mesial and distal), enamel is thinner than on lateral faces (lingual, buccal, and palatal). Moreover, enamel formation initiates later and finishes earlier in proximal faces than in lateral faces. The enamel cervix is higher on proximal faces, and mesial and distal margins are lower than cusps (Beynon et al. 1998). Therefore, the study of mesial and distal tooth faces underestimates the lateral and longitudinal development of the enamel. In addition, enamel thickness varies between tooth cusps (Macho and Berner 1993).

Lateral enamel thickness was measured using measure 3 in Beynon and Wood (1986) and measure 2 in Macho and Berner (1993). Macho and Berner (1993) suggested that differences in lateral enamel thickness are not significant between molars. The onset of enamel formation occurs at different times on cusps, but the longest interval in enamel formation initiation between the first and the last cusp of a tooth is never more than ten weeks in modern humans (Kronfeld 1935), and it was even shorter in Plio-Pleistocene hominids (Ramirez Rozzi 1997a). The offset of enamel formation on lateral faces occurs at different times. However, cervical enamel follows a horizontal orientation (Ramirez Rozzi 1993a), indicating that different times of enamel formation offset do not influence the height of the apparent dentine cap. Therefore, lateral development and longitudinal development of the enamel do not show significant differences between lateral tooth faces. Thus, teeth presenting broken surfaces on buccal as well as lingual/palatal tooth faces were included in this study.

Logarithmically transformed size and trait scales were used to compare molars from different periods and to assess whether changes followed a heterochronic pattern, with dissociation between growth and development— that is, whether the rate of change in size (the sum of metric traits) differs from the rate of change in shape (the ratio of trait to size) (Gould 1977). Lateral and longitudinal enamel development (Fig. 15.1), with their respective final products (enamel thickness and height of the apparent dentine cap), constitute a growth field (enamel/crown development). Age of this growth field is given by the total number of striae of Retzius (i.e., the time of crown formation). Lateral enamel thickness and the height of the apparent dentine cap are considered as traits. Size corresponds to the sum of these two traits and shape to the ratio between trait and size. Using growth, development, and age for a growth field, Gould's (1977) clock can be constructed to compare ancestor and descendant ontogenies.

Results

The plot of logarithmically transformed trait and size scales for enamel lateral thickness and the height of the apparent dentine cap of the lateral faces is shown in Figure 15.5. For height of the apparent dentine cap (Fig. 15.5*A*), ontogenetic trajectories are very close. Descendant ontogenies (for molars from Periods 2 and 3) are almost parallel to the isometric line and follow the ancestral ontogenies (that of molars from Periods 1 and 2, respectively). Thus, ancestral and descendant height of the apparent dentine cap are isometric, and the ancestor is at all sizes similar in shape to the descendant at any point along the descendant's trajectory. Size and shape are not dissociated, so this measurement does not distinguish paedomorphosis and peramorphosis.

In contrast, different ontogenetic trajectories can be discerned when enamel lateral thickness is examined (Fig. 15.5*B*). Comparison between molars from Periods 1 and 2 shows that enamel thickness increased while overall size decreased slightly in molars from Period 2. Descendant ontogeny does not follow ancestral ontogeny, as the isometric line passing through the ancestor is crossed by the descendant ontogeny trajectory in a higher position. This indicates that size and shape are dissociated, with the descendant attaining the ancestor's shape only after reaching a larger size. Thus, molars from Period 2 show a reduction in net change of shape relative to net change in size. This means that molars from Period 2 are paedomorphic relative to molars from Period 1 as the enamel had become proportionately thicker.

Analysis of molars from Period 3 shows that the descendant ontogenetic trajectory closely follows that of the ancestor, passing through ancestral shape (molars from Period 2). Size and shape are thus not dissociated. Even if molars from Periods 2 and 3 are significantly different in enamel lateral thickness and height of the apparent dentine cap of the lateral faces (Ramirez Rozzi 1997a, 1997b; Ramirez Rozzi et al. 1999), changes in size and shape are proportional. When molars from Period 3 are compared with those from Period 1, results are similar to those obtained for molars from Period 2, with molars from Period 3 having a paedomorphic development of enamel lateral thickness because change in shape was slowed relative to size (Fig. 15.5*B*).

Since the age of crown formation time is known for these teeth (Ramirez Rozzi 1995, 1997a), Gould's (1977) clock can be constructed to specify heterochronic process. Size and age are not significantly different between molars from Periods 1 and 2. Comparison between ancestor and descendant shows that, at a similar size and age, descendant shape has not attained ancestral adult shape (Fig. 15.5*C*), indicating a reduced shape transformation in relation to size change over a similar period of time, indicative of neoteny. Therefore, paedomorphosis has taken place by neoteny in molar crown develoment around 2.8 million years, resulting in a thicker lateral enamel.

In molars from Period 3, size and shape have increased over those from Periods 1 and 2. An increase in crown formation time is also recorded in molars from Period 3. Compared with those from Period 2, molars from Period 3 are peramorphic without size and shape dissociation. Peramorphosis was very probably attained by time hypermorphosis (Fig. 15.5*D*) around 2.4 million years ago.

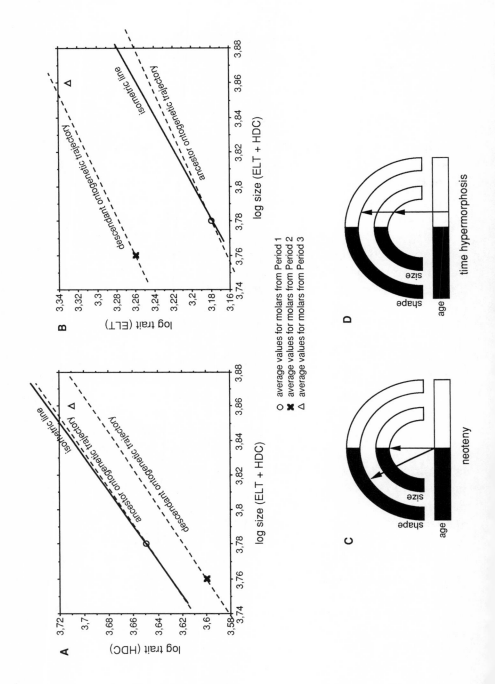

A

log trait (HDC)

isometric line

△

ancestor ontogenetic trajectory

descendant ontogenetic trajectory

log size (ELT + HDC)

B

log trait (ELT)

descendant ontogenetic trajectory

isometric line

△

ancestor ontogenetic trajectory

log size (ELT + HDC)

○ average values for molars from Period 1
✖ average values for molars from Period 2
△ average values for molars from Period 3

C

shape

size

age

neoteny

D

shape

size

age

time hypermorphosis

Discussion/Conclusions

Teeth analyzed here are considered to be in an ancestor-descendent relationship. *A. afarensis* is probably the ancestor of robust australopithecines, but it is uncertain whether teeth from Period 1 can be unquestioningly attributed to this species. Similarly, teeth from Periods 2 and 3 have been attributed to robust species, but to how many? If more than one, is *A. aethiopicus* ancestral to *A. boisei?* If enamel characteristics suggested as apomorphies in the robust lineages are accepted (Grine and Martin 1988), it would be difficult to view *A. aethiopicus* as ancestral to *A. boisei* because teeth from the former would have more apomorphic features than those from the latter (Ramirez Rozzi 1993a).

Although this relatively simple analysis of heterochrony is limited to only two traits, it yields useful observations. The analysis of lateral enamel

---

Fig. 15.5. Logarithmically transformed trait and size scales. Trait = height of the apparent dentine cap (*HDC*) or enamel lateral thickness (*ELT*) in molar lateral faces. Size = the sum of enamel lateral thickness and height of the apparent dentine cap in molar lateral faces. Ontogenetic trajectories are growth trajectories that correspond to trait/size relationship through development. They were determined from the first and last stages of crown formation. The first stage of crown development corresponds to the beginning of secretion of dentine by odontoblasts, and the final stage of crown development corresponds to crown completion. *Points* in this figure represent the average crown completion in molars for each period. An ontogenetic trajectory corresponds to each point.

A: When the height of the apparent dentine cap is considered as trait, ontogenetic trajectories are very close and almost parallel to the isometry line. Size and shape are thus not dissociated; the ancestor is at all sizes similar in shape to the descendant at any point along the descendant's trajectory. Changes in size and shape are proportional for this characteristic. *B:* If enamel lateral thickness is considered as trait, ontogenetic trajectories of molars from Periods 2 and 3 are distant from that of molars from Period 1 and are not parallel to the isometry line. Enamel became thicker in relation to size in molars from Period 2. Descendant ontogenetic trajectory (molars from Period 2) cross the isometry line that is passing through ancestral final stage, in a higher position. There is a reduction in the net change of shape relative to net change in size. Molars from Periods 2 and 3 have almost the same ontogenetic trajectories, indicating that there is no dissociation in size and shape between them. *C:* Age and size are similar for molars from Periods 1 (*solid*) and 2 (*arrows*), but shape was delayed (neoteny) in molars from Period 2. Molars from Period 2 show paedomorphosis by neoteny in relation to molars from Period 1. *D:* Molars from Period 3 change proportionally to molars from Period 2; however, size, shape, and age have increased in molars from Period 3. These molars are peramorphic relative to molars from Period 2, as a consequence of a time hypermorphosis.

Table 15.5 Results for Traits and Size

Specimen	LET	HDC	Size	Log LET	Log HDC	Log Size	Xlog LET	Xlog HDC	Xlog Size
B7-39c	1,630	4,280	5,910	3.21	3.63	3.77			
W7-559i	1,750	4,920	6,670	3.24	3.69	3.82			
W7-578	1,470	4,045	5,515	3.17	3.61	3.74			
L2-79	1,240	4,720	5,960	3.09	3.67	3.78	3.18	3.65	3.78
L183-11	2,240	3,990	6,230	3.35	3.60	3.79			
L849-1	2,040	4,180	6,220	3.31	3.62	3.79			
L9-11	1,570	3,680	5,250	3.20	3.57	3.72			
L296-1	1,500	4,000	5,500	3.18	3.60	3.74	3.26	3.60	3.76
L10-21	2,100	5,320	7,420	3.32	3.73	3.87			
Omo 57.5-320	2,410	7,230	9,640	3.38	3.86	3.98			
L209-18	2,580	5,320	7,900	3.41	3.73	3.90			
F22-1a	2,340	5,590	7,930	3.37	3.75	3.90			
F22-1b	2,480	5,180	7,660	3.39	3.71	3.88			
Omo 33-65	2,730	6,420	9,150	3.44	3.81	3.96			
Omo 33-3325	2,160	3,820	5,980	3.33	3.58	3.78			
Omo 33-3721	1,840	3,770	5,610	3.26	3.58	3.75			
Omo 33-6172	1,650	3,780	5,430	3.22	3.58	3.73			
Omo 57.6-244	1,680	5,410	7,090	3.23	3.73	3.85			
L398-266	1,930	4,830	6,760	3.29	3.68	3.83			
L465-112	1,670	4,630	6,300	3.22	3.67	3.80			
Omo 35-4024	2,560	5,620	8,180	3.41	3.75	3.91			
Omo 136-2	1,960	5,520	7,480	3.29	3.74	3.87			
Omo 136-3	2,420	6,100	8,520	3.38	3.79	3.93			
Omo 141-1	2,020	5,000	7,020	3.31	3.70	3.85	3.33	3.71	3.86

Note: Measurements are in microns. LET, lateral enamel thickness; HDC, height of the apparent dentine cap; Size = LET + HDC; XlogLET, average logarithm of the lateral enamel thickness for each Period.

336

thickness, for example, shows that heterochronic changes would have occurred with or without size/shape dissociation. When the height of the apparent dentine cap is considered as a trait, however, size and shape do not appear dissociated. In fact, different results are reached depending on which trait is considered. The height of the apparent dentine cap does not enable discernment of heterochrony, whereas heterochrony can be identified when lateral enamel thickness is used as a trait.

Lateral enamel thickness and the height of the apparent dentine cap correspond, respectively, to the lateral and longitudinal development of the enamel. As suggested earlier, they have changed through time in Plio-Pleistocene hominid molars as a result of an adaptation to a more abrasive, resistant diet. This study suggests that changes in molars of the robust australopithecine lineage from East Africa were produced by heterochrony. Enamel became thicker relative to size by the retention of ancestral juvenile shape in adult descendants around 2.8 million years ago. It became even thicker relative to a larger size by a longer duration of crown formation around 2.4 million years ago. Therefore, adaptation of robust australopithecine molars to drier environmental conditions occurred first by neoteny and later by time hypermorphosis.

Larger, robust teeth with thicker enamel are typically assigned to robust australopithecines. Yet new taxonomic attributions accepting the presence of *A. aethiopicus* (a robust australopithecine less specialized than *A. boisei*) and *Homo rudolfensis* (a more robust form than *Homo habilis*) show that it is inappropriate simply to attribute large teeth to robust australopithecines and small teeth to nonrobust australopithecines. In fact, it is now accepted that enamel thickness is not a characteristic with any taxonomic value in Plio-Pleistocene hominids; rather, there was a great amount of parallelism in heavy chewing Plio-Pleistocene hominid species (Macho and Thackeray 1992; Ramirez Rozzi 1992, 1993a, 1997a; Ramirez Rozzi et al. 1999). This study suggests that heterochronic changes would explain large teeth with thick enamel in robust australopithecines from East Africa. It would be interesting to know whether adaptation to heavy chewing in other Plio-Pleistocene hominid species (e.g., early *Homo*) resulted through similar evolutionary processes.

Dental Development Patterns

Many studies have focused on dental development in extant hominoids in recent years (see Anemone, Chap. 12), but their results are often dissimilar. Differences may arise from different methods of analysis (histologic vs. radiographic studies) or may simply illustrate a large variation within each taxon. However, patterns of dental development, such as crown and root formation time, as well as sequence and age of tooth eruption in modern humans and chimpanzees have become better understood (see Anemone, Chap. 12; Kuykendall, Chap. 13).

Dental development patterns in extant hominoids have been compared with patterns in Plio-Pleistocene hominid specimens that had not completed their dental development. These comparisons have led to discussions of whether development in Plio-Pleistocene hominids was more like that of modern humans or chimpanzees (see Kuykendall, Chap. 13). Smith (1986, 1989) developed an original method to observe whether Plio-Pleistocene hominid specimens followed a modern humanlike or a chimpanzee-like pattern of dental development. She plotted the dental stage reached by each tooth in an immature Plio-Pleistocene hominid specimen in charts showing the ages at which stages of dental development (onset and offset of crown formation, tooth eruption, and root completion) are attained in modern humans and in chimpanzees. Her chart corresponding to dental development in chimpanzees shows less spread in dental stages reached by teeth within a single individual than her modern human sample. According to these comparisons, dental development in Plio-Pleistocene hominids seems to have been closer to that of chimpanzees than that of modern humans.

However, Smith's comparisons do not provide information on dental patterns in Plio-Pleistocene hominids themselves. Indeed, her data from fossil hominids were limited to the comparison of developmental stages reached by each tooth, without taking into consideration information on crown formation time.

Previously, time of crown formation has been calculated in Plio-Pleistocene hominid teeth by counting incremental lines in the enamel. Crown formation time in molar teeth in Plio-Pleistocene hominids was close to that in modern humans and chimpanzees, whereas crown formation time in anterior teeth and premolars was shorter than in extant hominoids (see Anemone, Chap. 12). In extant hominoids the time of crown formation in premolars is longer than that in molars. However, Plio-Pleistocene hominids exhibit a similar number of incremental lines in premolars and molars, indicating that their crown formation time was similar for these two tooth classes (Beynon and Dean 1987; Ramirez Rozzi 1992, 1993c, 1995).

The aim of this final analysis is to propose a chart of dental development in Plio-Pleistocene hominids related to the time of crown formation.

Materials and Methods

Data for crown formation time of *A. africanus* and *A. afarensis* incisors comes from Bromage and Dean (1985), for crown formation time of *Australopithecus robustus* incisors and canines from Dean et al. (1993), and for premolar and molar crown formation times of robust australopithecines and early *Homo* from Beynon and Dean (1987), Ramirez Rozzi (1995, 1997a, 1998), and Ramirez Rozzi et al. (1997).

The time of crown formation was plotted on two charts representing tooth development and age of the individual. In one chart crown formation of Plio-Pleistocene hominids is assumed to initiate at the same age as in chimpanzees, and in the other at the same age as observed in modern humans. Time of root formation is not known in Plio-Pleistocene hominids.

In addition to the East African fossils, two South African juveniles were examined for comparison: STS 24 and SK 3978, from Sterkfontein and Swartkrans (South Africa), attributed to *A. africanus* and *A. robustus*, respectively. The teeth of STS 24 and SK 3978 had reached different stages of crown development (see Conroy and Vannier 1991a, 141; Conroy and Vannier 1991b, 130). Teeth from these two individuals were placed into two charts to observe how well their developmental stages match those of the other Plio-Pleistocene hominids examined here.

Results

Age of crown initiation in the anterior teeth is almost the same in modern humans and chimpanzees; thus, there are no differences between the two charts for these teeth (Fig. 15.6). However, modern humans are characterized by longer gaps between crown initiation in the cheek teeth than are chimpanzees, with the result that this pattern differs in the two charts. Relationships between the developmental stages of cheek teeth differ on the charts because of the influence of the short premolar crown formation times in Plio-Pleistocene hominids. M1 crown completion takes place when a third of the P3 crown and around a quarter of the P4 crown have formed in *both* modern humans and chimpanzees, and it occurs around three years

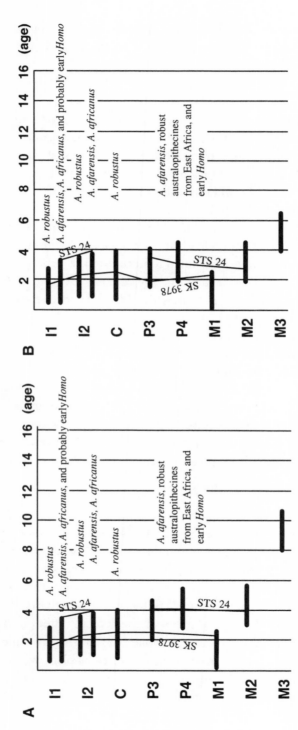

Fig. 15.6. Patterns of dental (crown) development in Plio-Pleistocene hominids. Time of crown formation in Plio–Pleistocene hominids (*bars*) is from Bromage and Dean (1985), Dean et al. (1993), Beynon and Dean (1987), Ramirez Rozzi (1995, 1997a, 1998), and Ramirez Rozzi et al. (1997). It was assumed that the age of onset of enamel formation was similar to that of modern humans (*A*) or chimpanzees (*B*). Dental developmental stages in STS 24 and SK 3978 are plotted in both charts. Incisor development is similar in both charts. I2, C, P, M1, and M2 in fossil hominids show less spread in the chart using age of onset of crown formation of modern humans. Only I1 formation is delayed in relation to other teeth. Eruption of M1 has just occurred in STS 24, suggesting that it took place at about four years of development age. No difference is observed between STS 24 (*A. africanus*) and SK 3978 (*A. robustus*).

of developmental age in both extant hominoid species (Reid et al. 1998). Assuming a similar age of crown initiation as in modern human teeth, M1 crown completion in Plio-Pleistocene hominids took place when approximately a quarter of the P3 crown had formed and the P4 was initiating crown formation (Fig. 15.6*A*). If age of crown initiation from chimpanzee teeth is used to create the chart of dental development in Plio-Pleistocene hominids, crown completion of M1 occurred when about half of the P3 and a third of the P4 crown had formed (Fig. 15.6*B*). Since age of M1 crown initiation is the same in modern humans and chimpanzees, it is assumed to be the same in Plio-Pleistocene hominids, so that M1 crown completion would be reached before three years of developmental age in these fossil taxa.

Plots of STS 24 and SK 3978 permit observations concerning the relationships within each tooth class and between anterior teeth and cheek teeth (Fig. 15.6). Developmental stages in anterior teeth in these two fossil specimens are similar in both charts. I1 development is delayed relative to I2 in STS 24 and in relation to I2 and C in SK 3978 regardless of whether it is following a chimpanzee or a modern human age of onset of crown formation. In contrast, differences exist between charts when the cheek teeth are considered. Developmental stages for P3, P4, and M2 in STS 24 and developmental stages for P3 and M1 in SK 3978 lie along a nearly straight line in the chart where the modern human age of crown initiation was used to estimate Plio-Pleistocene hominid dental patterns, whereas they deviate more when the chimpanzee age of crown initiation is used. P4 crown formation had not started in SK 3978 (Conroy and Vannier 1991a). The plot based on age of crown initiation in modern humans agrees with this fact, while the plot based on age of crown initiation of chimpanzees does not. This suggests, therefore, that the gap between crown initiation in Plio-Pleistocene hominid premolars, first molar, and second molar would be almost as long as in modern humans.

It has been suggested that the development of anterior teeth was delayed relative to cheek teeth in Plio-Pleistocene hominids, when compared with modern humans (Smith 1989; Conroy and Vannier 1991b). However, the plots of STS 24 and SK 3978 show only that I1 crown formation is delayed. Dental developmental stages of I2 and C lie in an almost straight line with those of cheek teeth when modern human standards are assumed (Fig. 15.6*A*). When chimpanzee standards are used (Fig. 15.6*B*), anterior teeth do not show a delayed developmental stage, but their development is spread around the same age as that of the cheek teeth.

M1 had just erupted in STS 24. This event happened at the age attributed to STS 24 based on dental developmental stages observed for P3, P4, and M2. Thus, M1 eruption took place at three to four years of developmental age in Plio-Pleistocene hominids. This study confirms an age at death of no greater than four years for STS 24 (Conroy and Vannier 1991a), suggesting that extended maturation appeared later in hominid evolution.

Discussion/Conclusions

Although these charts assume an age of crown initiation similar to that of modern humans or chimpanzees, they take into consideration the short time of premolar crown formation in Plio-Pleistocene hominids. Crown completion of M1 in Plio-Pleistocene hominids occurred around a quarter (Fig. 15.6A) or around a half of P3 (Fig. 15.6B) crown formation. This suggests a premolar crown formation time close to 2.5 years. But if a longer premolar crown formation (as long as that in humans or that in chimpanzees) is assumed, the proportion of premolar crown formed at M1 crown completion would be smaller. Alternatively, if a longer premolar crown formation time is assumed for STS 24, age attributed to premolars would be older, and the dental developmental stages in STS 24 would show less variation in Figure 15.6B but more variation in Figure 15.6A.

Previous studies on the same specimens based on the assumption of a longer time of premolar crown formation have suggested a chimpanzee-like pattern of tooth development in Plio-Pleistocene hominids (Conroy and Vannier 1991a; Smith 1986). Results from this study do not confirm this. Patterns of dental development in STS 24 and SK 3978 fit better when the age of crown initiation is considered using modern human standards. This does not mean that Plio-Pleistocene hominids are like modern humans because the age of crown initiation is the only standard assumed from extant hominoids. Indeed, the reason for a different result between previous studies and this one is that, in previous work, dental developmental stages of fossil specimens were plotted on chimpanzee and modern human dental development charts, whereas in the present study, they are plotted in a chart that represents patterns of dental development in Plio-Pleistocene hominids (i.e., with a short time of premolar crown formation).[3]

Some previous investigators assumed the presence of a chimpanzee-like pattern of dental development in Plio-Pleistocene hominids because gaps between the onset of crown formation were shorter than in humans. However, Beynon and Dean (1988) suggested that, in Plio-Pleistocene hominids, M2 crown formation began when M1 crown formation was completed, as in modern humans (Reid et al. 1998). Since M1 crown formation started around birth and took about 2.5 years to be completed in Plio-Pleistocene hominids (Beynon and Dean 1988; Ramirez Rozzi 1997a) and the time of crown formation of M1 in modern humans is between 2.5 and 3 years (Reid et al. 1998), the gap between M1 and M2 crown initiation in Plio-Pleistocene hominids was very close to that in modern humans. Thus, it is not surprising that STS 24 and SK 3978 better fit the chart where age of crown initiation similar to that of modern humans is assumed. Therefore, gaps between the onset of crown formation in cheek teeth were large in Plio-Pleistocene hominids, although probably not as large as in modern humans.

Some differences in dental developmental patterns between *A. africanus* and *A. robustus* have been reported (Conroy and Vannier 1991a, 1991b; Smith 1986), with the relationship between anterior teeth and cheek teeth more chimpanzee-like in the first but more humanlike in the second. Dental developmental stages in both STS 24 and SK 3978 better fit the chart assuming an age of crown initiation similar to that of modern humans (Fig. 15.6*A*). Although much variability seems to have existed in dental development in Plio-Pleistocene hominids (Conroy and Vannier 1991b), STS 24 and SK 3978 present similar patterns of dental development, suggesting that gracile and robust australopithecines show no differences in the time of crown formation and the age of crown initiation. Distinct dental developmental patterns between them probably result from differences in tooth emergence and root formation.

Assuming a relationship between anterior teeth and cheek teeth for Plio-Pleistocene hominids close to that in chimpanzees, Macho and Wood (1995) suggested two basic changes during hominid evolution: (1) a prolongation of anterior teeth development (slow crown and root extension rate) and (2) a retardation in the development of cheek teeth (larger gaps between onset of crown formation) relative to anterior teeth. However, the present study suggests that a large gap between premolar crown initiation and between first and second molar crown initiation was already present in Plio-Pleistocene hominids, although the gap was probably not as long in duration as in modern humans. Larger gaps in the onset of cheek teeth crown formation could have thus occurred subsequently (Macho and Wood 1995), but a significant change concerning cheek teeth during hominid evolution is the eventual appearance of a longer duration of premolar crown formation.

Acknowledgments

I thank Ken McNamara for inviting me to participate in this volume and for his editorial work on early drafts. I am grateful to David Beynon, Don Reid, Chris Dean, and Tim Bromage for interesting discussions on enamel microstructure and to all participants in the conference *African Biogeography, Climate Change, and Early Hominid Evolution* in Salima, Malawi, in 1995 for fruitful discussions on East African ecology, past and present. I thank B. Asfaw, Y. Coppens, and F. C. Howell for allowing me to examine material in their care.

Notes

1. The recognition of developmental stages of tooth formation is sometimes difficult from radiographs, as crowding can mask radiographic details. The determination of crown completion used in many studies on tooth development (i.e., Demirjian et al. 1973; Fanning 1961; Moorreess et al. 1963) is extremely difficult, not only because approximal faces are used (see below) but because x-rays enable visualization of enamel only after maturation is complete (i.e., about two weeks after enamel secretion). Since radiographic views of cheek teeth are usually taken from the lateral position, only approximal tooth faces (mesial and distal) are used. On these faces, enamel formation stops earlier than on buccal and lingual tooth faces; therefore, crown formation time is underestimated. Futhermore, x-ray transmission depends upon the thickness and degree of mineralization of the tissues traversed. The lack of enamel on dentine caps and on the cervix produces an insignificant x-ray absorption with the consequent loss of records for the first and the last enamel secreted (Beynon et al. 1998; Hess et al. 1932; Winkler 1995).

2. *Dentine cap* refers to the enamel covering the crown and, hence, capping the underlying dentine of the tooth crown.

3. Dental developmental stages used by Smith (1986) for STS 24 teeth are different from those defined by Conroy and Vannier (1991a). For example, the crown of I2 has not achieved formation in Smith's scheme, but it is completed and root formation has just begun according to Conroy and Vannier.

References

Baker, K.L. 1972. The fluorescent, microradiographic microhardness and specific gravity properties of tetracycline-affected human enamel and dentine. *Archives of Oral Biology* 17:525–536.

Beynon, A.D., and Dean, M.C. 1987. Crown-formation time of a fossil hominid premolar tooth. *Archives of Oral Biology* 32:773–780.

———. 1988. Distinct dental development patterns in early fossil hominids. *Nature* 335:509–514.

Beynon, A.D., and Reid, D. 1987. Relationships between perikymata counts and crown formation times in the human permanent dentition. *Journal of Dental Research* 66:889 (abstract).

Beynon, A.D., and Wood, B.A. 1986. Variations in enamel thickness and structure in East African hominids. *American Journal of Physical Anthropology* 70:177–195.

Beynon, A.D., Clayton, C.B., Ramirez Rozzi, F.V., and Reid, D.J. 1998. Radiographic and histological methodologies in estimating the chronology of crown development in modern humans and great apes: a review, with some applications for studies on juvenile hominids. *Journal of Human Evolution* 35:351–370.

Boyde, A. 1964. The structure and development of mammalian enamel. Ph.D. diss., University of London.

Boyde, A., and Fortelius, M. 1986. Development, structure and function of rhinoceros enamel. *Zoological Journal of the Linnean Society* 87:181–214.

Bowler, P. 1989. *Evolution: The History of an Idea.* Berkeley and Los Angeles: University of California Press.

Bromage, T.G. 1991. Enamel incremental periodicity in the pig-tailed macaque: a polychrome fluorescent labeling study of dental hard tissues. *American Journal of Physical Anthropology* 86:205–214.

Bromage, T.G., and Dean, M.C. 1985. Re-evaluation of the age at death of immature fossil hominids. *Nature* 317:525–527.

Bromage, T.G., Schrenk, F., and Zonneveld, F.W. 1995. Paleoanthropology of the Malawi Rift: an early hominid mandible from Chiwondo Beds, northern Malawi. *Journal of Human Evolution* 28:71–108.

Brown, F.H., and Feibel, C.S. 1988. "Robust" hominids and Plio-Pleistocene paleography of the Turkana basin, Kenya and Ethiopia. In F.E. Grine (ed.), *Evolutionary History of the "Robust" Australopithecines,* 325–341. New York: Aldine de Gruyter.

Brown, F.H., McDougall, I., Davies, I., and Maier, R. 1985. An integrated Plio-Pleistocene chronology for the Turkana basin. In E. Delson (ed.), *Ancestors: The Hard Evidence,* 82–90. New York: Alan R. Liss.

Conroy, G.C., and Vannier, M.W. 1991a. Dental development in South African australopithecines. Part 1: Problems of pattern and ontogeny. *American Journal of Physical Anthropology* 86:121–136.

Conroy, G.C., and Vannier, M.W. 1991b. Dental development in South African australopithecines. Part 2: Dental stage assessment. *American Journal of Physical Anthropology* 86:137–156.

Coppens, Y. 1972. Tentative de zonation du Pliocène et du Pléistocène d'Afrique par les grands mammifères. *Comptes Rendus de l'Académie des Sciences Paris* 274:181–184.

————. 1975. Evolution des mammifères, de leurs fréquences et de leurs associations, au cours du Plio-Pléistocène dans la basse vallée de l'Omo en Ethiopie. *Comptes Rendus de l'Académie des Sciences Paris* 281:1571–1574.

————. 1980. The difference between *Australopithecus* and *Homo:* preliminary conclusions from the Omo research expedition's studies. In L. Konigsson (ed.), *Current Argument on Early Man,* 207–225. New York: Pergamon Press.

Dean, M.C. 1987. Growth layers and incremental markings in hard tissues; a review of the literature and some preliminary observations about enamel structure in *Paranthropus boisei. Journal of Human Evolution* 16:157–172.

Dean, M.C., Beynon, A.D., Thackeray, J.F., and Macho, G.A. 1993. Histological reconstruction of dental development and age at death of a juvenile *Paranthropus robustus* specimen, SK 63, from Swartkrans, South Africa. *American Journal of Physical Anthropology* 91:401–419.

Demirjian, A., Goldstein, H., and Tanner, J.M. 1973. A new system of dental age assessment. *Human Biology* 45:211–227.

Denton, G.H. 1999. Cenozoic climate change. In T. Bromage and F. Schrenk (eds.), *African Biogeography, Climate Change, and Early Hominid Evolution,* 94–114. Oxford: Oxford University Press.

Fanning, E.A. 1961. A longitudinal study of tooth formation and root resorption. *New Zealand Dental Journal* 57:202–217.

Feibel, C.S., Brown, F.H., and McDougall, I. 1989. Stratigraphic context of fossil hominids from the Omo Group deposits: Northern Turkana basin, Kenya and Ethiopia. *American Journal of Physical Anthropology* 78:595–622.

Fitzgerald, C.M. 1998. Do enamel microstructures have regular time dependency? Conclusions from the literature and a large-scale study. *Journal of Human Evolution* 35:371–386.

Gould, S.J. 1977. *Ontogeny and Phylogeny.* Cambridge: Belknap Press of Harvard University Press.

Grine, F.E., and Martin, L.B. 1988. Enamel thickness and development in *Australopithecus* and *Paranthropus.* In F.E. Grine (ed.), *Evolutionary History of the "Robust" Australopithecines,* 3–42. New York: Aldine de Gruyter.

Harrold, F.B. 1991. The elephant and the blind men: paradigms, data gaps, and the Middle-Upper Paleolithic transition in Southwest France. In G.A. Clark (ed.), *Perspectives on the Past: Theoretical Biases in Mediterranean Hunter-Gatherer Research,* 164–182. Philadelphia: University of Pennsylvania Press.

Hess, A.F., Lewis, J.M., and Roman, B. 1932. A radiographic study of calcification of the teeth from birth to adolescence. *Dental Cosmos* 74:1053–1061.

Howell, F.C., Haesaerts, P., and de Heinzelin, J. 1987. Depositional environments, archeological occurrences, and hominids from Members E and F of the Shungura Formation (Omo Basin, Ethiopia). *Journal of Human Evolution* 16:665–700.

Kronfeld, R. 1935. First permanent molar: its condition at birth and its postnatal development. *Journal of the American Dental Association* 22 (July): 1131–1155.

Macho, G.A., and Berner, M.E. 1993. Enamel thickness of human maxillary molars reconsidered. *American Journal of Physical Anthropology* 92:189–200.

Macho, G.A., and Thackeray, J.F. 1992. Computed tomography and enamel thickness of maxillary molars of Plio-Pleistocene hominids from Sterkfontein, Swartkrans and Kromdraai (South Africa): an exploratory study. *American Journal of Physical Anthropology* 89:133–143.

Macho, G.A., and Wood, B.A. 1995. The role of time and timing in hominid dental evolution. *Evolutionary Anthropology* 4:17–31.

Moorreess, C.F.A., Fanning, E.A., and Hunt, E.E. 1963. Age variation of formation stages for ten permanent teeth. *Journal of Dental Research* 42:1490–1502.

Osborn, J.W. 1968. Directions and interrelationships of enamel prisms in cuspal and cervical enamel of human teeth. *Journal of Dental Research* 47:395–402.

Ramirez Rozzi, F.V. 1992. Le développement dentaire des hominidés du Plio-Pléistocène de l'Omo, Ethiopie. Paris: Thèse du Muséum National d'Histoire Naturelle.

———. 1993a. Teeth development in East African *Paranthropus. Journal of Human Evolution* 24:429–454.

———. 1993b. Modifications du développement dentaire des hominidés au cours du Plio-Pléistocène. *Comptes Rendus de l'Académie de Sciences Paris,* série II 317:1249–1254.

———. 1993c. Aspects de la chronologie du développement dentaire des hominidés Plio-Pléistocènes de l'Omo, Ethiopie. *Comptes Rendus de l'Académie de Sciences Paris,* série II 316:1155–1162.

———. 1994. Enamel growth markers of hominid dentition. *Microscopy and Analysis,* July (European edition), 21–23.

———. 1995. Time of formation in Plio-Pleistocene hominid teeth. In J. Moggi-Cecchi (ed.), *Aspects of Dental Biology: Palaeontology, Anthropology, and Evolution,* 217–238. Florence: International Institute for the Study of Man.

———. 1997a. *Les Hominidés du Plio-Pléistocène de l'Omo, Ethiopie. Caractérisation et modification au cours du temps de leur développement dentaire à partir de l'étude de la microstructure de l'émail.* Paris: CNRS.

———. 1997b. Développement dentaire des hominidés fossiles: le taux d'extension de l'émail dans les hominidés du Plio-Pléistocène. *Comptes Rendus de l'Académie des Sciences Paris* 325:293–296.

———. 1998. Can enamel microstructure be used to establish the presence of different species of Plio-Pleistocene hominids from Omo, Ethiopia? *Journal of Human Evolution* 35:543–576.

Ramirez Rozzi, F.V., Bromage, T., and Schrenk, F. 1997. UR 501, the Plio-Pleistocene hominid from Malawi: analysis of the microanatomy of the enamel. *Comptes Rendus de l'Académie des Sciences Paris* 325:231–234.

Ramirez Rozzi, F.V., Walker C., and Bromage, T. 1999. Early hominid dental development and climate change. In T. Bromage and F. Schrenk (eds.), *African Biogeography, Climate Change, and Early Hominid Evolution,* 349–363. Oxford: Oxford University Press.

Reid, D.J., Beynon, A.D., and Ramirez Rozzi, F.V. 1998. Histological reconstruction of dental development in four individuals from a medieval site in Picardie, France. *Journal of Human Evolution* 35:463–477.

Risnes, S. 1986. Enamel apposition rate and prism periodicity in human teeth. *Scandinavian Journal of Dental Research* 94:394–404.

———. 1990. Structural characteristics of staircase-type Retzius lines in human dental enamel analyzed by scanning electron microscopy. *Anatomical Record* 226: 135–146.

———. 1998. Growth tracks in enamel. *Journal of Human Evolution* 35:331–350.

Rose, J.C. 1979. Morphological variations of enamel prisms within abnormal striae of Retzius. *Human Biology* 51:139–151.

Shellis, R.P. 1984. Variations in growth of the enamel crown in human teeth and a possible relationship between growth and enamel structure. *Archives of Oral Biology* 29:697–705.

Shellis, R.P., and Poole, D.F.G. 1977. The calcified dental tissues of primates. In L.L.B. Lavelle, R.P. Shellis, and D.F.G. Poole (eds.), *Evolutionary Changes to the Primate Skull and Dentition,* 197–279. Springfield, Ill.: Charles C Thomas.

Smith, B.H. 1986. Dental development in *Australopithecus* and early *Homo. Nature* 323:327–330.

———. 1989. Dental development as a measure of life history in primates. *Evolution* 43:683–688.

———. 1991a. Standards of human tooth formation and dental age assessment. In M.A. Kelley and C.S. Larsen, *Advances in Dental Anthropology,* 143–168. New York: Wiley-Liss.

———. 1991b. Dental development and the evolution of life history in Hominidae. *American Journal of Physical Anthropology* 86:157–174.

Suwa, G. 1990. A comparative analysis of hominid dental remains from the Shungura and Usno formations, Omo Valley, Ethiopia. Ph.D. diss., University of California, Berkeley.

Suwa, G., White, T., and Howell, F.C. 1996. Mandibular postcanine dentition from the Shungura formation, Ethiopia: crown morphology, taxonomic allocations, and Plio-Pleistocene hominid evolution. *American Journal of Physical Anthropology* 101:247–282.

White, T.D. 1988. The comparative biology of "robust" *Australopithecus:* clues from context. In F.E. Grine (ed.), *Evolutionary History of the "Robust" Australopithecines,* 449–483. New York: Aldine de Gruyter.

Winkler, L.A. 1995. A comparison of radiographic and anatomical evidence of tooth development in infant apes. *Folia Primatologica* 65:1–13.

Chapter 16

Cranial Growth in *Homo erectus*

SUSAN C. ANTÓN

There are several compelling reasons why we might want to reconstruct the growth pattern of *Homo erectus*. The most basic of these is that evolution modifies the developmental pattern, so understanding ancestral development is critical to identifying how and when the descendant patterns of Neandertals and ourselves developed. Students of heterochrony argue that significant evolutionary changes can be accomplished by small changes in the timing of development (Gould 1977; Antón and Leigh 1998). And since an important suite of cranial characters are involved in the transition from *H. erectus* to *Homo sapiens*, it may be that studies of cranial development can provide clues to important adaptations in our lineage (*sensu* Antón and Leigh 1998).

Despite relatively few subadult fossils, there is a history of thinking that growth in *H. erectus* was accelerated relative to that in *H. sapiens* (e.g., Weidenreich 1943; Smith 1993). For Weidenreich, this view was based on ideas about the relationship between suture fusion and dental eruption in humans and great apes and on a general idea that *Sinanthropus* and *Pithecanthropus* were more apelike than humanlike in their cranial development (Weidenreich 1941, 1943; reviewed in Antón 1999). At the time of his writing, of course, *H. erectus* was the most primitive of the accepted fossil hominids, which otherwise included only Neandertals and fossil modern humans. More recently, Smith (1993) suggested that *H. erectus* achieved adulthood at the age of fifteen, which she recognizes as M3 eruption and equates with a human age of eighteen years. She bases her estimate of *H. erectus* adulthood on a developmental model that links primate life history with brain size and in which the *H. erectus* developmental pattern is approximately halfway between that of chimps and humans. Both Weidenreich's and Smith's models suggest, directly or indirectly, that at least early *H. erectus* may have lacked the adolescent growth spurt in linear height seen in *H. sapiens*.

The first step in any study of development must, of course, be the identification of subadult fossils. Unfortunately, as remarkable a find as is the nearly complete KNM-WT 15000 skeleton (the Nariokotome Boy), there is little other hard evidence on which to assess the growth pattern or tempo in *H. erectus*. And the evidence that remains is confused by a lack of systematic attention to many of the subadult remains, particularly those from Asia and Southeast Asia. Weidenreich suggested that a large number of cranial fossils were subadults, including six from Indonesia and five from China (Table 16.1). However, for Weidenreich any specimen with unfused sutures was a subadult. This has had some critical influence on the field, since many reference sources continue to use Weidenreich's age designations (e.g., Oakley et al. 1975; Santa Luca 1980; Wu and Poirier 1995) and thus many probable adults are excluded from analyses. Elsewhere I systematically examined the status of the Indonesian cranial material, resulting in my recognition of only two certain subadults, Mojokerto and Ngandong 2 (Antón 1997a, 1999; Antón and Franzen 1997). The Chinese subadults remain unreviewed.

Table 16.1   Cranial Remains Attributed to Subadult *Homo erectus*

Specimen	Weidenreich Age	Current Status	Element
Indonesia			
Mojokerto	18 mo	4–6 yr	Calvaria
Sangiran 3	5–8/8–9 yr	Young adult	Parietals/occipital ff
Ngandong 2	3–5 yr	8–10 yr	Frontal
Ngandong 5	Adolescent	Adult	Frontal/parietals
Ngandong 8	Adolescent	Older subadult/adult	Parietal fragment
Ngandong 9	Adolescent	Adult	Parietals
China			
Skull III	8–9 yr	Older subadult/adult	Calvaria
Skull IV	Subadult/adolescent	Adult	Parietal fragment
Skull VII	Adolescent/adult	Adult	Parietal fragment
Skull VIII	<3 yr	>6 yr	Occipital fragment
Skull IX	<6 yr	Older subadult/adult	Frontal frag + others
Africa*			
KNM-WT 15000	NA	11 yr (Smith 1993)	Skull + postcrania

*Does not include cranial specimens assigned to *Homo* aff. *erectus*.

My purpose here is to examine the purported subadult remains from Zhoukoudian, China, to compile a reliable subadult database and then to examine the metrical and morphological data that bear on the pattern and timing of growth in *H. erectus*. Even based on these few specimens, my data suggest that a reasonable argument might be made that an adolescent growth spurt did exist, even in early *H. erectus*.

Potential Cranial Subadults

Only twelve cranial specimens of *H. erectus* are reportedly subadult: Mojokerto, Ngandong 2, 5, 8, 9,[1] Sangiran 3, KNM-WT 15000, and Zhoukoudian Skull III, IV, VII, VIII, and IX. Mojokerto and Ngandong 2 are subadults of 4–6 and 8–11 cranial developmental years, respectively (Antón 1997a, 1999). In contrast, Sangiran 3 and Ngandong 5 and 9 are probably young adults, and Ngandong 8 is either an older subadult or young adult (Antón and Franzen 1997; Antón 1999; Table 16.2). As Smith (1993) has shown, KNM-WT 15000 is a subadult of approximately 11 dental developmental years. The Zhoukoudian remains from China have not been assessed since their original descriptions by Weidenreich (1943) and Black (1931).

In the 1920s and 1930s, Zhoukoudian (Choukoutien) Locality 1, some fifty kilometers southwest of central Beijing, yielded abundant fossilized remains popularly known as Peking Man; since then the area has yielded additional fossils. The hominid-bearing layers of Locality 1 range between 300,000 and 550,000 years old based on electron spin resonance, among other techniques (Grün et al. 1997). These remains, including the purported juveniles, were lost, perhaps by the U.S. Marines, during evacuations of China in World War II (Lanpo and Weiwen 1990). Thus, except for early work by Weidenreich and Black, all subsequent morphological work has been conducted on well-made casts of the fossils.

Although dubbed *Sinanthropus pekinensis,* the Chinese hominids were considered by most early workers to be conspecific with those from Java (e.g., Weidenreich 1943; von Koenigswald and Weidenreich 1939). Today, together with the Javanese fossils and specimens from Africa, they form the polytypic species *H. erectus* (e.g., Rightmire 1998). Some workers consider Asian and African specimens to represent two species, *H. erectus* and *Homo ergaster,* respectively (e.g., Wood 1994). While I attribute these differences to intraspecific geographic variation, the taxonomic designation of the Zhou-

Table 16.2 Evidence for Age-at-Death Determinations of *Homo erectus* Cranial Remains

Specimen	General Size	Thickness	Contours	Morphological Structures	Sutures
Subadult					
Mojokerto	Small	?	F/P-Round O-Flat	Incipient/absent	Unfused
Ngandong 2	Small	Thin	F-Round	Incipient	Unfused
Skull VIII	Small	Thin	?	Incipient	?
WT-15000	Small	Thin	F/P-Round O-Partially angled	Incipient	Unfused
Subadult/adult					
Skull III	Small	Low avg–thick	F-Round P/O-avg	Avg-thick SOT small	Unfused
Ngandong 8	?Average	Low average	?	Small	Unfused
Skull IX	Low average	Avg–thick	?	Avg-well	Unfused
Adult					
Sangiran	Average	Average	P-Avg	Average	Unfused
Skull IV	Average	Average	?	Average	Unfused
Ngandong 9	Average	Avg–thick	P-Avg	Average	Unfused
Ngandong 5	Average	Avg–thick	F/P-Avg	Average	Some fused
Skull VII	Average	Thick	?	Average	Some fused

Note: All comparisons are relative to adult ranges for *H. erectus* from a geographic region similar to that of the individual fossil. Average, within the range of adult development/size; low average, small end of adult range; thick, large end of adult range. F/P, frontal parietal; O, occipital; F, frontal; P, parietal; SOT, supraorbital torus; P/O, parietal/occipital.

koudian remains should have little effect on their age determination, since many of the comparisons I use to determine developmental age are relative to modern human standards or to the Zhoukoudian adults themselves, rather than to the greater *H. erectus* hypodigm (see below). The greater significance of this argument is to grouping of juveniles for the comparison of species-level developmental patterns and will be discussed below.

Five of the nineteen original Zhoukoudian cranial and facial fragments were considered subadult by Weidenreich (1943). Skull III is the most complete of these and was designated either an older subadult or possibly a young adult by Black (1931) and as an eight- or nine-year-old by Weidenreich (1943). Of the isolated cranial fragments, Skull VIII was designated "not older than 3 years" and Skull IX "approximately 6 years," with

Skulls IV and VII listed only as juvenile or adolescent. The criteria that can be assessed in each specimen are summarized in Table 16.2 and defined below.

In the assessment of cranial developmental status, the purported juveniles from Zhoukoudian are compared with *H. erectus* and *H. sapiens.* The adult *H. erectus* sample is as previously described for African and southeast Asian samples (Antón 1997a, 1999), but for thickness measures includes a larger number of Zhoukoudian adults (Table 16.3). The comparative *H. sapiens* sample includes adults and subadults from different geographic regions and focuses on Australia/Papua New Guinea because of the suggested connection between Australian *H. sapiens* and Asian *H. erectus* (e.g., Thorne and Wolpoff 1981).

Assessing Subadult Status: Cranial Criteria

Most of the purportedly subadult specimens from Asia and southeast Asia lack associated facial and dental remains. As a result, age determinations— even the seemingly simple distinctions between adult and subadult—become difficult. In the absence of teeth, it is essential to combine multiple lines of evidence for assessing skeletal maturation. I previously suggested that aspects of vault thickness, superstructure development, sutural development, and sagittal vault contours can be used together to assess the ontogenetic status of purported subadults (Antón 1997a, 1997b, 1999; Antón and Franzen 1997; Table 16.1).

In the Zhoukoudian specimens I evaluate three aspects of cranial vault thickness that should distinguish young adult from subadult *H. erectus:* (1) three layers of *fully* differentiated cranial bone; (2) fully developed, localized outer table hypertrophy of all cranial superstructures, including eminences, keels, and tori; and (3) values of absolute vault thickness within adult *H. erectus* ranges (Antón and Franzen 1997; Table 16.3). I also evaluate the relative development of their cranial superstructures by comparing vault thickness at the occipital and angular tori and at the bregmatic and parietal eminences with ranges for adults from Zhoukoudian (after Weidenreich 1943; Black 1931) and with development in accepted subadults (Antón 1997b). Incipient rather than full development of cranial superstructures is considered a strong indicator of subadult status.

In addition, I examine the development of sutural interdigitation (Antón and Franzen 1997) and bony sutural fusion (*sensu* Meindl and Love-

Table 16.3 Cranial Measurements for Zhoukoudian Subadults and Adults

							Thickness (mm)					
	III	IV	VII	VIII	IX	I	II	V	VI	X	XI	XII
Bregma	9.6 9.0b	—	—	—	—	—	9.0 8.0b	—	(9.9)	7.5	7.0	9.7
Mastoid angle	17.2 16.0b	—	17.4	—	—	14.0	13.5 13.0b	14.0	—	14.0	13.5	14.5
Occipital torus	20.4 17.0b	—	—	7.1	—	—	—	(12.3)	—	15.0	12.0	15.0
Parietal eminence	11.0	(10.7)	—	—	—	5.0?	11.0	—	11.2	12.5	16.0	9.0
Midsupraorbital torus	11.5	—	—	—	—	—	14.0	—	—	13.0	14.0	16.0
Frontal squama	10.0	—	—	—	—	13.0	10.0	—	(9.5)	7.0	11.0	7.0
Temporal fossa	4.8	—	—	—	—	—	6.5	—	4.6	(5.8)	4.6	5.5
Cranial capacity*	915	—	—	—	—	—	1,035	—	—	1,225	1,015	1,030

Note: Measurements are from Weidenreich (1943), except those marked *b*, which are from Black (1931). Numbers in parentheses are estimates.
*Measured in cubic centimeters.

joy 1985) in the Zhoukoudian specimens compared with adult and sub-adult *H. sapiens* and *H. erectus*. Although the age of bony sutural fusion is notoriously variable in humans, endo- and ectocranial vault suture fusion is rarely encountered during adolescence in nonpathological individuals (Meindl and Lovejoy 1985; Cohen 1986; Buikstra and Ubelaker 1994).

Finally, *H. erectus*, but not *H. sapiens*, vault contours provide a means of distinguishing between adult and subadult vaults (Antón 1997a, 1997b, 1999; Antón and Franzen 1997). The vault contours of the most complete Zhoukoudian "subadult" (Skull III) are evaluated in this light. Sagittal chords were measured using digital and spreading calipers and a steel tape for the frontal (glabella-bregma), parietal (bregma-lambda), and occipital (lambda-opistion) arcs. The relationship between cranial chords and arcs was evaluated for each bone using least-squares linear regression and re-duced major axis regression analyses (Systat, version 5). These data show that *H. erectus* has increased flattening of frontal contours and angulation of occipital contours with increasing developmental age. However, adult *H. sapiens* vaults, although larger than subadult vaults, are not significantly less round than subadult vaults. Thus, if Skull III is subadult it should have a more rounded frontal and less angulated occipital area than does adult *H. erectus*.

In the final assignation of cranial developmental age, it is important to consider that the male skull is usually thought to be both larger and more robust than the female and it may be possible to confuse a juvenile male for an adult female (see Antón 1999 for further arguments). In these cases the relative consistency of the results from multiple lines of evidence is particularly important. Here a specimen is considered to be in a certain age category (i.e., adult or subadult) if all lines of evidence indicate that age status and if there is no inconsistency within any of these lines, such as cranial thickness or superstructure development. A specimen that ex-hibits inconsistency among the different lines of evidence or within any given line is assigned to the transitional category of older subadult/young adult (see below). I am assessing cranial developmental maturation. Cra-nial developmental maturity may or may not correlate with reproductive maturity and may be achieved at different ages in males and females (Marshall and Tanner 1986; Bogin 1994). However, the assessment of cranial developmental maturity is important to fossil studies as it allows comparison among fully grown forms of each taxon and excludes those specimens that, for reasons of incomplete skeletal growth, have a differ-ent morphotype.

Description and Developmental Age of Zhoukoudian Remains

Below I provide general identifications of the subadult Zhoukoudian remains and describe details necessary for the assessment of age at death. Summaries of this information are provided in Tables 16.2 and 16.3. The reader is referred to Weidenreich's (1943) extensive anatomical descriptions and figures of each specimen for further information and to Black (1931) for additional information on Skull III.

Skull III (Locus E)

Skull III is a calvaria lacking only portions of the right orbital plate of the frontal bone, the body and right infratemporal portion of the sphenoid, the basilar and most of the exoccipital portions of the occipital, and a small portion of the left nuchal region of the occipital bone (Fig. 16.1). The tympanic portions of the temporal bones remain patent, with a linear groove extending medially from the external auditory meatus and dividing the tympanic plate into anterior and posterior moieties. This bilateral defect is probably the result of failed fusion of the tympanic ring and plate during development (Black 1931; see below).

In terms of general size, Skull III is somewhat smaller in virtually all di-

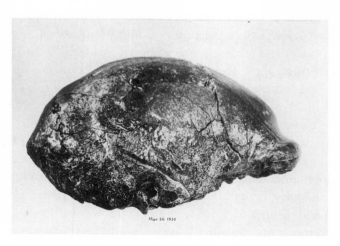

Fig. 16.1. Skull III, right lateral view of calvaria. Note the cleft visible on the lateral edge of tympanic. *Source:* Photo courtesy of Department of Library Services, American Museum of Natural History. Negative No. 273871.

mensions than the adult calvaria from Zhoukoudian. This is true for all length and most breadth and height measures when compared with those of the inferred male (larger) crania. It is likewise true for virtually all length measures in comparison with Skull XI (an inferred female). Skull III is close in size or slightly smaller than Skull XI in breadth and height measures. Skull III also has the smallest cranial capacity (915 cc) of any of the Zhoukoudian calvaria.

In addition to considering its overall small size, Weidenreich (1943) suggested that the small size of the mandibular fossa precluded an adult age for Skull III. He suggested that the fossa would only accommodate a subadult mandible, and one of the same size (Mandible F-I) was present in the vicinity of the Skull III find. Weidenreich considered F-I to be eight to nine years of age and used this to assign an age to Skull III. However, an age of fifteen or greater seems likely for this fragmentary specimen (see below, under "Assessing Subadult Status: Mandibular Criteria"). Although relative development of the mandibular fossa is useful in determining cranial developmental age in the childhood years (up to seven years, Antón 1997a), it is not clear whether mandibular fossa size in an older age range is a good indicator of adult versus subadult status. This is particularly true because Skull III differs from the adult means only in mandibular fossa depth, a measure that is reduced by the benign developmental defect of the tympanic plates mentioned above (length = 18 mm; breadth = 25 mm; depth = 11.5 mm vs. adult means of 18, 25, and 13.9, respectively).

The relative development of the cranial superstructures of Skull III presents a mixed message. In support of a developmental age of some eight or nine years, Weidenreich (1943) cited the relatively weak development of the frontal and occipital tori and faint nuchal markings in Skull III. The supraorbital torus, although well developed compared with young subadults such as Ngandong 2, is the most gracile and least thick (11.5 mm; adult range, 11–16 mm) of the Zhoukoudian calvaria. Likewise, the nuchal attachments are not well defined. However, the occipital torus is well developed and thick (17.0 mm; adult range, 12–17 mm; Table 16.3), as is the angular torus (16.0 mm; adult range, 13–16 mm). A sagittal keel is also present.

The general vault thicknesses of Skull III are within the adult range; however, they vary in terms of their relative thickness compared with adults from Zhoukoudian (Table 16.3). Some areas, such as the angular and occipital tori and the frontal squama including bregma, are among the thickest of the adult range for these measures. Other areas, such as the parietal eminence and temporal fossa region of the frontal bone, are among the thinnest of the adult range for these measures.

The unfused cranial sutures were the main reason that both Weidenreich (1943) and Black (1931) considered Skull III subadult. Although the sutures are unfused, they are thick. Black argued that the sutures indicated ongoing growth because they were not closely interlocking—that is, the interdigitations seemed too large for the "fingers of bone" they received. As the originals are no longer available, it is impossible to explore this possibility further, although confirmation of Black's observations would be strong support for a subadult age. I argued elsewhere that unfused sutures alone should not be the sole indicator of subadult status (Antón and Franzen 1997; Antón 1999). Thus, unless Black's observations could be confirmed, unfused sutures could be an equally good indicator of young adult status.

The cranial vault contours of Skull III also present a mixed message. In both parietal and occipital contours, Skull III plots among adult *H. erectus* specimens, including those from Zhoukoudian. However, Skull III frontal contours are significantly rounder than adult *H. erectus* frontal bones and plot along the subadult *H. erectus* trajectory (Fig. 16.2, *A* and *B*). This does not represent a regional difference in frontal contours, as all other adult Zhoukoudian frontals plot with adult *H. erectus* frontals from Indonesia and Africa.

In addition to sutures, Weidenreich (1943) suggested several other characters that he believed indicated that Skull III was some eight to nine years old at death. These included a large frontal sinus, cleft tympanic plates, and the presence of a lacrimal groove (depression). All of these are, he suggested, indicators of young age because they are not typically found in adults of the species. However, none of these are particularly compelling evidence of young age. According to Weidenreich's own measurements the frontal sinus of Skull III, while larger than that of adults from Zhoukoudian, is within the size range of the Indonesian hominid sinuses (see Weidenreich 1943, 164; Weidenreich 1951, 252). Likewise, there is no evidence that the juvenile Ngandong 2 has a larger frontal sinus than the Ngandong adults (Weidenreich 1951). As concluded by Black (1931), the cleft tympanic plates undoubtedly represent a minor developmental defect in which the tympanic plate fails to ossify completely. This condition is not infrequently encountered in adult modern humans, usually in the form of the Foramen of Huschke (tympanic dehiscence), although other forms are possible (De Stefano and Hauser 1989). In addition, in modern humans the tympanic plate is usually fused by the dental developmental age of 2.5

to 4 years (Weaver 1979; Sullivan and Weaver 1981). For the cleft to remain "normally" in an 8 or 9-year-old required Weidenreich to argue that in this feature *H. erectus* growth was retarded relative to that of modern humans (1943, 55), whereas in all other ways cranial growth was accelerated relative to humans. A lacrimal depression may be either present or absent in modern human adults. Its presence may be related to the relative development of the orbital versus squamous portions of the frontal bone and perhaps the supraorbital torus.

The age determination for Skull III is thus problematic, since some evidence points to a subadult age, whereas other evidence suggests an adult age. I am reluctant to place Skull III as a definitive adult simply because of this conflicting evidence. For example, in the case of Sangiran 3, which I have argued is an adult, the metrics are consistent in suggesting a relatively small adult. All linear distances and all thicknesses suggest a similarly sized, smallish individual quite consistent with Sangiran 2 (Antón and Franzen 1997). Likewise, all superstructure development in Sangiran 3 indicates a relatively gracile individual. In the case of Skull III, however, some metrics suggest that it is among the largest of the adult males, whereas others suggest that it is among the smallest of all adults. Some superstructures are the most robust of the entire sample (e.g., occipital torus), whereas others are the most gracile (e.g., supraorbital torus). This kind of inconsistency suggests the possibility that Skull III was a still-growing individual (perhaps male) that would have become a large adult, as it is now a large subadult. Recognizing this inconsistency both Black and Weidenreich considered Skull III a subadult male, although Weidenreich considered it no more than eight or nine years of age. However, if Skull III is a subadult, it must be a rather older subadult, as many adult structures are already in place. This is essentially the position Black eventually took and to which I am sympathetic.

I also find the cranial contour data rather compelling support of subadult status in Skull III. Other adult *H. erectus* frontals from distant geographic regions plot with one another and apart from subadult frontal bones, granted that the subadult sample is small (Fig. 16.2*A* and Antón 1997a, 1997b, among others). Subadult age could explain why Skull III does not plot with the adult frontal bones, since continued growth in its supraorbital torus would have changed the angulation of the frontal, presenting a flatter, more adult profile. This is consistent with what we know of human growth of the frontal bone, which, in the adolescent years, is largely accomplished by growth in the brow region (Björk 1955). While I prefer this explanation, the reverse is also possible, that the small supraorbital torus of

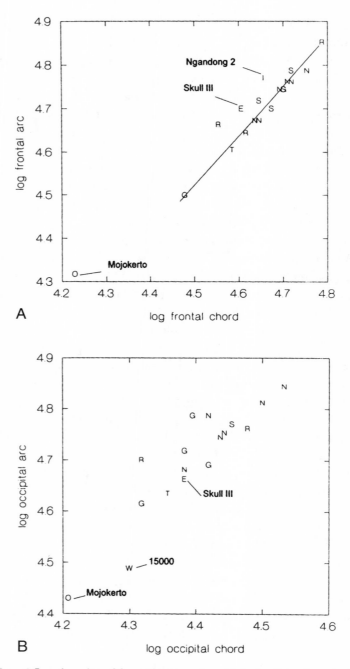

Fig. 16.2. *A:* Log-log plot of frontal contour data. Rounder frontals are toward the northwest of the plot. Skull III is more rounded than adult *H. erectus* specimens from Zhoukoudian, southeast Asia, and most from Africa. The roundest *R* is KNM-ER

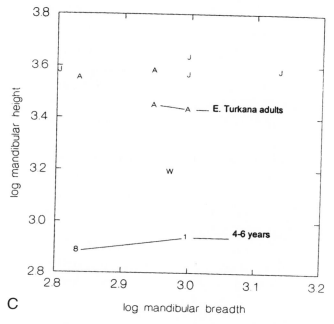

C

(*Fig. 16.2 continued*) 3733. *Unlabeled letters* are adults from Sangiran (*G*), Ngandong (*N*), Zhoukoudian (*S*), and Trinil (*T*). *B:* Log-log plot of occipital contour data. More angulated occipitals are toward the northwest of the plot. Skull III plots with adult *H. erectus.* Unlabeled letters are as for *A*. *C:* Log-log plot of mandibular breadth and height at M1. For KNM-WT 15000 to plot with adults from East Turkana, a 29 percent increase in its present mandibular height is required. Data points are subadults KNM-ER 820 (*8*), KNM-ER 1507 (*1*), KNM-WT 15000 (*W*), and adults from Africa (*A*, including KNM-ER 992 and 730 and Tighenif 1 and 2) and from Sangiran, Java (*J*).

Skull III is simply idiosyncratic adult variation, which explains its rounder frontal contours. The data, however, suggest that there is not such great variation in the rest of the adult *H. erectus* sample (Antón 1997a, 1999; Antón and Franzen 1997; Antón and Weinstein 1999). Whichever is the case, based on the above observations, Skull III must be placed in the older adolescent/young adult category, with my particular inclination being that this individual still had some growing left to do.

Skull IV (Locus G)

Skull IV is a fragment of a right anterior parietal bone, including portions of the coronal, sagittal, and squamosal sutures. Bregma and pterion are missing. The superior-inferior dimension of about 100 mm for this parietal fragment is within the general size range of adult parietal bones from Zhoukoudian. A sagittal keel is present. The temporal lines are well developed. Postmortem fractures reveal three, apparently fully developed, layers of cranial bone. As the original was not available for study, the absolute dimensions of each layer cannot be compared with those of adults, and it remains possible that these were not completely developed. Weidenreich's measurement of 10.7 mm for the parietal tuberosity is at the low end of the adult range for this measure, although this measure is likely to be small, since the true tuberosity is not preserved in this specimen (Table 16.3). The cranial sutures are unfused but fully interdigitated, and they seem to be fully developed in terms of thickness. Cranial contours cannot be assessed. All observations suggest that Skull IV is an adult (Table 16.2). However, this specimen is similar to the parietal bone of Skull III in some regards; thus, further information on cranial contours or superstructure development might make this designation more tentative, as is the case with Skull III.

Skull VII (Locus I)

Skull VII is a fragment of the right mastoid angle of the parietal bone (Fig. 16.3*A*). Portions of the lamdoidal and parietomastoid sutures are present. The angular torus is fully developed and robust. The thickness of the angular torus is the largest of the adults for this measure (17.4 mm, Table 16.3). Sections of the lambdoidal suture were apparently fused, since occipital remnants remain fused to the parietal bone. Cranial contours cannot be assessed. Weidenreich identified Skull VII as an adolescent in his summary tables but as an adult, because of the aforementioned suture fusion, in his text. All observations suggest that Skull VII is an adult (Table 16.2).

Skull VIII (Locus J)

Skull VIII is a fragment of occipital bone preserving the midline occipital torus and sagittal sulcus (Fig. 16.3*B*). No sutures are preserved. The occipital torus is partially developed on the midline, with a well-defined supratoral sulcus and less strongly defined nuchal attachments. Endocranially, the sagittal sulcus is well defined. Three layers of cranial bone are present, but the compact layers appear absolutely small relative to the thickness of the diploë. As the original was not available for study, the absolute dimen-

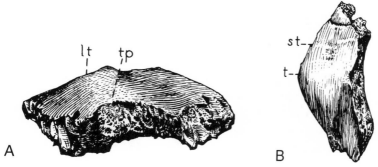

Fig. 16.3. *A:* Skull VII, posterior view of parietomastoid suture of right parietal bone drawn from the original. Note the well-developed angular torus (*tp*). *lt,* linea temporalis. Originally published as Figure 40 of Weidenreich 1943. *B:* Skull VIII, right posterolateral view of occipital fragment drawn from the original. Superior is toward *top.* Note the thin endo- and ectocranial tables. The occipital torus (*t*) and supratoral sulcus (*st*) are better developed than in the Mojokerto specimen. See text for dimensions. Originally published as Figure 45 of Weidenreich 1943. *Sources:* Photos courtesy of Department of Library Services, American Museum of Natural History, from Weidenreich archive. AMNH negative numbers 2A24734 and 2A24737, respectively. Photo of original drawing taken by J. Beckett.

sions of each layer cannot be measured. However, it is clear that the outer table hypertrophy typical of adult *H. erectus* is not present in this individual. As a result, the occipital torus and cerebellar fossae are absolutely thin in comparison with those of adult occipital bones from Zhoukoudian (Table 16.3). Cranial contours cannot be assessed. All observations suggest that Skull VIII is a subadult (Table 16.2).

Skull IX (Locus J)

Skull IX consists of small, unidentified cranial fragments and a fragment of anterior-inferior left frontal bone, including a portion of the orbital surface and zygomatic process of the frontal bone, temporal lines, temporal fossa, and coronal suture. The midline is not preserved. The anterior-posterior dimension from the orbital surface to the coronal suture near stephanion (42 mm) is at the low end of the size range for this measure in adult *H. erectus.* Weidenreich considered other measures, including the distance below the temporal lines, to be small for adult *H. erectus.* Because of the pattern of postmortem breakage, the region of the bregmatic and metopic eminences cannot be observed. However, the temporal lines are strongly developed. The thickness of the temporal fossa is at the large end of the adult range for this measure, whereas the thickness of the frontal squama is at the low end of the adult range for this measure. The coronal suture is unfused. Cranial contours cannot be assessed. Skull IX, like Skull III, presents mixed evidence, particularly with respect to cranial thickness. Based on these observations the specimen is considered either an older subadult or a young adult, with my inclination being to consider the specimen adult. However, both absolute size and frontal squama thickness could be used to argue for subadult status (Table 16.2).

Assessing Subadult Status: Mandibular Criteria

Assessing the age at death of mandibular and maxillary remains with teeth is less problematic than evaluating the status of cranial remains without teeth. While there are differences in opinion as to whether ape or human standards are more appropriate to suggest developmental ages for fossil hominids (e.g., Bromage and Dean 1985; Mann et al. 1990; Smith 1993; Anemone, Chap. 12, and Kuykendall, Chap. 13, this volume), the presence of unerupted or incompletely developed teeth obviates the problems of distinguishing between subadult and adult remains that were encountered with the cranial remains. Several workers have carefully reviewed the methods for assessing dental development (see Chaps. 12 and 13), and these will not be repeated here. Below I briefly review the specimens and the ages assigned to them.

Potential Mandibular Subadults

A total of nine subadult mandibles/mandibular fragments have been assigned to *H. erectus* or *H.* aff. *erectus*. Three subadult mandibles or partial mandibles are from Kenya: KNM-WT 15000, KNM-ER 820, and KNM-ER 1507 (Table 16.4). Based on human dental eruption standards, KNM-WT 15000 is between 10.5 and 11 years of age (Smith 1993), KNM-ER 820 is about 6.5 years of age, and KNM-ER 1507 is 4.0–6.1 years of age (Smith 1986). Perikymata counts suggest an age of 5.3 years for KNM-ER 820 (Bromage and Dean 1985), and KNM-ER 1507 has perikymata spacing and root development similar to that of KNM-ER 820 (Dean 1986). Six subadult mandibular fragments are from Zhoukoudian (Table 16.4). Based on human dental eruption standards, Weidenreich (1936) assigned

Mandibular Remains Attributed to Subadult *Homo erectus* or *Homo* aff. *erectus*

Specimen	Dental Eruption Age*	Element	Erupting	In Occlusion
China				
B-I mandible	8–9 yr (Weidenreich) 7–9 yr (Antón) 4.0–7.3 yr (Smith-hum) 4.9–5.7 yr (Smith-ape)	symphysis, rt corpus	M2 (in crypt)	Is, dc, dm1, dm2, M1
B-III mandible	8–9 yr (Weidenreich) 7 yr (Antón)	rt corpus, I1,2,dm,1,2,M1	MI (3/4 up)	Is (frag), dm1, dm2
B-IV mandible	5–6 yr (Weidenreich) 5 yr (Antón)	rt corpus di 1-dm2	M1 (in crypt)	All deciduous
B-V mandible	11 yr (Weidenreich) 10–11 yr (Antón)	symphysis + rt corpus	C (3/4 up), P3 (3/4 up)	Is, dm2
C-I mandible	8–9 yr (Weidenreich) 15 yr (Antón)	rt. corpus frag (M2)	M2 or M3 (1/4 up)	NA
F-I mandible	8–9 yr (Weidenreich) 15 yr (Antón)	rt. corpus frag M1 (M2)	M2 or M3 (1/2 up)	M1 or M2
Africa				
KNM-WT 1500	10.5–11 yr (Smith-mult) 5.5–8.2 yr (Smith-hum) 5.3–6.5 yr (Smith-ape) 5.3 yr (Bromage)	mandible	M3? (in crypt)	Is, C, Ps, M1, M2
KNM-ER 820		mandible-no left ramus or condyle, fragmtd right	permanent in crypt except M3	Is, dc, dm1, dm2
M1KNM-ER 1507	4.0–6.1 yr (Smith-hum) 4.8–5.5 yr (Smith-ape)	lft mandible corpus (Ldc-M2)	C, Ps, M2 all in crypt	dm1, dm2, M1

Note: At least one recent juvenile mandibular fragment has been found in Indonesia; however, it has not yet been published.

*Sources in parentheses: Antón, this study; Bromage, Bromage and Dean (1985); Smith (1989 or 1993; human, ape, or multiple standards); and Weidenreich (1936).

ages of between 5 and 11 years to these fossils. Only one (B-I) was assessed by Smith (1986), yielding an age of 4.0–7.3 years depending on the tooth examined.

My assessment based on eruption sequences using Ubelaker's (1984) criteria agrees broadly with Weidenreich, yielding an age-graded series from youngest to oldest of B-IV, B-III, B-I, B-V, F-I/C-I. However, the ages I assign to the locus B mandibles are somewhat younger than those of Weidenreich (Table 16.4). This systematic discrepancy is likely to be due to our use of different reference standards. Alternatively, Mandibles F-I and C-I may be older than suggested by Weidenereich. He argued that both retain M2 but not M3. However, the morphology of both "M2s" is more consistent with that of an M3: both are highly crenulated with reduced cusps and sometimes unidentifiable cusp patterns. If they retain unerupted M3s rather than M2s, both would be closer to fifteen dental developmental years or older. F-I is the mandible Weidenreich used to assign an eight- to nine-year-old age to Skull III.

Relative Ages of Accepted Subadults and Superstructure Development

Based on my reassessments of the cranial remains, there are four certain subadults and three older subadults/young adults of *H. erectus*. In addition, there are nine subadult mandibular remains of *H. erectus* or *H.* aff. *erectus*. These specimens span a relatively large developmental age range and include many parts of the cranium; thus, they may inform us about the pattern of development in *H. erectus*. Thickness measures for the cranial specimens are compared with adults from their region in Tables 16.3 and 16.5.

From youngest to oldest the cranial remains are Mojokerto, Skull VIII/Ngandong 2, WT-15000, followed by the subadult/adults Ngandong 8, Skull III, and Skull IX. In previous studies Mojokerto was assigned a developmental age of four to six years (Antón 1997a) and Ngandong 2 a range of greater than six and less than 11 years, with a more likely age of eight to ten years (Antón 1999). Skull VIII is certainly older than Mojokerto based on the greater degree of definition in its occipital torus, and Skull VIII has a well-defined supratoral sulcus that Mojokerto lacks. The relative ages of Skull VIII and Ngandong 2 cannot be assessed, since they do not preserve similar anatomical structures. However, both are likely to be younger than KNM-WT 15000, although this is more certain for Ngandong 2 than

Table 16.5 Cranial Thickness Measures for KNM-WT 15000 and Ngandong Juveniles Compared with Adults from the Same Area

			Thickness (mm)			
	WT 15000*	ER 3733[†]	ER 3883[†]	Ng 2	Ng 8	Ng Adult range
Bregma	7.9	—	8.0	7.2[‡]	—	8.0–11.0[§]
Asterion	10.1	5.0	9.0	—	9.7[‡]	—
Parietal eminence	5.0	7.5	10.0	—	—	7.9–11.8[§]
Mid-supraorbital torus	10.4	9.0	14.0	5.8[†]	—	11.0–14.2[†]

Note: Metrics are from *Walker and Leakey (1993), [†]Wood (1991), [‡]Antón (1999), [§]Gauld (1996).

Skull VIII, as both cranial contours and relative morphological development can be compared for the former specimen. As I argue above, Skull III provides some suggestive evidence of subadult status, and if this is the case then certainly Ngandong 8 would be younger than Skull III, as the angular torus in Ngandong 8 is not particularly well developed.

Trying to integrate the mandibular remains with the cranial remains in terms of an age series cannot be precise. However, based on human eruption standards, KNM-ER 820, 1507, and Zhoukoudian B-1 (Smith age) and B-IV fall in the developmental age range suggested for Mojokerto, whereas Mandibles B-I (Weidenreich age), B-III, and B-V are approximately of the age range suggested for Ngandong 2/Skull VIII, and, of course, KNM-WT 15000 is associated with the cranial remains of that individual. If both Mandibles C-I and F-I retain M3 rather than M2, then they are the oldest of the juvenile mandibles, perhaps corresponding to Skull III in age.

On the basis of these series and comparison of the relative development of different structures in the more complete individuals, we can evaluate cranial superstructure development in *H. erectus*. The cranial superstructures are localized thickenings of ectocranial bone and diploë that develop as the cranial tables differentiate throughout the subadult years (Weidenreich 1940; Hublin 1978, 1986; Antón 1997b). The occipital torus and metopic eminence are the first to develop, as only these structures are present in the youngest specimen, Mojokerto. Mojokerto's occipital torus is more developed than its metopic eminence, suggesting that the occipital torus is the first of the superstructures to differentiate. The bregmatic eminence, sagittal keel, and angular torus are not present in Mojokerto, suggesting that the metopic eminence is the second structure to develop. Although Mojokerto's supraorbital gutter (ophryonic) region has already taken on the characteristic *H. erectus* form, the supraorbital torus itself is undeveloped in all preserved areas.

The bregmatic eminence and supraorbital torus are both incipient in Ngandong 2. Because the bregmatic eminence is contiguous with the metopic eminence, it is tempting to suggest that all three structures develop before a sagittal keel. However, the lack of a parietal bone in this specimen precludes our knowing whether a sagittal keel or angular torus was also beginning to form at this stage. The midline swelling in the region of the supraorbital torus in Ngandong 2 indicates that this structure begins to develop after the metopic eminence and at about the same time as the bregmatic eminence.

By the age of Skull III, all of the superstructures, except perhaps the supraorbital torus, are fully developed, suggesting that the supraorbital torus may be the last to achieve adult form. This delay in development of the supraorbital torus is consistent with human data that show that the superciliary region develops relatively late as part of adolescent facial development (Brodie 1953). The order of acquisition among the sagittal keel, angular torus, bregmatic eminence, and supraorbital torus is not discernable from the subadults available. Either all these structures appear together, as seems to be the case for the bregmatic eminence and supraorbital torus, or the sagittal keel and angular torus appear later than the other structures.

Comparative Growth Patterns in *H. erectus* and *H. sapiens*

We can now consider the time of appearance during development of the cranial vault contours that differentiate adult *H. erectus* from adult *H. sapiens*. When these differences appear may provide important clues as to how and why our derived neurocranial shape evolved (Antón and Leigh 1998). The developmental timing may also provide clues as to the presence or absence of an adolescent growth spurt in *H. erectus*.

As discussed previously, the adult *H. erectus* vault has a flatter frontal and more angulated occipital bone than do subadult *H. erectus* vaults, and these contours get progressively closer to the adult form as we look at developmentally older individuals (Antón 1997a, 1997b, 1999; Antón and Leigh 1998). Developmentally young *H. erectus* specimens, such as Mojokerto, have rounder frontal bones than their older counterparts, such as Ngandong 2. Moreover, Ngandong 2 has a rounder frontal contour than KNM-WT 15000 (estimated) or Skull III. Similarly, occipital contours become increasingly angulated in older youths. Thus, modern human adult vaults retain the appearance of the juvenile ancestor, and it may be that associated benefits of the juvenile form, perhaps even behavioral plasticity, should be considered as a possible source of the selective advantage conferred by this change in vault shape. Any behavioral correlates with vault shape remain to be tested in primates, although selection for certain behaviors seems to have also resulted in more juvenile cranial form in domesticated foxes (Trut 1991; Trut et al. 1991).

Despite the increasingly adult shape of the vault with increasing age of the juvenile, there is still a sizable gap between the shape of the vault of KNM-WT 15000 and that of adults from the same region, including KNM-ER 3733 and 3883 (Figs. 16.2*B* and 16.4). In other words, the vault of the eleven-year-old Nariokotome Boy has not achieved the size or proportions of adult *H. erectus*. Particularly in the occipital region, the Nariokotome Boy would have required a great deal of growth before achieving maturity (adult shape). This may suggest that there was an adolescent growth spurt yet to be undergone by the Nariokotome Boy and thus a growth spurt in *H. erectus*. However, because of the small number of subadult vaults the data are not conclusive.

In both of the above arguments, interpretations of results are confounded by taxonomic arguments surrounding African and Asian specimens of this sample. If these two subsamples represent separate species, then pooling them to discuss developmental patterns, without first establishing that their specific-level growth patterns are similar, is unwarranted. Given the small size of subadult samples, I cannot compare patterns of cra-

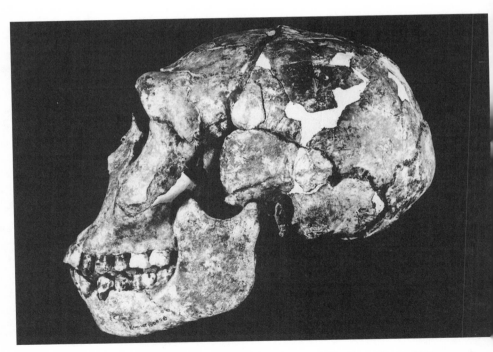

Fig. 16.4. KNM-WT 15000, left lateral view of the skull. Note the small degree of angulation in the occipital region, and compare with the position of this specimen in Figure 16.3. *Source:* Photo courtesy of Dr. Alan Walker.

nial contour growth between regions and thus have no solution that will satisfy those who do not consider African and Asian specimens conspecific. However, as luck and accidents of preservation would have it, the discussions of growth patterns below are largely limited to evidence only from Africa, thus obviating the taxonomic issue.

An adolescent growth spurt—that is, a dramatic increase in skeletal growth due to an increase in the velocity of growth around the time of sexual maturity—occurs in most human groups (Bogin 1994). Because adolescence is a time when individuals learn critical adult roles, the time of origin of this life stage could have important implications for hominid evolution, particularly relating to cultural complexity. The problem of recognizing such a spurt in the fossil record, however, would be manifold, even with a much larger sample than is currently at hand, because of the difficulties of recognizing changes in growth velocity in cross-sectional samples; changes in velocity known to occur in living humans are artificially depressed in cross-sectional samples (Tanner 1978; Leigh 1996). However, some studies have attempted to discern the presence or absence of an adolescent growth spurt in *H. erectus* by accepting as proxies for changes in growth velocity either (1) the percentage of adult growth achieved in the postcranial skeleton at a certain dental developmental age or (2) the development of certain morphological modifications undergone during adolescent growth.

Bogin and Smith (1996) argue that human adolescence and the growth spurt may appear relatively late in the hominid lineage, after the earliest *H. erectus*. This late acquisition of adolescence is suggested to them by the relationship between the preadolescent dental stage of KNM-WT 15000 (eleven years) and his stature (1.6 meters), which is comparable to that attained "even in tall populations" at ages of thirteen to fifteen years, which is well into an adolescent period (Smith 1993, 206). Smith (1993) suggests reasons why dental and postcranial developmental ages might be inconsistent: the growth pattern of *H. erectus* might differ from that of humans in not having a growth spurt, indicating that maturation was accelerated in *H. erectus* relative to *H. sapiens,* or KNM-WT 15000 might represent idiosyncratic (or even pathological) variation in his developmental modes between dental and skeletal maturation (i.e., he might be an outlier, although encompassed by human standards).

In contrast, Tardieu (1998) suggests that a growth spurt was in fact present in *H. erectus,* albeit one that was less pronounced and of shorter duration than that in modern humans. She bases her conclusions on discrete changes in the morphology of the distal femoral epiphysis, which, she suggests, indicate the presence of an adolescent growth spurt. These markers include a strong projection of the lateral lip of the femoral trochlea, anteroposterior lengthening of the distal femoral epiphysis, and an increase in the radius of curvature of the lateral condyle. She finds these markers in KNM-WT 15000, the same specimen that Smith (1993) uses to argue that there is *not* a growth spurt in early *H. erectus*. Although these studies disagree as to whether an adolescent growth spurt existed in early *H. erectus*, both seem to suggest that the postcranial skeleton of KNM-WT 15000 indicates a slightly more advanced developmental age than does the dentition.

If *H. erectus* lacked an adolescent growth spurt, then all skeletal systems usually affected by this spurt should reflect its absence and show relatively greater attainment of adult size/shape at earlier dental ages. Smith (1993) suggests that this is the case in terms of limb length and epiphyseal development in the Nariokotome Boy, whereas Tardieu (1998) argues that it is *not* the case in terms of his femoral morphology, and I suggest that it does not seem to be the case in terms of his cranial contours. But should we be able to see a growth spurt, if present, in the cranial skeleton?

Arguably, the adolescent growth spurt in modern humans affects all skeletal tissue, causing increases in nearly all linear skeletal dimensions, including those in the cranium, particularly mandibular height (Marshall and Tanner 1986; Smith 1993). Although the correlation between accelerated postcranial growth and growth of the face and vault is not well understood, peak facial growth velocities occur within a few months of peak growth velocities in stature, although they may be less extreme (e.g., Nanda 1955; Thompson et al. 1976; Mitani 1977; Baughan et al. 1979; Ekstrom 1982). Mandibular height is the most strongly affected facial measure, increasing some 25 percent between twelve and twenty years of age (Goldstein 1936; Marshall and Tanner 1986). While this increase reflects a relatively wide window of time and does not necessarily imply the strongly constrained period of peak growth seen in the adolescent spurt of the postcranial skeleton, the percentage of adult cranial growth achieved at a particular developmental age should give us some indication of whether a larger percentage of growth has occured than in a comparably aged modern human. Thus, further exploration of the apparent inconsistency in growth pattern between the postcranial and dental development of KNM-WT 15000 is possible by assessing relative development of the cranium and dentition. That is, are relative dental development and cranial development inconsistent with one another in the same way that postcranial and dental development differ?

Measures of Facial Height

To investigate whether cranial data other than vault contours might also suggest the presence of a growth spurt, I consider measures of facial height. Although adult human cranial capacity is achieved relatively early in childhood (Scammon 1930; Tanner 1988), facial height reaches adult size relatively late in development (Goldstein 1936; Brodi 1953; Enlow 1975; Moore 1982). If the 25 percent increase in mandibular height between twelve and twenty years of age is indicative of a facial growth spurt, then the presence or absence of such a pattern of increase in the face of *H. erectus* might signal whether a similar spurt existed in that species.

Restricted by specimen preservation, I consider two measures of facial height, upper facial height and mandibular height. For upper facial height (nasion-prosthion), there is only a single subadult data point, KNM-WT 15000. This data point can be compared with adult *H. erectus* from the same geographic region to see how much growing, if any, might have been left to KNM-WT 15000. Mandibular corpus height at M1 allows the inclusion of more subadult data points, including KNM-WT 15000, KNM-ER 820, and 1507, as well as several adults from Kenya (Table 16.6). None of

Table 16.6 Facial Metrics of Juvenile and Adult *Homo erectus* or *Homo* aff. *erectus*

Specimen	Age	Facial Metrics (mm)*		
		Height at M1	Breadth at M1	Nasion–Prosthion
KNM-ER 820	6.5 yr	18.0	17.0	—
KNM-ER 1507	—	19.0	20.0	—
KNM-WT 15000	11.0 yr	24.4	19.5	<78.0
KNM-ER 992	Adult	31.0	20.0	—
KNM-ER 730	Adult	31.5	19.0	—
Javanese *H. erectus*	Adult	36.5	20.0	—
Chinese *H. erectus*	Adult	28.5	15.5	—
KNM-ER 3733	Adult	—	—	83.0

*Measurements from Wood (1991), except for KNM-WT 15000, which are from Walker and Leakey (1993). Walker and Leakey (1993) provide for alveolare-nasion a slightly longer measure, by a few millimeters, than that for prosthion-nasion.

the juvenile Zhoukoudian remains is complete enough to measure mandibular height.

Both datasets suggest that KNM-WT 15000 had some facial height left to achieve before adulthood (Table 16.6). We accept all the caveats of using cross-sectional data to look at growth and the likelihood that such data will underestimate growth (Boas 1892); however, the facial height of KNM-WT 15000 is about 10–12 percent smaller than that of KNM-ER 3733. Given that KNM-WT 15000 is male and most authors consider KNM-ER 3733 to be female, we may infer that an adult male would present even greater differences in facial height. The mandibular data are still more compelling. The eleven-year-old KNM-WT 15000 mandible is approximately five millimeters taller than the roughly five-year-old KNM-ER 820 and 1507 mandibles and seven millimeters smaller than the adult mandibles of KNM-ER 730 and KNM-ER 992. However, mandibular breadth does not vary significantly among the specimens (Fig. 16.2C), a finding consistent with modern human growth, in which breadth dimensions are least responsive to the adolescent growth spurt (Goldstein 1936; Savara and Tracey 1967). To achieve adult mandibular height, KNM-WT 15000 would have needed to increase its current mandibular height by about 29 percent. This degree of increase is similar to or slightly greater than that seen in postcranial length in modern humans from eleven years to adulthood (18–31%) and much greater than the 14–18 percent increase in chimpanzees between comparable developmental ages (Gavin 1953; Smith 1993). It is also similar to the increases in mandibular height achieved by humans between twelve and twenty years of age (Goldstein 1936; Buschang et al. 1983; Marshall and Tanner 1986). Thus, unlike his postcranial lengths, the size and shape of the cranial skeleton of KNM-WT 15000 are consistent with a dental developmental age that is preadolescent and thus with the presence of an adolescent growth spurt.

There is no a priori reason for suggesting that gross patterns of skeletal growth in the face should be more strongly correlated with dental maturation than are patterns of postcranial growth. Bogin (1988) suggested that the facial skeleton, although controlled by a different system, will harmonize with the dentition, thus adjusting skeletal growth to dental development over evolutionary time. However, there is little evidence for the synchrony of dental and facial skeletal maturation, although correlations between dental maturation and other indicators of biological maturity are quite strong (Falkner 1957; Björk and Helm 1967; Billewicz et al. 1973; Robinow 1973; Dermirjian 1986). Thus, the dental development of KNM-

WT 15000 is the most likely of the three systems to reflect the degree of developmental maturity, suggesting that either the cranial or the postcranial skeleton of KNM-WT 15000 is exhibiting an idiosyncratic variation of the *H. erectus* growth pattern.

At present, there are few data to assess which of these patterns might be atypical. However, if the cranial and dental data can be viewed as independent, I am inclined to believe that it is the postcranial data that are aberrant. Although this cannot be definitively shown to be the case, we have sufficient data to urge caution in discussions of the origin of the adolescent growth spurt. At best, we cannot distinguish unequivocally whether early *H. erectus* lacked an adolescent growth spurt. In my view, the majority of data suggest the presence of such a growth spurt.

Conclusions

Clearly, more work is needed on the origin of the adolescent stage of development. Such work is, however, hampered because both dental developmental age and some measure of skeletal growth must be assessed in the same specimen and over a number of juveniles. Given this requirement and the paucity of associated individuals in the fossil record, evaluation of the pattern of adolescent growth may be most possible through analysis of the cranial skeleton. Such evaluation requires a more thorough understanding of the correlation between facial growth and physiological, dental, and postcranial maturity. It also requires a systematically assessed subadult dataset, such as that now available for Asian *H. erectus*.

Acknowledgments

I am grateful to Nancy Minugh-Purvis and Kenneth McNamara for the invitation to join this volume and for their insights on this topic. A. Walker graciously supplied Figure 16.4. I thank the following individuals for access to specimens in their care: T. Jacob, Gadjah Mada University, Yogyakarta, Indonesia; F. Aziz, Quaternary Research Lab, Geological Research and Development Center, Bandung, Indonesia; J. L. Franzen, Senckenberg Museum, Frankfurt; J. de Vos, P. Storm, and R. van Zelst, Rijksmuseum, Leiden; C. Stringer and R. Kruszynski, Natural History Museum, London; I. Tattersall and J. Grand, American Museum of Natural History;

D. Dechant, Atkinson Collection, University of the Pacific, School of Dentistry, San Francisco; W. Maples, C. A. Pound Human Identification Laboratory, University of Florida. W. R. Leonard and C. C. Swisher III provided valuable discussion. This study was partially funded by the Department of Sponsored Research, University of Florida, and National Science Foundation grant SBR 9804861.

Note

1. Many authors use Ngandong and Solo specimen numbers interchangeably. However, only Ngandong and Solo numbers 1 through 3 are the same. From number 4 onward Ngandong numbers are one greater than Solo numbers—that is, Ngandong 5 is Solo 4 or Skull IV and so on (Oakley et al. 1975; Santa Luca 1980). Weidenreich's (1951) Solo skull numbers for suggested subadults correlate with the Ngandong system as follows: Ngandong 2 (Solo 2), Ngandong 5 (Solo 4), Ngandong 8 (Solo 7), Ngandong 9 (Solo 8). Care should be taken when comparing specimen numbers between systems, since many authors simply substitute "Ngandong" for "Solo" without adjusting the numbers.

References

Antón, S.C. 1997a. Developmental age and taxonomic affinity of the Mojokerto child, Java, Indonesia. *American Journal of Physical Anthropology* 102:497–514.

———. 1997b. Cranial growth in *Homo erectus:* the development of frontal recession and occipital angulation. *American Journal of Physical Anthropology* 24 (suppl): 67 (abstract).

———. 1999. Cranial growth in *Homo erectus:* how credible are the Ngandong juveniles? *American Journal of Physical Anthropology* 108:223–236.

Antón, S.C., and Franzen, J.L. 1997. The occipital torus and developmental age of Sangiran-3. *Journal of Human Evolution* 33:599–610.

Antón, S.C., and Leigh, S.R. 1998. Paedomorphosis and neoteny in human evolution. *Journal of Human Evolution* 34:A2 (abstract).

Antón, S.C., and Weinstein, K.J. 1999. Artificial deformation and fossil Australians revisited. *Journal of Human Evolution* 36:195–209.

Baughan, B., Demirjian, A., Levesque, G.Y., and Lampalme-Chaput, L. 1979. The pattern of facial growth before and during puberty, as shown by French-Canadian girls. *Annals of Human Biology* 6:59–76.

Billewicz, W.Z., Thomson, A.M., Baber, F.M., and Field, C.E. 1973. The development of primary teeth in Chinese (Hong Kong) children. *Human Biology* 45:229–241.

Björk, A. 1955. Cranial base development: a follow-up x-ray study of the individual variation in growth occurring between the ages of 12 and 20 years and its relation to brain case and face development. *American Journal of Orthodontics* 41:198–225.

Björk, A., and Helm, S. 1967. Prediction of the age of maximum pubertal growth in body height. *Angle Orthodontist* 37:134–143.

Black, D. 1931. On an adolescent skull of *Sinanthropous pekinensis* in comparison with an adult skull of the same species and with other hominid skulls, recent and fossil. *Palaeontologica Sinica* Series D 7(2):1–114.

Boas, F. 1892. The growth of children. *Science* 19:281–282.

Bogin, B. 1988. *Patterns of Human Growth*. Cambridge: Cambridge University Press.

———. 1994. Adolescence in evolutionary perspective. *Acta Paediatrica* 406 (suppl): 29–35.

Bogin, B., and Smith, B.H. 1996. Evolution of the human life cycle. *American Journal of Human Biology* 8:703–716.

Brodie, A.G. 1953. Late growth changes in the human face. *Angle Orthodontist* 23: 146–157.

Bromage, T.G, and Dean, M.C. 1985. Re-evaluation of the age at death of immature fossil hominids. *Nature* 317:525–527.

Buikstra, J.E., and Ubelaker, D.H. 1994. Standards for data collection from human skeletal remains. *Arkansas Archeological Survey Research Series* No. 44.

Buschang, P.H., Baume, R.M., and Nass, G.G. 1983. Craniofacial growth maturity gradient for males and females between 4 and 16 years of age. *American Journal of Physical Anthropology* 61:373–381.

Cohen, M.M. 1986. *Craniosynostosis: Diagnosis, Evaluation and Management*. New York: Raven Press.

De Stefano, G.F., and Hauser, G. 1989. *Epigenetic Variants of the Human Skull*. Stuttgart: E. Schweizerbart'sche Verlagsbuchhandlung.

Dean, M.C. 1986. The dental developmental status of six East African juvenile fossil hominids. *Journal of Human Evolution* 16:197–213.

Demirjian, A. 1986. Dentition. In F. Falkner and J.M. Tanner (eds.), *Human Growth: A Comprehensive Treatise*, 2:269–298. New York: Plenum Press.

Ekstrom, C. 1982. Facial growth rate and its relation to somatic maturation in healthy children. *Swedish Dental Journal* 11 (suppl): 5–99.

Enlow, D.H. 1975. *Handbook of Facial Growth*. Philadelphia: W. B. Saunders Co.

Falkner, F. 1957. Deciduous tooth eruption. *Archives of Disease in Childhood* 32: 386–391.

Gauld, S.C. 1996. Allometric patterns of cranial bone thickness in fossil hominids. *American Journal of Physical Anthropology* 100:411–426.

Gavin, J. 1953. Growth and development of the chimpanzee: a longitudinal and comparative study. *Human Biology* 25:93–143.

Goldstein, M.S. 1936. Changes in dimensions and form of the face and head with age. *American Journal of Physical Anthropology* 22:37–89.

Gould, S.J. 1977. *Ontogeny and Phylogeny*. Cambridge: Belknap Press of Harvard University Press.

Grün, R., Huang, P-H., Wu, X., Stringer, C.B., Thorne, A.G., and McCulloch, M. 1997. ESR analysis of teeth from the paleoanthropological site of Zhoukoudian, China. *Journal of Human Evolution* 32:83–91.

Hublin, J-J. 1978. Le torus occipital transverse et la structures associées: evolution dans le genre *Homo.* Ph.D. diss., Université de Paris VI.

———. 1986. Some comments on the diagnostic features of *Homo erectus:* fossil man, new facts—new ideas. *Anthropos (Brno)* 23:175–187.

Lanpo, J., and Weiwen, H. 1990. *The Story of Peking Man: From Archaeology to Mystery.* Oxford: Oxford University Press.

Leigh, S.R. 1996. Evolution of human growth spurts. *American Journal of Physical Anthropology* 101:455–474.

Mann, A., Lampl, M., and Monge, J. 1990. Patterns of ontogeny in human evolution: evidence from dental development. *Yearbook of Physical Anthropology* 33:111–150.

Marshall, W.A., and Tanner, J.M. 1986. Puberty. In F. Falkner and J.M. Tanner (eds.), *Human Growth: A Comprehensive Treatise,* 2:171–209. New York: Plenum Press.

Meindl, R.S., and Lovejoy, C.O. 1985. Ectocranial suture closure: a revised method for the determination of skeletal age at death based on the lateral-anterior suture. *American Journal of Physical Anthropology* 68:57–66.

Mitani, H. 1977. Occlusal and craniofacial growth changes during puberty. *American Journal of Orthodontics* 72:76–84.

Moore, K.L. 1982. *The Developing Human,* 3d ed. Phildelphia: W. B. Saunders Co.

Nanda, R.S. 1955. The rates of growth of several facial components measured from serial cephalometric roentgenograms. *American Journal of Orthodontics* 14: 658–673.

Oakley, K.P., Campbell, B.G., and Molleson, T.I. 1975. *Catalogue of Fossil Hominids. Part 3: Americas, Asia, Australasia.* London: British Museum of Natural History.

Rightmire, G.P. 1998. Evidence from facial morphology for similarity of Asian and African representatives of *Homo erectus. American Journal of Physical Anthropology* 106:61–85.

Robinow, M. 1973. The eruption of the deciduous teeth (factors involved in timing). *Journal of Tropical Pediatrics and Environmental Child Health* 19:200–202.

Santa Luca, A.P. 1980. The Ngandong fossil hominids: a comparative study of a far eastern *Homo erectus* group. *Yale University Publications in Anthropology* 78: 1–175.

Savara, B.S., and Tracey, W.E. 1967. Norms of size and annual increments for five anatomical measures of the mandible from 3 to 16 years of age. *Archives of Oral Biology* 12:469–486.

Scammon, R.E. 1930. The measurement of the body in childhood. In J.A. Harris, J.M. Jackson, D.E. Paterson, and R.E. Scammon (eds.), *The Measurement of Man,* 173–215. Minneapolis: University of Minnesota Press.

Smith, B.H. 1986. Dental development in *Australopithecus* and early *Homo*. *Nature* 323:327–330.

———. 1989. Growth and development and its significance for early hominid behavior. *Ossa* 14:63–96.

———. 1993. The physiological age of KNM-WT 15000. In A. Walker and R. Leakey (eds.), *The Nariokotome* Homo erectus *Skeleton*, 195–220. Cambridge: Harvard University Press.

Sullivan, N.C., and Weaver, D.S. 1981. An inter-observer analysis of tympanic plate growth stages. *Homo* 33:210–213.

Tanner, J.M. 1978. *Fetus into Man*. Cambridge: Harvard University Press.

Tanner, J.M. 1988. Human growth and constitution. In G.A. Harrison, J.M. Tanner, D.R. Pilbeam, and P.T. Baker (eds.), *Human Biology: An Introduction to Human Evolution, Variation, Growth, and Adaptability*, 337–435. Oxford: Oxford University Press.

Tardieu, C. 1998. Short adolescence in early hominids: infantile and adolescent growth of the human femur. *American Journal of Physical Anthropology* 107: 163–178.

Thompson, G.W., Popovich, F., and Anderson, D.L. 1976. Maximum growth changes in mandibular length, stature and weight. *Human Biology* 48:285–293.

Thorne, A.G., and Wolpoff, M.H. 1981. Regional continuity in Australasian Pleistocene hominid evolution. *American Journal of Physical Anthropology* 55:337–349.

Trut, L.N. 1991. Intracranial allometry and morphological changes in silver foxes (*Vulpes vulpes*) under domestication (in Russian). *Genetika* 27:1440–1450.

Trut, L.N., Dzerhzhinsky, F.J., and Nicolsky, V.S. 1991. A principal component analysis of changes in cranial characteristics appearing in silver foxes (*Vulpes vulpes*) under domestication (in Russian). *Genetika* 27:1605–1611.

Ubelaker, D.H. 1984. *Human Skeletal Remains: Excavation, Analysis, Interpretation*, rev. ed. Washington, D.C.: Taraxacum.

von Koenigswald, G.H.R., and Weidenreich, F. 1939. The relationship between *Pithecanthropus* and *Sinanthropus*. *Nature* 144:926–929.

Walker, A., and Leakey, R. 1993. The skull. In A. Walker and R. Leakey (eds.), *The Nariokotome* Homo erectus *Skeleton*, 63–94. Cambridge: Harvard University Press.

Weaver, D.S. 1979. Application of the likelihood ratio test to age estimation using the infant and child temporal bone. *American Journal of Physical Anthropology* 50:263–270.

Weidenreich, F. 1936. The mandibles of *Sinanthropus pekinensis:* a comparative study. *Palaeontologia Sinica*, Series D, no. 7: 1–164.

———. 1940. The torus occipitalis and related structures and their transformations in the course of human evolution. *Bulletin of the Geological Society of China* 19:479–559.

———. 1941. The brain and its role in the phylogenetic transformation of the human skull. *Transactions of the American Philosophical Society* 31:321–442.

————. 1943. The skull of *Sinanthropus pekinensis:* a comparative study on a primitive hominid skull. *Palaeontologia Sinica,* Series D, no. 10: 1–298.

————. 1951. Morphology of Solo man. *American Museum of Natural History Anthropological Papers* 43:1–290.

Wood, B. 1991. *Koobi Fora Research Project,* vol. 4, *Hominid Cranial Remains.* Oxford: Clarendon Press.

————. 1994. Taxonomy and evolutionary relationships of *Homo erectus.* In J.L. Franzen (ed.), *100 Years of* Pithecanthropus, *the* Homo erectus *Problem. Courier Forschungsinstitut Senckenberg* 171:159–165.

Wu, X., and Poirier, F.E. 1995. *Human Evolution in China: A Metric Description of the Fossils and a Review of the Sites.* Oxford: Oxford University Press.

Chapter 17

Peramorphic Processes in the Evolution of the Hominid Pelvis and Femur

CHRISTINE BERGE

The notion of human evolution in terms of heterochrony is rooted in the fetalization theory of Bolk (1926), a theory based on the resemblance in shape between the skull of a juvenile chimpanzee and the skull of an adult human. Bolk assumed that humans, as compared with apes, retained juvenile traits by growth retardation. In contrast to information on the skull, Bolk and his contemporaries found no reliable arguments to strengthen the neotenic theory in the postcranium (see list of Bolk's traits in Gould 1977 and discussion of pelvic traits in Berge 1998). Schultz (1927) was among the first to express doubts about Bolk's theory because, as he wrote, body proportions in humans do not correspond to a fetal condition of primates—"just the opposite seems to be the case" (Schultz 1927, 61). He also argued that postcranial growth in humans is not particularly slow in terms of relative changes in shape.

Six decades later, this discussion, as exemplified by many of the chapters in this book, is ongoing. Can we still consider humans to be "essentially" neotenic? (Gould 1977, 365). In recent publications the neotenic hypothesis is always described as a global process, even though it is admitted that humans did not evolve through a single heterochronic process (see Alba, Chap. 2). I believe that such oversimplification may explain paradoxical diagnoses, such as "peramorphic paedomorphosis" (Godfrey and Sutherland 1995, 1996). Indeed, current literature depicts the human morphology resulting from neoteny as a retarded shape (underdeveloped shape of the skull) or paedomorphosis, combined with an increase in body size (overdeveloped size) or peramorphosis. Thus, such heterochrony (peramorphic paedomorphosis) implicitly suggests that changes in shape are not related to size (nonallometrical traits). However, according to Shea (1989, Chap. 4), many cranial features cited as evidence of human neoteny have to be reconsidered because they probably do not correspond to such shape-size dissociation (see also Penin and Berge 2001).

Similarities in pelvic shape in juvenile humans and adult chimpanzees suggest heterochronic processes that are the opposite of neoteny (Berge 1993). The iliac blade of a human fetus is flat and laterally positioned, as is the case in apes. At the beginning of childhood, the ventral portion of the ilium is rapidly modified by additional traits (acetabulocristal buttress, cristal tubercle, internal iliac fossa) leading to important perturbations of growth allometries compared with apes (Berge 1993, 1995a, 1995b, 1996, 1998). Other pelvic traits, such as the drastic shortening of the human ilium compared with apes, suggest that accelerative processes occurred at the beginning of the hominid lineage (Berge et al. 1986; Chaline et al. 1986). However, a hypothetical apelike trait in human ancestors, such as a long ilium, never appears in human ontogeny, even in the earlier stages. Thus, fossils from the human lineage are essential to reconstruct intermediate steps because they may represent "heterochronoclines" (i.e., progressive heterochronic changes in shape in adults through evolutionary time) (McNamara 1982; McKinney and McNamara 1991). For example, the anterior portion of the ilium is gradually modified from *Australopithecus* to *Homo sapiens,* and this change in shape resembles that seen in modern humans during ontogeny (Berge 1998).

My purpose here is to specify different types of heterochronies that have been previously suspected and partially demonstrated in pelvic and in femoral morphologies (Berge 1993, 1995a, 1998; Tardieu 1997, 1998). The ancestral pelvic and femoral morphotype can be considered to be that of the gracile australopithecines (*Australpithecus africanus, Australopithecus afarensis*), which was already involved in hominization (in terms of bipedal specializations) but retained many primitive apelike traits or intermediate ones (Chopra 1962; Zihlman 1971; Lovejoy et al. 1973; McHenry 1975, 1984, 1994; Jungers 1982, 1988; Tardieu 1981, 1983; Stern and Susman 1983; Berge 1984, 1991, 1993; Sigmon 1986; McHenry and Berger 1998). This study uses graphic methods and clock models defined in Gould (1977) and Alberch et al. (1979) in an attempt to estimate and compare the pelvic and femoral growth trajectories in ancestors (australopithecines) and descendants (modern humans). The methods use very simple morphometric data (long bone lengths, indices of proportion) to represent morphological changes in size and shape. For this reason some morphological traits, which have been described in the literature as being different in australopithecines and modern adult humans, cannot be interpreted in terms of heterochrony, either because it is difficult to measure them in early stages of growth or because the measured traits (indices of proportion) do not vary during ontogeny.

Materials and Methods

Materials

Modern Humans. The skeletal sample of modern humans (Table 17.1) comprises 134 adult and juvenile pelves and 107 adult and juvenile femora, coming from many different origins (all the continents). The material of known age and sex is mainly European. Only 46 pelves and femora are associated (36 of known age and sex). Two specimens of particular interest are added because they represent extreme values of adult body size: a very small female pygmy (Baminga) of known stature (1.37 meters), whose pelvic length is similar to that of adult australopithecines, and a French male giant from the nineteenth century (named Hugo), of known age (35 years) and stature (2.29 meters).

Table 17.1   Number of Specimens Studied

	Pelvis	Femur	Pelvis + Femur
Modern *Homo sapiens*			
Juveniles	93 (73)	63 (45)	32 (30)
Subadults	11 (11)	9 (9)	3 (3)
Adults	30 (6)	18 (6)	11 (3)
Total	134 (90)	90 (60)	46 (36)
Fossil *Australopithecus*			
Juvenile *A. africanus*	MLD 7		
Subadult *A. africanus*	Sts 14		
Adult *A. afarensis*	AL 288	AL 288	AL 288
Total	3	1	1

Note: Numbers in parentheses, number of specimens with known ages. For definitions of growth stages, see text.

The specimens are classified in Table 17.1 according to three stages of growth: (1) juveniles (*sensu largo*) have unfused pelvic and femoral growth plates (separated ilium, ischium, and pubis); (2) subadults have partly fused growth plates (incompletely fused iliac crest, ischial tuberosity, and sacral vertebrae); and (3) adults have fully fused pelvic and femoral growth plates. The modern specimens come from the Orfila collection of the Institut d'Anatomie, Centre Universitaire des Saints-Pères, Paris; collections of the Laboratoire d'Anthropologie of the Muséum National d'Histoire Naturelle, Paris; and collections of the Laboratoire d'Anthropologie Biologique of Université Paris 7.

Fossil Hominids. The fossil sample (Table 17.1) comprises the casts of the ilium MLD 7 (*A. africanus*), the pelvis Sts 14 (*A. africanus*), and the pelvis and femur AL 288, "Lucy" (*A. afarensis*). The complete pelvis and left femur of AL 288 are Schmid's reconstructions (Schmid 1983). The pelvis has been reconstructed from the left hip bone and sacrum by mirror molding, and the femur by joining the two components of the diaphysis. The damaged part of the distal epiphysis has been corrected on the basis of a very similar fossil (AL 129-1a) coming from the same site.

According to the three stages of growth used in this study, MLD 7 is a juvenile ilium (Dart 1949, 1958; Robinson 1972). Sts 14 is a subadult pelvis that preserves some immature traits at the level of the iliac crest, ischial tuberosity, and sacrum (Berge and Gommery 1999). AL 288 is an adult pelvis and femur (Table 17.1).

The casts come from the Musée de l'Homme (Paris), Laboratoire de Paléontologie des Vertébrés et Paléontologie Humaine (Université Paris 6), and Anthropologisches Institut und Museum der Universität (Zürich).

Methods

Gould (1977) and Alberch et al. (1979) define ontogenetic trajectories of specimens or species by their coordinates in a three-dimensional space defined by three independent axes, or "vectors"—the vectors of size, shape, and age. Complete ontogenetic trajectories are hard to obtain not only in human specimens, whose given ages are often unreliable, but above all in fossil hominids, whose ages are unknown. Thus, three sorts of graphic methods were used: (1) age-size graphs, or growth curves of pelvic and femoral lengths in humans (known ages), which allow us to locate adult and juvenile fossils (estimated ages); (2) shape-size graphs that approximate the best ontogenetic changes in shape in humans and australopithecines; (3) Gould's clocks, which best illustrate heterochronic processes between adult ancestors (*Australopithecus*) and adult descendants (*H. sapiens*). As previously suggested by McKinney and McNamara (1991) and Godfrey and Sutherland (1996), Gould's clocks may be used to study specific morphological traits instead of the entire organism. To avoid any confusion, I use heterochronic methods and terms that follow those in Gould (1977) and Alberch et al. (1979).

Measurements. Nine distances were measured (Fig. 17.1). The pelvic measurements are the following: PEL, maximum pelvic length; AIL, anterior iliac breadth; SAP, posterior iliac breadth; ACE, acetabular diameter. The pelvic measurements are described in Berge (1993, 1998). For the measurement of AIL, the position of landmark *T* in the cristal tubercle must be specified in juveniles as the location of maximum pelvic breadth (distance *T-T* in the complete pelvis). The femoral measurements are FEL, maximum femoral length; FEH, femoral head diameter (maximum diameter taken vertically); NEC, biomechanical neck length (projected distance between

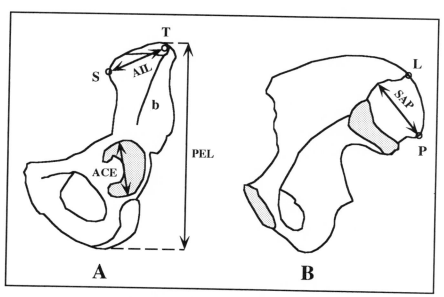

Fig. 17.1. Pelvic measurements. *A*, external face of the os coxae; *B*, internal face of the os coxae. *ACE*, acetabular diameter, measured in the prolonged ischium axis; *AIL*, anterior iliac breadth, measured from the anterior superior iliac spine (*S*) to the cristal tubercle (*T*); *PEL*, pelvic length, maximum dimension; *SAP*, posterior iliac breadth, maximum dimension; *b*, acetabulo-cristal buttress; *L*, spina limitans; *P*, posterior superior iliac spine.

center of femoral head and lateral point on the greater trochanter taken perpendicular to the long axis of the femoral shaft); SAE, sagittal diameter of the distal epiphysis (maximum anteroposterior diameter of the lateral condyle); TRE, tranverse diameter of the distal epiphysis (maximum transverse diameter of the posterior aspect of both condyles). NEC was defined in Lovejoy (1975). Other femoral measurements are measurements 11, 1, 16, and 12 in McHenry and Corruccini (1978), except the two last dimensions, which are close to measurements 2 and 3 in Tardieu (1997, 1998).

Shape Vectors. Shape vectors were studied in both shape-size graphs and clocks. They are indices of proportion, or ratios, as in Gould (1977). Three of these ratios were previously studied in human growth. In the pelvis, the relative breadth of the anterior part of the ilium (AIL/PEL) and the relative breadth of the posterior part of the ilium (SAP/PEL) were allometrically studied in Berge (1993, 1995a, 1998). In the femur, the shape of the outline of the distal femoral epiphysis (SAE/TRE) was studied by Tardieu (1997, 1998). The other ratios correspond to anatomical traits that have often been cited in the literature as being different in *Australopithecus* and *Homo* (see McHenry and Temerin 1979; Stern and Susman 1983): femoral length in proportion to the pelvic length (FEL/PEL), acetabular diameter in proportion to the pelvic length (ACE/PEL), femoral head diameter in proportion to the femoral length (FEH/FEL), and femoral neck in proportion to the femoral length (NEC/FEL).

Size Vectors. The material consists of numerous isolated pelves and femora. Size vectors used to build shape-size graphs are either the pelvic length or the femoral length.

Age Vectors. Growth curves allow us to represent the age vectors of humans (known ages) and australopithecines (estimated ages). The adult age in Gould's clocks is the age of the adult size of pelvic bones and femora, rather than the age of sexual maturity. The age of adult femora is well known in humans because it corresponds to the age of adult stature, given by growth curves as being about sixteen years of age for females and twenty-one years for males (see, e.g., Bogin 1999; Feldesman 1992). In the present sample, we observe that the adult stage of growth, when the pelvic bones and femur are completely united, is reached at the age of sixteen to eighteen in females and twenty-one to twenty-six in males.

Adult age has been estimated in australopithecines by comparing adult and juvenile pelvic bones with different degrees of fused pelvic growth plates, compared with modern humans and African apes (Curgy 1965). This gives an approximate age of eight to ten years for the juvenile ilium MLD7 (i.e., a prepubertal stage of growth) (Berge 1993, 1998). In contrast, Sts 14 is a subadult (Berge and Gommery 1999), and AL 288 is fully adult. An approximate adult age of twelve years has been suggested to explain the small change in pelvic length between the juvenile MLD7 and the subadult Sts 14 (see below).

Alternatively, Tardieu (1997, 1998) extrapolated a similar adult age for *Australopithecus* from adult femoral morphology (combination of morphological data taken from AL 288 and AL 129, as in the present study).[1] However, many recent investigators admit a short maturity for *Australopithecus* on the basis of dental development (see Ramirez Rozzi, Chap. 15).

Shape-Size Graphs. Seven shape-size graphs allow comparison of changes in shape relative to pelvic and femoral size during human growth and at the end of australopithecine growth. Regression lines or logarithmic curves (the choice depends on which fits best as R^2 reaches high values) are calculated for humans and transposed to the fossils.

Gould's Clocks. Six clocks correspond to six morphological traits studied in shape-size graphs. One trait—relative femoral neck length—has been excluded (low R^2). Otherwise, it would be very close to the clock corresponding to femoral head diameter. The six clocks allow us to compare adult stages in the ancestral pattern (*Australopithecus*) and descendant pattern (*H. sapiens*). The size and shape vectors are the hands of the clock moving on the two dials, and the age vector (adult ages) runs along the horizontal axis. The three vectors of the ancestor are on a vertical line, whereas the three vectors of the descendant move independently on the right side (positive side: increased values) or the left side (negative side: decreased values). Calibration of the vectors of age, size, and shape is explained in the text.

Results

Growth Curves in Modern Humans versus Australopithecines

In Figure 17.2, the growth curves of the human pelvis and femur are calculated for specimens of known ages (mainly Europeans). The growth curve for pelvic length (Fig. 17.2*A*) is used to confirm the estimated age of about twelve for the adult stage of *Australopithecus*. As is explained above, ages of the fossils have been previously estimated using comparisons with similar growth stages in modern apes and humans. MLD 7 is presumably

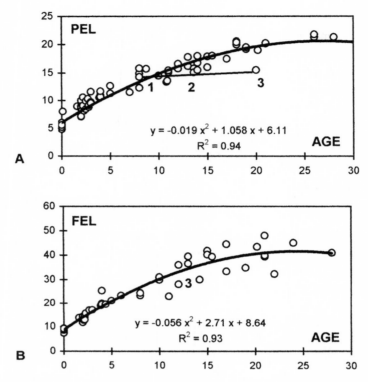

Fig. 17.2. Growth curves of the pelvic and femoral lengths in hominids. *PEL*, pelvic curve; *FEL*, femoral curve; X-axis (in years): known ages in humans and estimated ages in fossils (see under "Methods"); Y-axis (in centimeters): *PEL*, pelvic length; *FEL*, femoral length. *1:* MLD 7 (juvenile *Australopithecus africanus*). *2:* Sts 14 (subadult *A. africanus*, see text). *3:* AL 288 (adult *Australopithecus afarensis*).

a juvenile aged eight to ten years (Berge 1998), Sts 14 a subadult aged about twelve years (Berge and Gommery 1999), and AL 288 an older adult. MLD 7 has been connected with the MLD 8 juvenile ischium of a similar stage of growth (probably the same individual according to Dart 1949, 1958). Figure 17.2*A* indicates that the MLD 7 juvenile has almost the same pelvic length (133 mm) as the subadult Sts 14 (138 mm), whereas the adult AL 288 is slightly bigger (144 mm). On the other hand, we may observe that the juvenile australopithecine (MLD 7) also has the same pelvic size as juvenile humans of similar age (eight to ten years), whereas adolescent and adult australopithecines (Sts 14, AL 288) have the same pelvic size as juvenile humans of around twelve years (in a range of ten to fourteen years). In contrast, modern humans continue to grow in pelvic size after twelve years, with a noticeable growth spurt in pelvic length until the adult stage (i.e., completely fused growth plates) is reached at around sixteen to eighteen years in females and twenty-one to twenty-six years in males.

The growth curve for femoral length (Fig. 17.2*B*) is similar to that of the pelvis, apart from larger variability in adult femoral length (very large dispersion of specimens after ten years of age in the curve). The femoral length of the adult AL 288 equals that of a juvenile human of around twelve years of age (in the same range of age variation as for its pelvis).

Ontogenetic and Phylogenetic Changes in Pelvic Size and Shape

The three pelvic traits (Fig. 17.3) indicate two different types of heterochronies. The fossils are either fully separated from the modern human trajectory (Fig. 17.3, *A* and *B*) or they are superimposed on it (Fig. 17.3*C*). It can be observed in Figure 17.3*A* that the relative breadth of the ventral ilium (AIL/PEL) increases with pelvic size in modern humans. The curved trajectory indicates that this change in shape occurs at the beginning of growth. In contrast, the fossils have a very small ventral portion of the ilium (AIL/PEL). The adult pelvic shape is that of the modern neonates (same coordinates of AIL/PEL in Y-axis), whereas their pelvic size is that of older modern juvenile specimens before the adolescent growth spurt at around twelve years of age (Fig. 17.2*A*). Another australopithecine trait resembles that of a human neonate: the very small dorsal region of the ilium (Berge 1996). In Figure 17.3*B*, the fossils plot far from modern humans of similar pelvic size (juveniles of about twelve years old), with SAP/PEL coordinates similar to those of fetuses and neonates. In contrast, Figure 17.3*C* shows that the relative size of the australopithecine acetabulum (ACE/PEL) is within the modern human range.

Ontogenetic and Phylogenetic Changes in Femoral Size and Shape

The four femoral traits depicted in Figures 17.4 and 17.5 indicate a single type of heterochrony. In the four graphs, the femur of the adult *Australopithecus* (AL 288) is superimposed on the human trajectory but displaced toward juveniles in terms of size and shape. The slopes of regression lines indicate that two femoral traits are particularly modified in human ontogeny and hominid phylogeny. First, femoral length increases more

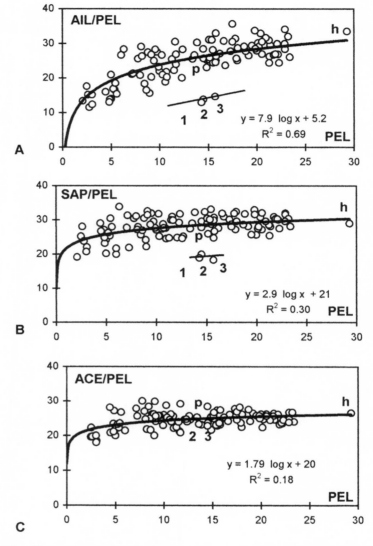

Fig. 17.3. Shape-size graphs of pelvic morphology. X-axis in centimeters; Y-axis: ratios (in percentage). *AIL/PEL,* relative breadth of the anterior part of the ilium; *SAP/PEL,* relative breadth of the posterior part of the ilium; *ACE/PEL,* relative size of the acetabulum; *AIL,* anterior iliac breadth; *ACE,* acetabular diameter; *SAP,* posterior iliac breadth; *PEL,* maximum pelvic length; *1,* MLD 7 (juvenile *A. africanus*); *2,* Sts 14 (subadult *A. africanus*); *3,* AL 288 (adult *A. afarensis*); *p,* adult pygmy; *h,* adult giant specimen Hugo (see text).

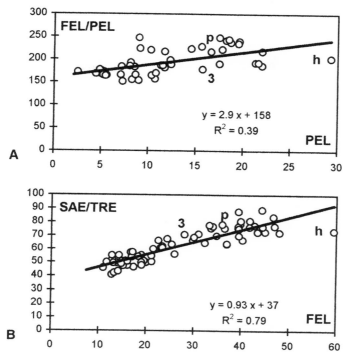

Fig. 17.4. Shape-size graphs of femoral morphology. X-axis in centimeters; Y-axis: ratios (in percentage). *FEL/PEL*, relative length of the femur (in proportion to the pelvic length); *SAE/TRE*, sagittal to transverse diameter of distal femoral epiphysis; *FEL*, femoral length; *PEL*, pelvic length; *SAE*, sagittal diameter of the distal femoral epiphysis; *TRE*, transverse diameter of the distal femoral epiphysis; *3*, AL 288 (adult *Australopithecus*); *p*, adult pygmy; *h*, adult giant specimen.

rapidly than does pelvic length (Fig. 17.4*A*). The relatively short femur of the adult australopithecine corresponds to body proportions observed in juvenile humans of similar pelvic size (juveniles of about twelve years of age, Fig. 17.2*A*). Second, as previously demonstrated by Tardieu (1997, 1998), the distal epiphysis of the human femur changes in shape during growth and becomes more sagittally developed at the level of the internal condyle. In Figure 17.4*B*, the shape of the outline of the australopithecine distal femoral epiphysis (SAE/TRE) is similar to that of juvenile humans of similar femoral size (i.e., juveniles of twelve years of age) (see also Fig. 17.2*B*).

In contrast, the femur of early *Homo* (KNMER 1472) is already closer to that of adult modern humans in both size and shape. Contrary to expected results, the relative size of the femoral head and the relative length

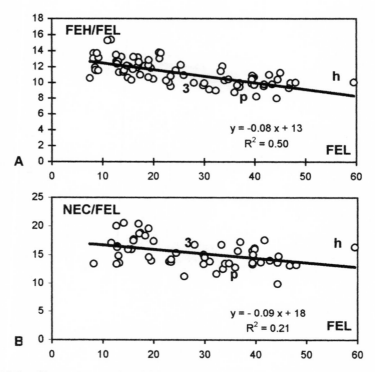

Fig. 17.5. Shape-size graphs of femoral morphology. X-axis in centimeters; Y-axis: ratios (in percentage). *A*, relative size of the femoral head; *B*, relative length of the femoral neck. *FEH/FEL*, proportion of the distal femoral epiphysis; *FEH*, femoral head diameter; *FEL*, femoral length; *NEC*, biomechanical neck length; *3*, AL 288 (adult *Australopithecus*); *p*, adult pygmy; *h*, adult giant specimen.

of the femoral neck show very little variation during ontogeny and phylogeny (Fig. 17.5). It seems that, in proportion to femoral length, these two dimensions of the proximal femur scale down slightly in human ontogeny and hominid phylogeny.

Pelvic and Femoral Clocks

Godfrey and Sutherland (1996) made the point that clock dials have to be calibrated. Calibration of the age vector is simple. As explained above, it corresponds to adult ages (in terms of pelvic and femoral sizes) given by growth curves (Fig. 17.2)—twelve years in ancestors and sixteen to twenty-one years in descendants. Size and shape dials are calibrated from the shape-size graphs. We may consider two scenarios: (1) In Figure 17.3*C*

(ACE/PEL) and Figures 17.4 and 17.5 (FEL/PEL, SAE/TRE, and FEH/ FEL), the fossils are superimposed on the growth trend for modern humans. Calibrations are calculated from the modern human regression lines (or curves). They are values that approximate the best mean values of adult shape and size in ancestors and descendants. (2) In Figure 17.3, *A* and *B* (AIL/PEL, SAP/PEL), the fossils are further off the human trends. A short portion of the ancestral growth curve may be reconstructed in fossils in reference to the end of the human growth period. Shape and size values, respectively, are calculated from the ancestral and descendant regression curves. A hypothetical shape value for ancestors who would be the size of descendants is also calculated from the ancestral growth curve. Such calculations are used to estimate the dissociation in shape between ancestors and descendants. For example, in the AIL/PEL clock (Fig. 17.6), a shape ratio of sixteen was calculated from the ancestral curve given in Figure 17.3*A* (AIL/PEL) for a pelvic size of 20 cm. The dissociation in shape between ancestors and descendants is the difference between sixteen and twenty-eight in the dial of shape.

We may describe the pelvic and femoral clocks as follows (Fig. 17.6). In all clocks, the three vectors are positively displaced. There are two sorts of peramorphoses: (1) Four clocks (ACE/PEL, FEL/PEL, FEH/FEL, SAE/ TRE) resemble the classical example of peramorphosis given in Gould (1977, 258, Fig. 39). There is no dissociation of size and shape vectors in descendants compared with ancestors. The overdeveloped size and shape in descendants is related to the prolongation of growth duration. In other words, the shape of descendants is a consequence of larger size (allometry) via hypermorphosis. (2) The clocks corresponding to two pelvic traits (AIL/ PEL, SAP/PEL) are more complex. I interpret them as the combination of two different types of peramorphoses: first, as was the case above, a hypermorphic process that leads to an increase in size and shape (undissociated shape) with a prolonged period of growth; second, an acceleration of shape that clearly dissociates from the size vector. Shape-size dissociation is particularly marked in both clocks.

Discussion

As was previously discussed in Berge (1998), the pelvic morphology of modern humans and fossil hominids is already strongly derived from a hypothetical apelike ancestor. Pelvic specializations in humans led to an on-

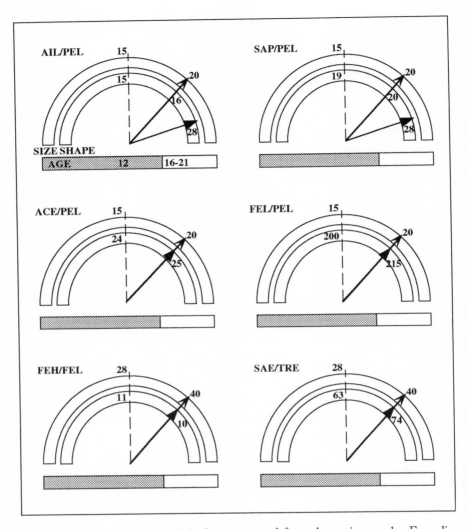

Fig. 17.6. Pelvic and femoral clocks constructed from shape-size graphs. For calibration of size and shape dials and age axis, see text and Figures 17.3, 17.4, and 17.5. *AIL/PEL*, relative anterior iliac breadth; *SAP/PEL*, relative posterior iliac breadth; *ACE/PEL*, relative size of the acetabulum; *FEL/PEL*, relative femoral length; *FEH/FEL*, relative femoral head; *SAE/TRE*, proportion of the distal femoral epiphysis.

togenetic allometric pelvic pattern that is fundamentally different from that of living apes (Berge 1993, 1995a, 1995b). In my opinion the gracile australopithecines (*A. africanus* and *A. afarensis*) may represent a good ancestral prototype for human pelvic and femoral morphology (Fig. 17.7*B*) because the fossils share many traits (in spite of specific variations) that are

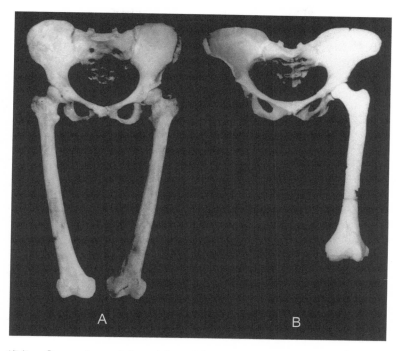

Fig. 17.7. Comparison of the pelvis and femur in adult hominids. *A, Homo sapiens* (female pygmy); *B, Australopithecus afarensis* (AL 288), Schmid's reconstruction (1983).

intermediate between those of apes and modern humans (see below). Another argument is given by the heterochronic study, since some of the peculiar australopithecine pelvic traits (the shape of the ventral and dorsal parts of the ilium) are still visible in early stages of human growth. One may object that the ancestral morphotype is defined here by a small number of fossil specimens (two adults and a juvenile). However, numerous fragmentary fossils, not included in the present study, confirm a common pelvic and femoral morphotype for gracile australpithecines (see, e.g., McHenry and Temerin 1979).

In a variety of descriptions, australopithecine pelvic and femoral morphology is viewed either as being very close to modern human morphology or as having peculiar traits. The first view amplifies the likeness with human morphology as irrefutable evidence of bipedalism (Dart 1949, 1958; Robinson 1972; Lovejoy et al. 1973; Walker 1973; Lovejoy 1975; Johanson and Coppens 1976; Wolpoff 1976; McHenry and Corruccini 1978; McHenry and Temerin 1979, for example). Some years later, however, the fossils were described as having peculiar traits, which has been seen as evidence of locomotor differences (Stern and Susman 1983; Jungers and Stern 1983; Berge 1994). Some of these traits may recall an apelike morphology. For example, the adult AL 288 has been described as having very short lower limbs, as in the hindlimbs of apes (Jungers 1982; Jungers and Stern 1983; McHenry 1994). The articular surfaces (acetabulum and femoral head), studied in AL 288 and in other fragmentary fossils, have been described as smaller than their modern human counterparts (Zihlman 1971; McHenry 1975, 1984, 1986; Corrucini and McHenry 1978; Sigmon 1986; Jungers 1988). The distal epiphysis in australopithecines has been described as being more transversally developed (i.e., as having a rectangular outline as in apes) and not a square one as in modern humans (McHenry and Corruccini 1978; Tardieu 1981, 1983, 1997, 1998; Stern and Susman 1983; McHenry 1984, 1986; Bacon and Baylac 1995, 1996).

In contrast, some peculiar traits in australopithecines are not shared with apes. The femoral neck, measured in AL 288 and in other fossils from the same site, is considered proportionally longer than in either modern humans or apes (Wolpoff 1978; McHenry 1984; McHenry and Corrucini 1978; Corrucini and McHenry 1980; Stern and Susman 1983).

Pelvic traits suggest rather an intermediate morphology between apes and humans. In addition, the australopithecine pelvis is characterized by a ventral situation of the acetabulocristal buttress and of the iliac tubercle—in other words, a small anterior iliac breadth (Zihlman 1971; Stern and Susman 1983; McHenry 1986; Sigmon 1986; Berge 1993)—and a small but humanlike auricular surface—that is, a small posterior iliac breadth (SAP) (McHenry 1984, 1986; Sigmon 1986; Berge 1984, 1991, 1993).

An important consideration in examining these results is that we need to clearly distinguish studies of adults (australopithecines and modern humans) from those including juveniles. The present results confirm some of the peculiar traits described above. As compared with adult modern humans, adult australopithecines have a relatively short femur in proportion to pelvic length, proportionally small anterior and posterior iliac breadths, and a rectangular outline of the distal femoral epiphysis. Other traits described above as being different from modern human ones are here within the range of adult modern humans. In AL 288, the articular surfaces (acetabulum and femoral head) do not seem to be particularly small, nor does the femoral neck appear particularly long by comparison with the sample of adult modern humans.

In the light of the present results, I believe that three main reasons may explain the contradictory descriptions in the literature. The first reason is that australopithecine traits are gradually amplified from gracile species (*A. africanus* and *A. afarensis*) to robust ones (*Australopithecus robustus* and *Australopithecus boisei*). This indicates a common allometric pattern (all species included) that is globally different from that of modern humans (see McHenry and Corruccini 1975, 1978; Berge 1984). Thus, as noted by McHenry and Temerin (1979), some authors have a tendency to generalize observations from one species to another. For example, the small femoral head and the long femoral neck are clearly visible in robust species but not in gracile ones (McHenry and Corruccini 1978; McHenry and Temerin 1979).

The second reason, as previously suggested by Aiello and Dean (1990), is the use of different standards of reference in morphological comparisons, more specifically the use of femoral length as a size reference. However, in the present study, the choice of bone lengths (pelvis and femur) as references of size is suitable because it best represents the measurement of diaphyseal growth in length. One may suggest that multivariate methods may be used to calculate a size reference and shape changes for heterochronic studies (McKinney and McNamara 1991; Eble, Chap. 3, this volume). In previous studies, I used multivariate analyses that gave similar results in terms of heterochronies, but in a qualitative way and in a single organ, the pelvis (Berge 1993, 1998). In the present study, shape-size graphs and Gould's clocks seem to be better instruments for comparing several heterochronies in two different organs and in a quantitative way.

The third reason that may explain contradictory opinions is the choice of the modern human comparative sample, classically made of a homogeneous population: adult pygmies, Europeans (or Americans of European origin), and, in some studies, Amerindians. The present human sample, which comes from all continents, shows quite a range of adult body proportions. Africans, African pygmies, and many Europeans have relatively long lower limbs and femora by comparison with other people, such as Asians and some Europeans. African pygmies have extremely long femora in proportion to the pelvic length. We may observe in graphs that Lucy (AL 288) is markedly further from the female pygmy in terms of shape than from other people. More precisely, AL 288 and the female pygmy have quite similar pelvic lengths (respectively, 150 and 160 mm), similar acetabular diameters (36 and 38 mm), similar femoral neck lengths (47 and 48 mm), but extremely different femoral lengths (280 and 350 mm). When compared with short-legged people, Lucy (AL 288) does not seem to have an extremely short femoral length, contrary to what Jungers (1982) and

Jungers and Stern (1983) maintained. They considered that such proportioning is far from allometric (contra Wolpoff 1983).

Alternatively, we can also examine the present results when juveniles are included. Interestingly, most of the differences in femoral and pelvic proportions in *Australopithecus* and *Homo* almost disappear when the fossils are compared with younger modern human specimens—that is, when heterochronic processes are taken into account. In short, the pelvis of *Australopithecus* closely resembles that of a human neonate (Figs. 17.3, *A* and *B*, and 17.6), whereas the femur tends to resemble that of a juvenile about twelve years of age (Tardieu 1998; Figs. 17.4 to 17.6).

Two kinds of accelerative processes are revealed by shape-size graphs and Gould's clocks. First is an accelerative process of change in shape that involves pelvic shape and, more specifically, the ventral and dorsal portions of the ilium. The clocks of AIL/PEL and SAP/PEL clearly illustrate size-shape dissociation, whereas the shape-size trajectories indicate that the change in shape occurs rapidly at the beginning of growth. Such dissociation in size and shape could result from two accelerative processes. The first is the predisplacement of morphological traits, arising earlier in descendants than in ancestors (see Berge 1998). The other is the change in the allometric pattern of growth that leads to a change in the ilium shape during growth (Berge 1995a, 1995b). Second, a general process of hypermorphosis arises at the end of the adolescent period. Delayed growth in humans gives rise to associated changes in size and shape mostly visible in femoral morphology. The femur becomes longer in proportion to the pelvis, and the distal epiphysis becomes more sagittally developed. Thus, the term *global acceleration*, proposed by McKinney and McNamara (1991) as synonymous with hypermorphosis, seems to be particularly suitable here to describe the growth spurt in femoral size and shape in humans.

One of the questions raised in this study is whether the hypermorphic process (i.e., the prolonged and accelerated growth in the adolescent period) is specific to modern humans and whether it is possible to interpret allometries in terms of heterochrony. Tardieu (1997, 1998) studied the change in shape of the distal femoral epiphysis in fossil hominids and in modern human growth. Although there are no juvenile fossils, she estimated that the morphological differences between *Australopithecus* and *Homo* correspond to a change in the hominid growth pattern (lengthening of the adolescent growth period in *Homo*). I believe that juvenile fossils are essential to prove that the australopithecine growth pattern lacked an adolescent growth spurt, as has been demonstrated from the comparison of adult and juvenile pelvic bones (Fig. 17.2*A;* see also Berge 1993, 1995a, 1998).

In the case of the femur, when adult fossils are compared with modern humans, allometries of size, growth, and evolution are combined in the same change in shape. For example, when two fossils from the same site (Hadar) but of very different sizes (AL 129-1a and AL 333-4) are compared, we do not know whether the differences in femoral shape correspond to two different growth patterns (*Australopithecus* and *Homo*), as proposed by Tardieu (1983) and Senut and Tardieu (1985), or to allometrical variations in a single pattern (*A. afarensis*), as proposed by Stern and Susman (1983) and Tardieu (1997, 1998). Bacon and Baylac (1995, 1996) used "Procrustes" analyses to compare these two fossils with modern humans and other primates. They concluded that there is no conclusive way of deciding whether the fossils correspond to the same pattern.

In an attempt to discuss the hypermorphic process among modern humans, we may now consider the example of pygmies, who are relatively close in the graphs to the australopithecines in terms of size (pelvic length) but more distant in terms of shape (pelvic and femoral traits). When compared with other humans, African pygmies (Baminga) are described as having a shorter adult size due to less sensitivity to growth hormones (Cavalli-Sforza 1986; Froment 1993). Are the pygmies smaller than other peoples because they stop growing earlier (i.e., with no hypermorphic process as in apes— Cavalli-Sforza's hypothesis), or are they smaller because they grow slower at any period of growth (i.e., they are neotenic—Froment's hypothesis)?

Pasquet et al. (1995; pers. comm.) calculated growth curves of the statures of 462 female and 446 male African pygmies of known ages. The male and female growth curves indicate that the African pygmies reach adult size at the same age as other peoples (sixteen years in females and twenty-one years in males), with a smooth growth spurt at the end of the adolescent period. This example suggests that the adolescent growth spurt in body size is common to all modern humans but encompasses some variations.

In general, as explained above, size and shape are associated in the hypermorphic process. However, in some specimens or peoples distinct shape-size dissociation may be observed by comparison with the general human trend. Thus, pygmies grow more in shape than in size. At adulthood, they gain adult traits of shape (a proportionally very long femur like other African peoples, a sagittally developed distal femoral epiphysis), but no adult traits in body size (short pelvic length, small acetabulum and femoral head, gracile bones, and short trunk). In contrast, the giant specimen changes more in size than in shape (Figs. 17.3 to 17.5). It is common to suppose that pyg-mies, who are close to the australopithecines in terms of body size, are also close to them in terms of body shape (Jungers and Stern 1983). Actually, in light of the present study, they are strikingly different in shape, thus rein-forcing the hypothesis of different patterns of growth in *Australopithecus* and *Homo*. The example of pygmies also suggests that the hypermorphic changes in body size and shape in humans are probably more complex in terms of genetic and developmental systems than they seem at first sight.

Conclusions

Two very different peramorphoses may explain the changes in pelvic and femoral shape in hominid evolution. One is an accelerative process of shape (shape-size dissociation) arising in ilial morphology. This process takes place at the beginning of childhood. The second one is a hypermorphic process corresponding to a global acceleration in size and shape (shape-size asso-ciation) mainly visible in femoral morphology. This accelerative process is directly related to the prolonged and accelerated modern human growth during the adolescent period. In terms of functional adaptations, the per-amorphic processes lead humans to attain longer and more sagittally posi-tioned lower limbs, facilitating higher speed and the sagittal movements characteristic of human bipedalism.

Ackowledgments

I thank J. Langaney, J.-P. Lassau, J. Repérant, and F. Demoulin for allow-ing me to study the pelvic and femoral material and A. Jérome for stylistic corrections. Special thanks are due to J.-P. Gasc, K. J. McNamara, and the anonymous referee for constructive remarks and critical evaluations of the manuscript. This research is supported by the Centre National de la Re-cherche Scientifique (UMR 85 70).

Note

1. Such extrapolation is discussed under "Results" because growth duration cannot be deduced from adult morphology (the three vectors of the ontogenetic

trajectory—age, size, and shape—are, by definition, independent; see Gould 1977; Alberch et al. 1979; Devillers 1989; Klingenberg 1998).

References

Aiello, L., and Dean, C. 1990. *An Introduction to Human Evolutionary Anatomy.* London: Academic Press.

Alberch, P., Gould, S.J., Oster, G.F., and Wake, D.B. 1979. Size and shape in ontogeny and phylogeny. *Paleobiology* 5:296–317.

Bacon, A.M., and Baylac, M. 1995. Landmark analysis of distal femoral epiphysis of modern and fossil primates with particular emphasis on *Australopithecus afarensis* (AL 129–1 and AL 333–4). *Comptes Rendus de l'Académie des Sciences* 321 (series IIa): 553–560.

———. 1996. Allometry of the distal femur in the Hominoidea *Papio* and *Australopithecus afarensis:* a morphometric approach. *Cahiers d'Anthropologie et Biométrie Humaine* 3–4:497–508.

Berge, C. 1984. Multivariate analysis of the pelvis for hominids and other extant primates: implications for the locomotion and systematics of the different species of Australopithecines. *Journal of Human Evolution* 13:555–562.

———. 1991. Size and locomotion-related aspects of hominid and anthropoid pelves: an osteometrical approach. *Human Evolution* 6:365–376.

———. 1993. *L'Evolution de la Hanche et du Pelvis des Hominidés: Bipédie, Parturition, Croissance, Allométrie.* Cahiers de Paléoanthropologie. Paris: CNRS.

———. 1994. How did the australopithecines walk? A biomechanical study of the hip and thigh of *Australopithecus afarensis. Journal of Human Evolution* 26: 259–273.

———. 1995a. The pelvic growth in extant and extinct hominids: implications for the evolution of body proportions and body size in humans. *Anthropologie* 33: 47–56.

———. 1995b. La croissance du pelvis de l'homme comparée à celle des pongidés africains: implications dans l'évolution de la bipédie. *Anthropologie et Préhistoire* 106:45–56.

———. 1996. The evolution and growth of the hominid pelvis: a preliminary thinplate spline study of ilum shape. In L.F. Marcus, M. Corti, A. Loy, G.J.P. Naylor, and D.E. Slice (eds.), *Advances in Morphometrics,* 441–448. New York: Plenum Press.

———. 1998. Heterochronic processes in human evolution: an ontogenetic analysis of the hominid pelvis. *American Journal of Physical Anthropology* 105:441–459.

Berge, C., and Gommery, D. 1999. The sacrum of Sterkfontein Sts14 Q (*Australopithecus africanus*): new data on the growth and on the osseous age of the specimen (homage to R. Broom and J. T. Robinson). *Comptes Rendus de l'Académie des Sciences* 329:227–232.

Berge, C., Marchand, D., Chaline, J., and Dommergues, J.L. 1986. La bipédie des hominidés: comparaisons des itinéraires ontogénétiques des dimensions fémoro-pelviennes des pongidés, australopithèques et hommes. In J. Chaline and B. Laurin (eds.), *Ontogenèse et Evolution*, 63–79. Paris: CNRS.

Bogin, B. 1999. *Patterns of Human Growth*, 2nd ed. Cambridge: Cambridge University Press.

Bolk, L. 1926. On the problem of anthropogenesis. *Proceedings of the Section of Sciences, Koninklijke Akademie van Wetenschappen te Amsterdam* 29:465–475.

Cavalli-Sforza, L.L. (ed.). 1986. *African Pygmies*. New York: Academic Press.

Chaline, J., Marchand, D., and Berge, C. 1986. L'Évolution de l'homme: Un modèle gradualiste ou ponctualiste? *Bulletin de la Société Royale Belge Anthropologie et Préhistoire* 97:77–97.

Chopra, S.R.K. 1962. The innominate bone of the australopithecines and the problem of erect posture. *Biblioteca Primatologica* 1:93–102.

Corruccini, R.S., and McHenry, H.M. 1978. Relative femoral head size in early hominids. *American Journal of Physical Anthropology* 49:145–148.

———. 1980. Hominid femoral neck length. *American Journal of Physical Anthropology* 52:397–398.

Curgy, J.J. 1965. Apparition et soudure des points d'ossification des membres chez les mammifères. *Mémoire du Muséum National Histoire Naturelle* 32:12–307.

Dart, R.A. 1949. Innominate fragments of *Australopithecus prometheus*. *American Journal of Physical Anthropology* 7:301–333.

———. 1958. A further adolescent australopithecine ilium from Makapansgat. *American Journal of Physical Anthropology* 16:473–479.

Devillers, C. 1989. Ontogenie et phylogénie: Haeckel ou Garstang? *Geobios Mémoire Special* 12:133–144.

Feldesman, M.R. 1992. Femur/stature and estimates of stature in children. *American Journal of Physical Anthropology* 87:447–459.

Froment, A. 1993. Adaptation biologique et variation dans l'espèce humaine: le cas des pygmées d'Afrique. *Bulletin et Mémoires de la Société Anthropologie* 5:417–448.

Godfrey, L.R., and Sutherland, M.R. 1995. What's growth got to do with it? Process and product in the evolution of ontogeny. *Journal of Human Evolution* 29:405–431.

———. 1996. Paradox of peramorphic paedomorphosis: heterochrony and human evolution. *American Journal of Physical Anthropology* 99:17–42.

Gould, S.J. 1977. *Ontogeny and Phylogeny*. Cambridge: Belknap Press of Harvard University Press.

Johanson, D.C., and Coppens, Y. 1976. A preliminary anatomical diagnosis of the first Plio-Pleistocene hominid discoveries in the Central Afar, Ethiopia. *American Journal of Physical Anthropology* 45:217–234.

Jungers, W.L. 1982. Lucy's limbs: skeletal allometry and locomotion in *Australopithecus afarensis*. *Nature* 297:676–678.

———. 1988. Relative joint size and hominoid locomotor adaptations with implications for the evolution of hominid bipedalism. *Journal of Human Evolution* 17:247–265.

Jungers, W.L., and Stern, J.T. 1983. Body proportions, skeletal allometry and locomotion in the Hadar hominids: a reply to Wolpoff. *Journal of Human Evolution* 12:673–684.

Klingenberg, C.P. 1998. Heterochrony and allometry: the analysis of evolutionary change in ontogeny. *Biological Reviews* 73:79–123.

Lovejoy, C.O. 1975. Biomechanical perspectives on the lower limb of early hominids. In R.H. Tuttle (ed.), *Primate Functional Morphology and Evolution*, 291–326. The Hague: Mouton.

Lovejoy, C.O., Heiple, K.G., and Burstein, A.H. 1973. The gait of *Australopithecus*. *American Journal of Physical Anthropology* 38:757–779.

McHenry, H.M. 1975. A new pelvic fragment from Swartkrans and the relationship between the robust and gracile australopithecines. *American Journal of Physical Anthropology* 43:245–262.

———. 1984. The common ancestor. In R.L. Susman (ed.), *The Pygmee Chimpanzee*, 201–230. New York: Plenum Press.

———. 1986. The first bipeds: a comparison of the *A. afarensis* and *A. africanus* post-cranium and implications for the evolution of bipedalism. *Journal of Human Evolution* 15:177–191.

———. 1994. Early hominid postcrania. In R.S. Corruccini and R.L. Ciochon (eds.), *Phylogeny and Function. Integrative Paths to the Past: Paleoanthropological Advances in Honor of F. Clark Howell*, 251–268. Englewood Cliffs, N.J.: Prentice Hall.

McHenry, H.M., and Berger, L.R. 1998. Body proportions in *Australopithecus afarensis* and *A. africanus* and the origin of the genus *Homo*. *Journal of Human Evolution* 35:1–22.

McHenry, H.M., and Corruccini, R.S. 1975. Multivariate analysis of early hominid pelvic bones. *American Journal of Physical Anthropology* 43:263–270.

———. 1978. The femur in early human evolution. *American Journal of Physical Anthropology* 49:473–488.

McHenry, H.M., and Temerin, L.A. 1979. The evolution of hominid bipedalism: evidence from the fossil record. *Yearbook of Physical Anthropology* 22:105–131.

McKinney, M.L., and McNamara, K.J. 1991. *Heterochrony: The Evolution of Ontogeny*. New York: Plenum Press.

McNamara, K.J. 1982. Heterochrony and phylogenetic trends. *Paleobiology* 8:130–142.

Pasquet, P., Froment, A., and Koppert, G. 1995. Variations staturales liées à l'âge chez l'adulte en Afrique: Étude semi-longitudinale de la sénescence et recherche de tendances séculaires en fonction du milieu au Cameroun. *Cahiers d'Anthropologie et Biométrie Humaine* 13:233–245.

Penin, X., and Berge, C. 2001. Étude des hétérochronies par superposition procruste: application aux crânes de primates Hominoidea. *Comptes Rendus de l'Académie des Sciences* 324:87–93.

Robinson, J.T. 1972. *Early Hominid Posture and Locomotion.* Chicago: University of Chicago Press.

Schmid, P. 1983. Eine Reconstruktion des Skelettes von A.L.288-1 (Hadar) und deren Konsequenzen. *Folia Primatologica* 40:283–306.

Schultz, A.H. 1927. Studies on the growth of gorilla and other higher primates with special reference to a fetus of gorilla, preserved in the Carnegie Museum. *Memoirs of the Carnegie Museum* 11:1–88.

Senut, B., and Tardieu, C. 1985. Functional aspects of Plio-Pleistocene hominid limb bones: implications for taxonomy and phylogeny. In E. Delson (ed.), *Ancestor: The Hard Evidence,* 193–201. New York: Alan R. Liss.

Shea, B.T. 1989. Heterochrony in human evolution: the case for neoteny reconsidered. *American Journal of Physical Anthropology* 32:69–101.

Sigmon, B.A. 1986. Evolution in the hominid pelvis. *Paleontologica Africana* 26:25–32.

Stern, J.T., and Susman, R.L. 1983. The locomotor anatomy of *Australopithecus afarensis. American Journal of Physical Anthropology* 60:279–317.

Tardieu, C. 1981. Morphofunctional analysis of the articular surfaces of the knee-joint in primates. In A.B. Chiarelli and R.S. Corrucini (eds.), *Primate Evolutionary Biology,* 68–80. Berlin: Springer Verlag.

———. 1983. *L'Articulation du Genou des Primates Catarrhiniens et des Hominidés Fossiles.* Cahiers de Paléoanthropologie. Paris: CNRS.

———. Femur ontogeny in humans and great apes: heterochronic implications for hominid evolution. *Comptes Rendus de l'Académie des Sciences* 325:899–904.

———. 1998. Short adolescence in early hominids: infantile and adolescent growth of the human femur. *American Journal of Physical Anthropology* 107: 163–178.

Walker, A. 1973. New *Australopithecus* femora from East Rudolf, Kenya. *Journal of Human Evolution* 2:545–555.

Wolpoff, M.H. 1976. Fossil hominid femora. *Nature* 264:812–813.

———. 1978. Some implications of relative biomechanical neck length in hominid femora. *American Journal of Physical Anthropology* 48:143–148.

———. 1983. Lucy's little legs. *Journal of Human Evolution* 12:443–453.

Zihlman, A.L. 1971. The question of locomotor differences in *Australopithecus. Proceedings of the 3rd International Congress on Primatology, Zürich 1970,* 1: 54–66.

Chapter 18

Heterochrony and the Evolution of Neandertal and Modern Human Craniofacial Form

Frank L'Engle Williams, Laurie R. Godfrey, and Michael R. Sutherland

In this chapter we examine craniofacial ontogeny in Neandertals and modern humans, and we contrast the degree and direction of their differences with those distinguishing chimpanzees (*Pan troglodytes*) and bonobos (*Pan paniscus*). Three regions of the skull are treated separately in this analysis to maximize the use of fragmentary Neandertal fossil materials. These include the calotte (frontal, parietal, and occipital bones), the face, and the mandible. We calculate growth allometries on logarithmically transformed axes and Euclidean distances in multidimensional shape-space, and we use principal components analysis (PCA) of craniofacial shapes to address the following questions: (1) How do patterns of allometric growth compare, and what are the relative amounts of growth of craniofacial traits in the various taxa? (2) Are the shape differences manifested in adult Neandertals and modern humans already present during infancy, or do they emerge later? (3) To what extent do Neandertals and modern humans differ in their degree of departure from infant shapes and morphologies? (4) Can adult modern human craniofacial form be described as paedomorphic, or juvenilized, with respect to Neandertals? In other words, are modern human adults closer in shape to Neandertal infants or juveniles than they are to Neandertal adults? Are modern human adults closer in shape to Neandertal infants than Neandertal adults are to modern human infants? (5) Finally, which variables tend to distinguish developmental stages of Neandertals

405

and modern humans, as summarized by the first and second axes of PCA space? For each of the above questions, we address the parallel case of the comparison of chimpanzees and bonobos.

The first fossil hominid whose ontogeny became a focus of paleoanthropological research was *Australopithecus*. This was because the earliest-discovered australopithecine skull was that of an immature individual (the Taung child; see Anemone, Chap. 12, and Kuykendall, Chap. 13). Debate over whether Taung more closely resembled modern humans or chimpanzees helped to spur the development of methods to determine absolute biological age of fossil hominids by, for example, using incremental markers on teeth (see Anemone, Chap. 12; Kuykendall, Chap. 13; and Ramirez Rozzi, Chap. 15). With the discovery of the Nariokotome juvenile, the ontogeny of *Homo erectus* came into the spotlight (see Antón, Chap. 16).

Meanwhile, general heterochronic models for human and ape evolution were posited and debated, often without consideration of the fossil record. Humans were described as "paedomorphic" (i.e., juvenilized in comparison to their ancestors) and indeed "neotenic" with reference to chimpanzees and other nonhuman primates (de Beer 1940; Montagu 1989; Gould 1977; Privratshy 1981).[1] McKinney and McNamara (1991, 295) offered the opposing view that humans are predominantly "peramorphic" (i.e., hyperadult in comparison to their ancestors) and that this condition arose via "sequential hypermorphosis" (delayed offset, or cessation, of all developmental phases; see McNamara, Chap. 5). This subject continues to stimulate lively debate (see Shea, Chap. 4).

The first-discovered Neandertal (Engis 2), found in 1829 in a cave near Liège, Belgium,[2] was an older infant (Fig. 18.1; see Fraipont 1936; Tillier 1983). Since the early 1800s, many additional immature Neandertals have been found, and today Neandertals unequivocally provide the most complete (albeit fragmentary) craniofacial growth series in the hominid fossil record. Nevertheless, most comparisons of Neandertal and modern human craniofacial form have focused on adults (e.g., Boule 1912; Hrdlička 1927; Brace 1964; Howells 1967; Wolpoff 1986; Rak 1986; Stringer and Andrews 1988; Smith et al. 1989; Bräuer and Smith 1992; Trinkaus and Shipman 1992; Stringer and Gamble 1993; Wolpoff and Caspari 1997).

Only rarely have researchers compared the ontogenetic trajectories of Neandertals and modern humans (Tillier 1983, 1989; Dean et al. 1986; Minugh-Purvis 1988; Stringer et al. 1990). Minugh-Purvis (1988) and Tillier (1989) observed that some morphological differences distinguish-

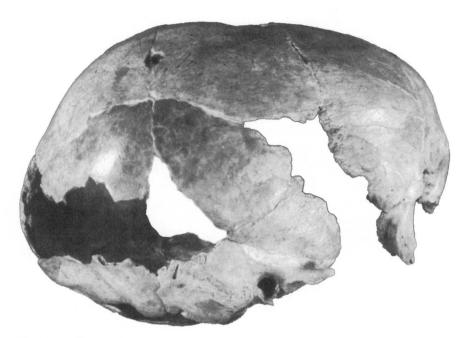

Fig. 18.1. Engis 2. This infant cranium was found in association with a palatal fragment. The frontal bone is damaged, both parietals are fragmentary, the right temporal bone is nearly complete, and the left parietal articulates with a virtually complete occipital bone. Much of the basicranium is preserved. Most of the anterior dentition was found in situ, although the deciduous molars and M1 were found in isolation. It is not possible to tell whether M1 had begun to erupt. *Source:* Photograph courtesy of the Laboratoire de Paléontologie, Université de Liège, and the Institut Royal des Sciences Naturelles de Belgique.

ing adult Neandertals and modern humans are manifested early in ontogeny, while others appear much later. Dean et al. (1986) and Stringer et al. (1990) argued on the basis of the low perikymata count and large brain size of the Devil's Tower child from Gibralter that Neandertals grew and developed more quickly than modern humans (but see Skinner 1997). Minugh-Purvis (1988) conducted the most thorough analysis of Neandertal craniofacial ontogeny, contrasting aspects of Neandertal ontogeny with that of early and recent modern humans. She found that, while Neandertals grew and developed at different rates than modern humans, early modern humans (including Skhul and Qafzeh, as well as Upper Paleolithic samples) do the same; indeed, they are more similar developmentally to

Neandertals than they are to modern humans. Similarly, Tompkins (1996) found that Neandertal dental development resembles that of early moderns, but that Upper Pleistocene and recent human populations differ in their relative timing of dental eruption and rates of dental crown calcification. In contrast, Lieberman (1997, 1998) found differences from birth onward in the craniofacial development of Neandertals and both recent and early modern humans, and he saw this as a rationale for distinguishing Neandertals from modern humans at the species level.

It is imperative that data concerning Neandertal development be brought to bear on the question of the craniofacial evolution of modern humans. To the extent that Neandertals preserve developmental trajectories that are ancestral for both Neandertals and modern humans, the heterochronic shifts affecting the evolution of modern humans can be inferred. (Conversely, to the extent that modern humans preserve ancestral developmental trajectories, the heterochronic shifts affecting Neandertal evolution can be inferred.) Thus, Neandertals offer paleoanthropologists the best opportunity to address heterochronic patterns and processes in the evolution of modern humans.

Recent attempts to address ontogenetic change in Neandertal craniofacial morphology have used a variety of mathematical tools. Williams (1996, 1997) explored the fit of von Bertalanffy's growth curves to univariate data collected for fossil samples and plotted multivariate size and shape parameters for ontogenetic series of Neandertals and modern humans. Krovitz et al. (1997) used three-dimensional digitization of cranial landmarks and Euclidean distance matrix analysis to capture shape differences between crania of immature Neandertals and modern humans (applying tools developed earlier by Lele and Richtsmeier 1991, Corner et al. 1992, Richtsmeier et al. 1993, and Richtsmeier and Walker 1993). Computer images using computed tomographic (CT) scans are currently being developed to reconstruct the internal skull morphology of immature Neandertals.

To test for heterochronic shifts in the evolution of modern human craniofacial morphology, we compared ontogenetic series of modern and Neandertal crania. The extent to which modern humans can be considered "paedomorphic" in shape with respect to Neandertals was evaluated by analyzing shape differences between modern humans and Neandertals at different life-history stages. These differences were compared with those distinguishing bonobos from chimpanzees.

Materials

Our Neandertal samples comprise the remains of 39 individuals housed at museums in England (Natural History Museum, London); France (Musée de l'Homme, Paris; Musée d'Antiquités Nationales, Saint Germain-en-Laye; Museum National de Préhistoire, Les Eyzies de Tayac; and Université de Poitiers, Poitiers); Belgium (Institut Royal des Sciences Naturelles de Belgique, Brussels; Université de Liège, Liège; and Direction de l'Archéologie, Ministère de la Région Wallonne, Namur); Hungary (Termeszettudomanyi Museum, Budapest); Israel (Tel Aviv University, Tel Aviv); Italy (Pigorini Museum and Instituto di Paleontologia Umana, Rome); and Croatia (Croatian Natural History Museum, Zagreb). Of the 39 individuals, 17 are in good to excellent condition and state of completeness, while 22 are fragmentary or poorly preserved. Thirty-eight are original fossils; 1 is an excellent plaster cast (Teshik-Tash, housed at Tel Aviv University). Thirteen of the 39 were used in our analyses of the calotte; 11 in the analyses of the face; and 16 in the analyses of the mandible (Table 18.1). Only 24 of these 39 had complete datasets for at least one region. Of these, 4 infants, 4 juveniles, and 12 adults figured

Table 18.1 Neandertal Specimens That Figured Heavily in Our Analyses, Listed by Craniofacial Region

Calotte	Face	Mandible
Pech de l'Azé	Pech de l'Azé	Pech de l'Azé
Roc de Marsal	Roc de Marsal	Roc de Marsal
Engis 2*	Engis 2	Teshik-Tash
Subalyuk 2	Subalyuk 2	Tabun C1
La Quina 18	Devil's Tower*	La Ferrassie
Teshik-Tash	La Quina 18	La Chapelle-aux-Saints
La Ferrassie	Teshik-Tash	Spy 1
La Quina 5	Guattari 1	Malarnaud*
La Chapelle-aux-Saints	Amud 1	Circeo 2 and 3
Tabun C1	Krapina 47	Archi 1
Spy 1 and 2	Tabun C1	Amud 1
Guattari 1		Amud 7*
Amud 1		Kebara 2
		Krapina 59
		Sclayn

*Individuals not sufficiently complete to be included in our Euclidean distance analysis.

heavily in our Euclidean distance analyses (see below). Table 18.2 gives a breakdown of our Neandertal sample by developmental stage (infant, juvenile, subadult, and adult).

The modern human sample (n = 272) was obtained from museums and universities in the United States (Department of Anthropology, American Museum of Natural History, New York, and Anatomy Department, Johns Hopkins University, Baltimore) and Europe (Vakgroep Anatomie en Embryologie, Rijksuniversiteit Groningen, the Netherlands; Centrum voor Fysische Antropologie/Anatomie Museum, Rijksuniversiteit Leiden; and Nationaal Natuurhistorisch Museum, Leiden, the Netherlands; Rijksinstituut voor Ouderheidskunde Bodemonderzoek, Amersfoort, the Netherlands; and Institut Royal des Sciences Naturelles de Belgique, Brussels, Belgium). This sample includes individuals from archeological, historical, and recent populations and from different geographic locations. Many specimens are from Europe (n = 60); others come from southeast Asia (n = 18), the Americas (n = 20), the Middle East (n = 11), Sub-Saharan Africa (n = 14), and Papua New Guinea (n = 24). Eight additional adults had no locality data. Samples from medieval Belgium (n = 54) and two collections of infants and children with no locality data (n = 60) were also included. The latter includes seventeen fetuses, neonates, and very young infants that are housed at the Centrum voor Fysische Antropologie/Anatomie Museum, Leiden.

The *P. troglodytes* sample (n = 149) was obtained from the Powell-Cotton Museum (Kent, U.K., n = 87); the Nationaal Natuurhistorisch Museum (Leiden, the Netherlands, n = 32); the Anatomie Museum (Rijksuniversiteit Leiden, the Netherlands, n = 1); and the Museum of Comparative Zoology (Harvard University, Cambridge, Mass., n = 7); and the Muséum National d'Histoire Naturelle (Paris, France, n = 21). Nine of these individuals were captive, fifteen had no locality information, and all other individuals were wild-caught in West Africa.

The *P. paniscus* sample (n = 72) was largely acquired from the Koninklijk Museum voor Midden-Afrika (Tervuren, Belgium, n = 69). Data on additional individuals were obtained from the Nationaal Natuurhistorisch Museum (Leiden, the Netherlands, n = 2) and the Muséum National d'Histoire Naturelle (Paris, France, n = 1). One individual was captive; all others were wild-caught in the "Belgian Congo" (today, Republic of Congo, formerly Zaire).

Table 18.2 Neandertal Specimens Used in This Study, Arranged by Life Cycle Stage

Life Cycle Stage:	Infant	Juvenile	Subadult	Adult
Age Range (yr):	2.5–5.5	6–11	14–16	19–45
Range Midpoint (yr):	4.0	8.5	15	32
Dental Stage:	Pre-M1	Post-M1 and Pre-M2	Post-M2 and Pre-M3	Post-M3
	Pech de l'Azé	La Quina 18	Malarnaud	La Ferrassie
	Roc de Marsal	Teshik-Tash	Krapina 48, 54	La Quina
	Subalyuk 2	Krapina 46, 47	Krapina 55, 56	Krapina 3, 57, 58
	Engis 2	Krapina 49, 53		Krapina 59, 63, 66
	Devil's Tower	Sclayn SCLA 4A-1		La Chapelle-aux-Saints
	Archi 1			Tabun C1
	Châteauneuf-sur Charente			Spy 1 and 2
	Amud 7 (4 mo)			Guattari 1
				Circeo 2 and 3
				Amud 1
				Kebara 2
				Subalyuk
Total sample (n)	8	7	5	19

411

Methods

To examine heterochronic patterns in modern humans with respect to Neandertals and in bonobos with respect to chimpanzees, we first plotted craniofacial variables against age. These plots allowed us to select variables for shape analysis and to assess intra- and interpopulation variability for modern humans. Whenever known ages were available (i.e., for thirty-four infant and juvenile modern humans housed at the Anatomie Museum, Rijksuniversiteit Leiden, and for one adult and two juvenile chimpanzees housed at the collection of the Muséum National d'Histoire Naturelle, Paris), they were used. Other individuals were aged according to dental eruption patterns, stage of cranial bone development (e.g., of the tympanic region), extent of cranial suture closure (e.g., basilar, palatal, and neurocranial), degree of dental wear, and, whenever possible, long bone epiphyseal plate closure. In aging the Neandertal sample, the procedures and developmental profiles used to age modern humans were followed.

We used the dental eruption schedules developed by Buikstra and Ubelaker (1994) for modern humans and by Dean and Wood (1981), based in part on Nissen and Riesen (1964), for wild-caught *Pan*. Neandertals were assumed to follow a humanlike dental eruption schedule, although Stringer et al. (1990) and Wolpoff (1979) suggested that Neandertals may have exhibited more rapid dental eruption. *P. troglodytes* dental eruption standards were used to age immature *P. paniscus* for two reasons. First, all ape species seem to follow similar dental eruption patterns (Dean and Wood 1981; Winkler et al. 1996). Second, precise standards for *P. paniscus* are lacking in the literature. Because the dental developmental sequence of the individuals in our sample more closely corresponds to that reported by Dean and Wood (1981) than to those reported by other authors (e.g., Anemone et al. 1996; Kuykendall and Conroy 1996), we adopted Dean and Wood's aging schedule.

Our observation that *Pan* exhibits much more variability in the sequence of dental eruption than do modern humans corroborates observations made earlier by Dean and Wood (1981) and Winkler et al. (1996) for apes and by Garn et al. (1965) for modern humans. Judging from recorded dispersions of ages at eruption for individual teeth in both human and chimpanzee samples (see Smith et al. 1994), it seems unlikely that this difference is due to a lower variance in dental eruption ages in humans; rather, it is likely to be due to the greater length of time between the eruption of individual human teeth. The fact that Neandertals exhibit minimal sequence variability, as do humans but not chimpanzees, strongly suggests that Neandertals exhibit a more human- than chimplike dental eruption schedule (see also Tompkins 1996). To assign minimum ages to individual *Pan*, we constructed aging criteria that accommodate dental eruption sequence variability (Williams 2001).

Trait-age plots for our ontogenetic series of geographically mixed modern humans (n = 197) were compared with those derived for samples from Papua (n = 24) and medieval Belgium (n = 55) to assess inter- and intrapopulation variability in growth curves. The latter two series represent our most distinct modern human populations. Individuals from Papua and medieval Belgium generally fall clearly within the range of variation represented by the larger, geographically mixed sample (Fig. 18.2). Furthermore, using the geographically mixed sample does not introduce a large amount of error, as among-population variability barely exceeds within-population variability.[3]

Two criteria were applied in selecting traits for our intertaxon comparisons. First, for all taxa we required that each trait demonstrate a monotonic increase in size during the postnatal growth phase of ontogeny (Fig. 18.3). (Traits, such as foramen magnum breadth and length, that exhibit very little ontogenetic change in size were excluded from further analysis.) Second, any trait that was poorly represented in our Neandertal sample (Table 18.2) was also excluded from further analysis.

Using the above criteria, twelve traits were selected for analysis of ontogenetic change in size and shape (Table 18.3). Three shape-spaces were constructed to describe the three regions (the calotte, face, and mandible) that are well represented in our Neandertal samples. A shape-space can be constructed by converting trait measurements to proportions (such that they always sum to 1) and then plotting these proportions (from 0 to 1) on orthogonal axes (Godfrey et al. 1998). Figure 18.4 shows a three-dimensional shape-space representing traits X (from the origin, 0,0,0, to 1,0,0), Y (from the origin to 0,1,0), and Z (from the origin to 0,0,1). In three-dimensional shape-space, all shape paths are constrained to lie in a single plane. The shape paths of three species, labeled *a* (ancestor), *b*, and *c* (two hypothetical descendants), are shown. Arrows indicate directions of ontogenetic change. Descendant *b* follows its ancestor's shape path, whereas descendant *c* does not. The shape paths do not reveal the sizes of species at any shape or their rates of development (age at any shape). Only species

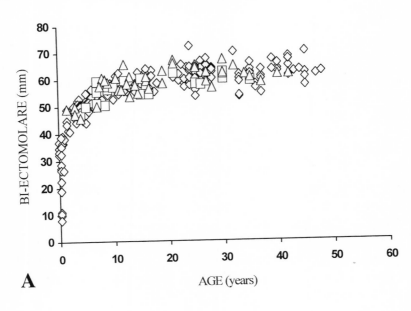

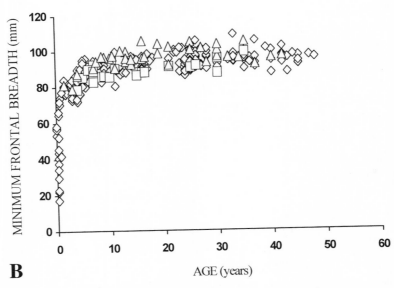

Fig. 18.2. Comparison of growth trajectories (trait by age) for modern human populations. *A*, bi-ectomolare; *B*, minimum frontal breadth. *Diamonds*, mixed sample; *triangles*, Papuan; *squares*, medieval Belgians.

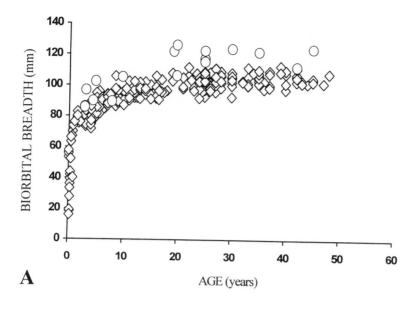

A

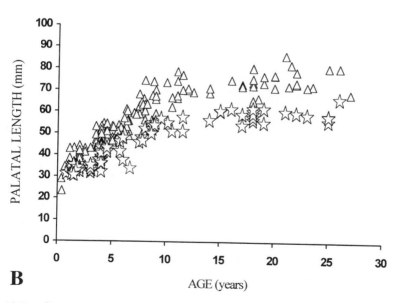

B

Fig. 18.3. Comparison of growth trajectories (trait by age) by genus. *A:* Biorbital breadth in *Homo. Circles,* Neandertals; *diamonds,* modern humans. *B:* Palatal length in *Pan. Stars,* bonobos; *triangles,* chimpanzees.

Table 18.3 Craniofacial Traits Used in This Study

Calotte	*Face*	*Mandible*
Maximum cranial length (glabella-opisthocranion)	Bi-ectomolare (breadth across the most lateral points on the alveolar margins of the maxilla)	Bigonial breadth
		Mandibular length (gonion-gnathion)
Maximum cranial breadth (bi-parietal)	Palatal length (prosthion-staphylion)	Mandibular symphyseal height (infradentale-gnathion)
Minimum frontal breadth (maximum postorbital constriction)	Maximum nasal aperture breadth (bi-alare)	Height of the mandibular corpus (at the mental foramen)
Biorbital breadth (at fronto-maxillary suture)		Thickness of the mandibular corpus (at the mental foramen)

b is a true paedomorph for the growth field represented by these three traits.

For this study, we constructed a four-dimensional shape-space for the four traits selected to describe the calotte, a three-dimensional shape-space for the three traits selected to describe the face, and a five-dimensional shape-space for the five traits selected to describe the mandible of Neandertals, modern humans, chimpanzees, and bonobos. "Size" variables were constructed for each region, following Godfrey and Sutherland (1996) and Gould (1977). These were defined as the sums of the linear measurements describing each region. Thus, the sum of four linear traits describing the calotte was used to approximate calotte size. Face size was the sum of the three traits describing the face, and mandible size was the sum of the five traits describing the mandible. To calculate growth allometries, the twelve craniofacial traits were log-transformed and individually regressed against their appropriate log-transformed "size" variable. We tested the null hypothesis of isometry (i.e., the slope, or allometric coefficient k, of the linear regression on logarithmic scales of trait on regional size = 1.00) using 95 percent confidence limits.

After the twelve selected craniofacial traits were converted to proportions, Euclidean distances between individuals were calculated for each of the three shape-spaces. We used a resampling procedure with 100 trials, followed by a t-test, to measure the statistical significance of the differences between interindividual Euclidean distances calculated for different group comparisons. Thus, we randomly selected infant-adult pairs within the

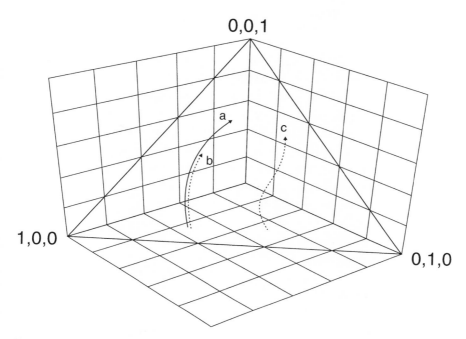

Fig. 18.4. Schematic representation of shape paths in three-dimensional shape-space. See text for explanation. *Source:* Modified from Godfrey et al. 1998.

same "taxon" and cross-"taxon" pairs holding developmental stage constant. (When the total population of available interindividual distances was less than 100, we used only 30 resampling trials to avoid artificially inflating the degrees of freedom available for the *t*-test.)

We restricted our Neandertal "infant" sample to Pech de l'Azé, Roc de Marsal, Archi 1, and Subalyuk 2 to limit life cycle stage shape variability. These Neandertal infants ranged in age from approximately 2.75 to 3.75 years. To make our modern human "infant" sample comparable, we selected the same age subset of these data. The modern human infant sample thus comprised eleven calottes, nine faces, and twelve mandibles. Our Neandertal "juvenile" sample (La Quina 18, Teshik-Tash, Krapina 47, and Scalyn SCLA 4A-1) consisted of individuals ranging in dental age from approximately seven to eleven years. Neandertal and modern human "adults" were identified by their having M3 in full occlusion. The adult sample comprised 7 Neandertal calottes, 3 Neandertal faces, and 12 Neandertal mandibles, as well as 113 modern human calottes, 107 modern humans faces, and 101 modern human mandibles.

Similar procedures were also applied to chimpanzee and bonobo infants and adults. Bonobo adults were compared with chimpanzee subadults to test for paedomorphosis. The chimpanzee infant sample comprised 25 calottes, 26 faces, and 25 mandibles; the bonobo infant sample comprised 14 calottes, 13 faces, and 14 mandibles. Our chimpanzee adult sample comprised 40 calottes, 41 faces, and 41 mandibles; the bonobo adult sample included 22 calottes, 23 faces, and 21 mandibles. The chimpanzee subadult sample included 24 calottes, 24 faces, and 23 mandibles.

Principal components analyses were performed separately for *Homo* and for *Pan*, for each of the regional suites of craniofacial proportions. For each PCA, the entire ontogenetic series for *Homo* or for *Pan* was included, and life cycle stage centroid scores on the first and second principal component axes were calculated for infants and for adults. This effectively allowed us to examine ontogenetic shape change in a series of shape-spaces of reduced dimensionality. The individuals used to calculate our modern human and Neandertal "infant" centroids included all specimens falling within the dental age range of 2.5–5.5 years. (Engis 2 and Devil's Tower were thus included, for this analysis, in the Neandertal "infant" category.) Chimpanzees and bonobos younger than 3.1 years were considered infants.

Results

Growth Allometries

Patterns of Growth Allometry in Modern Humans and Neandertals. Neandertals and modern humans differ in their patterns of craniofacial growth and development in several ways. In comparison with modern humans, Neandertals grew faster and achieved larger adult sizes for most craniofacial traits (e.g., see Fig. 18.3). Larger adult size for craniofacial traits is not universal, however. Adult ascending ramus height and bigonial breadth, for example, are not significantly different in Neandertal and modern human adults. Neandertal adults have larger faces (with generally broader palates and broader nasal apertures) and longer as well as broader crania. They also tend to have longer and more robust mandibles.

There are some striking differences between *patterns* of shape change (i.e., directions of allometric change) in Neandertals and modern humans. Neandertal crania increase in breadth with negative allometry ($k = 0.724$); as a consequence, Neandertal adults display relatively narrower skulls than do Neandertal infants (Table 18.4). Modern humans show little such change ($k = 0.977$). Nasal aperture breadth increases with weak negative allometry in modern humans ($k = 0.922$); this is not the case for Neandertals ($k = 1.19$) (Table 18.5). There is little difference between the patterns of allometric change among modern humans and Neandertals for the mandible (Table 18.6).

There are also differences between Neandertals and modern humans in the *strengths* of their growth allometries. Even for suites of traits that show comparable patterns of allometric change in Neandertals and modern humans, Neandertals tend to show stronger allometric signals. This difference holds regardless of the statistical significance of the difference between allometric coefficients.[4] For example, both modern humans and Neandertals are positively allometric for biorbital breadth and mandibular symphyseal height, and both show negative allometry for bigonial breadth, but Neandertals exhibit statistically significantly stronger positive and stronger negative allometries for each of these. The same directionality holds (without statistical significance) for mandibular corpus height, bi-ectomolare, and

Table 18.4 Comparison of Growth Allometries of the Calotte in Modern Humans and Neandertals and Significance of the Difference between Their Allometric Coefficients

	Modern Humans			*Neandertals*			
Trait	*k*	*SE*	*Allometry*	*k*	*SE*	*Allometry*	*Signif.*
Max. cranial length	1.05	0.009	Positive	0.931	0.083	Isometry	NS
Max. cranial breadth	0.977	0.012	Isometry	0.724	0.087	Negative	$p < 0.05$
Min. frontal breadth	0.917	0.007	Negative	0.952	0.086	Isometry	NS
Biorbital breadth	1.03	0.011	Positive	1.54	0.109	Positive	$p < 0.05$

Table 18.5 Comparison of Growth Allometries of the Face in Modern Humans and Neandertals and Significance of the Difference between Their Allometric Coefficients

	Modern Humans			*Neandertals*			
Trait	*k*	*SE*	*Allometry*	*k*	*SE*	*Allometry*	*Signif.*
Bi-ectomolare	0.822	0.012	Negative	0.784	0.097	Negative	NS
Palatal length	1.31	0.020	Positive	1.21	0.160	Isometry	NS
Nasal breadth	0.922	0.026	Negative	1.19	0.134	Isometry	$p < 0.05$

Table 18.6 Comparison of Growth Allometries of the Mandible in Modern Humans and Neandertals and Significance of the Difference between Their Allometric Coefficients

Trait	Modern Humans			Neandertals			Signif.
	k	SE	Allometry	k	SE	Allometry	
Bigonial breadth	0.979	0.009	Negative	0.750	0.078	Negative	$p < 0.05$
Mandibular length	0.994	0.008	Isometry	1.12	0.077	Isometry	NS
Symphyseal height	1.04	0.016	Positive	1.30	0.115	Positive	$p < 0.05$
Corpus height	1.19	0.013	Positive	1.41	0.181	Positive	NS
Corpus thickness	0.783	0.024	Negative	0.562	0.161	Negative	NS

mandibular corpus thickness. Modern humans tend to grow with weaker allometry, particularly in the calotte and in the mandible. Differences between Neandertal infants and adults are particularly pronounced in the supraorbital and facial regions (compare Figs. 18.5 and 18.6).

Patterns of Growth Allometry in Chimpanzees and Bonobos. In general,

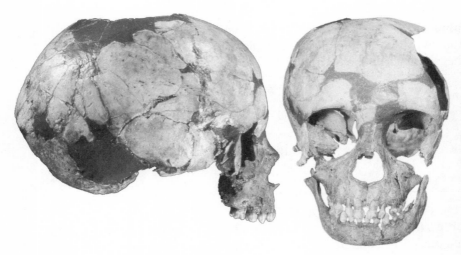

Fig. 18.5. Roc de Marsal. Most of the neurocranium, face, and mandible of this infant were found intact. The frontal, parietal, and occipital bones have been reconstructed. The complete set of deciduous teeth have erupted, but M1 is unerupted. The mandible is anteriorly squared off, like those of many adult Neandertals, and the neurocranium is elongated posteriorly, a feature present in most Neandertals regardless of age (and in very young modern humans). Supraorbital relief has already begun to develop. *Source:* Photograph courtesy of the Musée National de Préhistoire, Les Eyzies de Tayac.

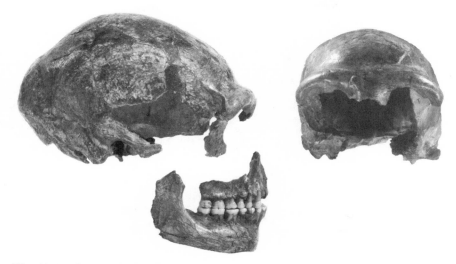

Fig. 18.6. Spy 1. This adult is represented by an almost complete calvarium and mandible but only a portion of the palate. Both parietal and temporal bones are nearly complete, as is the frontal. The occipital is missing its basal portion. The robusticity of the supraorbital torus and mandible, increased mandibular height, decreased gonial angle, and extreme elongation of the calvarium seen here typify adult Neandertals. *Source:* Photograph courtesy of the Institut Royal des Sciences Naturelles de Belgique.

craniofacial traits grow faster and reach larger adult sizes in *P. troglodytes* than in *P. paniscus* (Fig. 18.3). *P. troglodytes* adults tend to possess larger adult skulls, faces, and mandibles than do *P. paniscus.* However, chimpanzees and bonobos differ little in their craniofacial growth allometries (Tables 18.7, 18.8, and 18.9). Allometric patterns are virtually identical for the calotte and face; only in the mandible are differences apparent, and even here they are slight. For both the calotte and the face, very strong allome-

Table 18.7 Comparison of Growth Allometries of the Calotte in *Pan troglodytes* and *Pan paniscus* and Significance of the Difference between Chimpanzee and Bonobo Allometric Coefficients

Trait	*Pan troglodytes*			*Pan paniscus*			
	k	SE	Allometry	k	SE	Allometry	Signif.
Max. cranial length	0.981	0.025	Isometry	1.02	0.037	Isometry	NS
Max. cranial breadth	0.501	0.031	Negative	0.401	0.050	Negative	NS
Min. frontal breadth	0.530	0.036	Negative	0.609	0.055	Negative	NS
Biorbital breadth	1.96	0.047	Positive	2.03	0.073	Positive	NS

Table 18.8 Comparison of Growth Allometries of the Face in *Pan troglodytes* and *Pan paniscus* and Significance of the Differences between Chimpanzee and Bonobo Allometric Coefficients

Trait	*Pan troglodytes*			*Pan paniscus*			Signif.
	k	SE	Allometry	k	SE	Allometry	
Bi-ectomolare	0.736	0.014	Negative	0.684	0.021	Negative	$p < 0.05$
Palatal length	1.26	0.015	Positive	1.32	0.027	Positive	NS
Nasal breadth	0.965	0.036	Isometry	1.02	0.079	Isometry	NS

Table 18.9 Comparison of Growth Allometries of the Mandible in *Pan troglodytes* and *Pan paniscus* and Significance of the Difference between Chimpanzee and Bonobo Allometric Coefficients

Trait	*Pan troglodytes*			*Pan paniscus*			Signif.
	k	SE	Allometry	k	SE	Allometry	
Bigonial breadth	0.926	0.021	Negative	0.904	0.034	Negative	NS
Mandibular length	1.07	0.017	Positive	1.19	0.032	Positive	$p < 0.05$
Symphyseal height	1.13	0.027	Positive	1.06	0.039	Isometry	NS
Corpus height	0.965	0.028	Isometry	0.863	0.039	Negative	$p < 0.05$
Corpus thickness	0.734	0.028	Negative	0.579	0.041	Negative	$p < 0.05$

tries are present for almost every trait examined. For the mandible, bonobos and chimpanzees display isometry or weak allometry. Perhaps surprisingly, whenever bonobos and chimpanzees exhibit significant differences in their allometric coefficients, it is *P. paniscus* that displays the stronger (positive or negative) craniofacial growth allometries. For example, mandibular corpus thickness grows with stronger negative allometry in bonobos than in chimpanzees, and mandibular length grows with slightly stronger positive allometry.

There are certain consistencies in craniofacial growth allometries across all four taxa (compare Tables 18.4, 18.5, and 18.6 with Tables 18.7, 18.8, and 18.9). Biorbital breadth grows with positive allometry in relation to other traits of the calotte; this is undoubtedly characteristic of all primates, as it reflects the general positive growth allometry of the face relative to the neurocranium. Biorbital breadth grows with greater positive allometry in *Pan* than in *Homo* and with greater positive allometry in Neandertals than in modern humans. This undoubtedly reflects the trend in recent human evolution toward reduced prognathism and diminished relative splanchno-cranial size.

Again across the board, the bi-ectomolare grows with negative allometry relative to other traits in the face (i.e., the palate becomes relatively narrower). *Pan* exhibits stronger negative allometry than does *Homo*. In the mandible, bigonial breadth and especially corpus thickness grow with negative allometry relative to other mandibular traits; here, Neandertals exhibit the strongest negative allometries. Mandibular length is positively allometric in *Pan* and isometric in *Homo*. There is no consistency in the relative degree of similarity of Neandertals or of modern humans to *Pan*. For example, Neandertals are more similar than are modern humans to *Pan* in the degree to which they exhibit negative allometry of neurocranial breadth vis-à-vis other traits of the calotte; compare Tables 18.4 and 18.7. They are less similar than are modern humans to *Pan* in the pattern of allometry for mandibular corpus height (vis-à-vis the other traits of the mandible); compare Tables 18.6 and 18.9.

Euclidean Distances in Shape-Space

Figure 18.7 shows the scatter of points for each of our four taxa in facial shape-space. Their developmental trajectories (or shape paths) are depicted as arrows running through the range midpoints for infants and for adults. Table 18.10 describes the mean Euclidean distances for pairwise comparisons of individuals drawn either from different life cycle stages or across "taxa" for Neandertals and modern humans. Table 18.11 describes the same for pygmy chimpanzees and common chimpanzees. For both *Homo* and *Pan,* ontogenetic shape modifications across the life span, as well as shape differences between closely related "taxa," tend to be greater in the face than in the calotte.

The first rows of Tables 18.10 and 18.11 show cross-taxonomic comparisons of individuals belonging to the same life cycle stages. This comparison addresses the developmental source of shape differences that characterize adults. Are strong shape differences between adults of closely related taxa already present during infancy? Or are infants of closely related taxa far more alike in their cranial shapes? As Table 18.10 shows, only in the face are adult Neandertals and adult modern humans more different in shape than infant Neandertals and modern humans. For the calotte and mandible, proportions change after infancy (albeit not as dramatically as in the face), but the Euclidean shape distance between the two groups stays

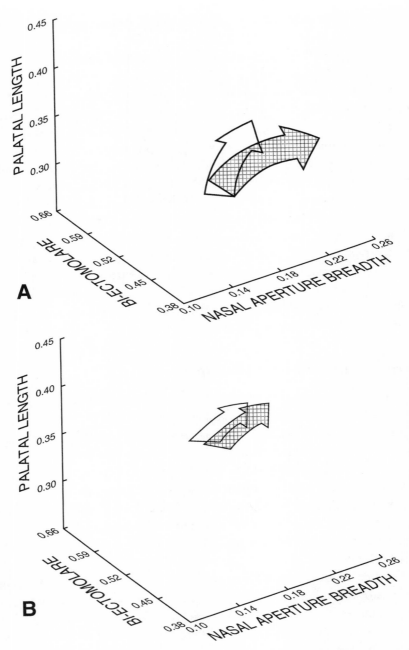

Fig. 18.7. Ontogenetic paths for the four taxa in three-dimensional facial shape-space. *Arrows* show directionality of shape change, from infancy to adulthood. *A:* Comparison of modern human (*open arrow*) and Neandertal (*hatched arrow*) shape paths. *B:* Comparison of bonobo (*open arrow*) and common chimpanzee (*hatched arrow*) shape paths.

roughly constant. Table 18.11 (row 1) shows that the same applies to the mandible in *Pan;* the degree of shape difference between bonobos and chimpanzees does not change significantly from infancy to adulthood. This suggests that substantial shape differences between taxa are manifested at a very young age and that they increase little over the life span. In general, Neandertals differ from modern humans in craniofacial proportions as much, or more, than pygmy chimpanzees differ from common chimpanzees.

The second rows of Tables 18.10 and 18.11 contrast the degree of shape change, from infancy to adulthood, in Neandertals and modern humans (Table 18.10) and in bonobos and chimpanzees (Table 18.11). Neandertals change more in the face and mandible than do modern humans, whereas modern humans change more in the calotte. The shape change in the calotte and face is greater for both species of *Pan* than for *Homo;* however, bonobos and common chimpanzees differ little in the degree to which they undergo postnatal change in facial proportions.

The third through fifth rows of Tables 18.10 and 18.11 test the hypotheses that modern humans are paedomorphic with respect to Neandertals and that bonobos are paedomorphic with respect to common chimpanzees. If modern humans were paedomorphic with respect to Neandertals, then we would expect modern human adults to be more similar in shape to Neandertal infants than Neandertal adults are to modern human infants. For calotte shape, modern human adults are closer to Neandertal infants than Neandertal adults are to modern human infants, but the same does not hold for either the face or the mandible. If modern humans were strongly paedomorphic with respect to Neandertals, then we might also expect modern human adults to be more like Neandertal infants than they are like Neandertal adults. If modern humans were more weakly paedomorphic, they might resemble Neandertal juveniles or subadults. Modern humans are decidedly unlike Neandertal infants, but they are significantly more like Neandertal juveniles than they are like Neandertal adults (particularly in the face and mandible). We conclude that modern humans are, indeed, weakly paedomorphic with respect to Neandertals. We will return to this question in the discussion.

A similar case can be made for paedomorphosis in pygmy chimpanzees (vis-à-vis common chimpanzees), but here, too, the signal is not strong. *P. paniscus* adults are significantly closer to *P. troglodytes* infants than *P. troglodytes* adults are to *P. paniscus* infants. This is true for every region of the cranium. However, bonobo adults resemble common chimpanzee adults far

Table 18.10 Euclidean Mean Distances between Individuals Belonging to Selected Life Cycle Stages, Generated via Resampling Modern Humans and Neandertals, 100 Trials (30 trials for row 1)

Comparison	Calotte	Face	Mandible
Across taxa			
Modern human infants and Neandertal infants	$t = -1.39, p = 0.17$, NS 0.022	$t = -2.53, p = 0.01$ 0.051	$t = -0.01, p = 0.99$, NS 0.047
Modern human adults and Neandertal adults	0.024	0.069	0.047
Across life cycle stages			
Modern human infants and modern human adults	$t = 3.15, p = 0.00$ 0.035	$t = -5.19, p = 0.00$ 0.057	$t = -8.34, p = 0.00$ 0.038
Neandertal infants and Neandertal adults	0.029	0.078	0.068
Across taxa and life cycle stages	$t = -12.37, p = 0.00$	$t = 2.00, p = 0.04$	$t = 2.26, p = 0.02$
Modern human adults and Neandertal infants	0.025	0.077	0.060
Neandertal adults and modern human infants	0.48	0.066	0.053
Are modern humans very paedomorphic?	$t = 0.36, p = 0.72$, NS	$t = 1.58, p = 0.11$, NS	$t = 4.37, p = 0.00$
Modern human adults and Neandertal infants	0.025	0.077	0.060
Modern human adults and Neandertal adults	0.024	0.069	0.047
Are modern humans slightly paedomorphic?	$t = -2.65, p = 0.01$	$t = -11.07, p = 0.00$	$t = -9.3, p = 0.00$
Modern human adults and Neandertal juveniles	0.019	0.037	0.011
Modern human adults and Neandertal adults	0.023	0.066	0.047

TABLE 10.11 Euclidean Mean Distances between Individuals Belonging to Selected Life Cycle Stages, Generated via Resampling *Pan troglodytes* and *Pan paniscus*, 100 Trials

Comparison	Calotte	Face	Mandible
Across taxa	$t = -4.26, p = 0.00$	$t = 3.19, p = 0.00$	$t = 1.38, p = 0.17$, NS
P. troglodytes infants and P. paniscus infants	0.020	0.047	0.037
P. troglodytes adults and P. paniscus adults	0.027	0.036	0.034
Across life cycle stages	$t = 4.54, p = 0.00$	$t = -0.84, p = 0.59$, NS	$t = -3.45, p = 0.00$
P. troglodytes infants and P. troglodytes adults	0.065	0.086	0.040
P. paniscus infants and P. paniscus adults	0.057	0.089	0.049
Across taxa and life cycle stages	$t = -13.06, p = 0.00$	$t = -11.43, p = 0.00$	$t = -3.8, p = 0.00$
P. troglodytes adults and P. troglodytes infants	0.051	0.070	0.041
P. troglodytes adults and P. paniscus infants	0.76	0.115	0.052
Are bonobos very paedomorphic?	$t = 14.63, p = 0.00$	$t = -9.62, p = 0.00$	$t = 3.15, p = 0.00$
P. paniscus adults and P. troglodytes infants	0.051	0.070	0.041
P. paniscus adults and P. troglodytes adults	0.027	0.036	0.034
Are bonobos sightly paedomorphic?	$t = -2.2, p = 0.03$	$t = -1.02, p = 0.31$, NS	$t = -0.19, p = 0.84$, NS
P. paniscus adults and P. troglodytes subadults	0.021	0.036	0.037
P. paniscus adults and P. troglodytes adults	0.024	0.039	0.037

more than they resemble common chimpanzee infants. They are slightly closer in the form of the calotte to *P. troglodytes* subadults than they are to *P. troglodytes* adults.

Principal Component Analysis of Shape Change

Table 18.12 summarizes variable loadings (or correlations of original trait ratios with taxon scores) for the first two principal component axes of all analyses reported in this section. As described under "Methods," we analyzed separately each of the three regions (calotte, face, and mandible) for both *Homo* and *Pan*.

Calotte. PCA 1 (*Homo* analysis) explains 56.4 percent of the total variance (Fig. 18.8*A*). It is a contrast vector distinguishing individuals with relatively broad and short crania and small faces (and strongly negative scores) from individuals with relatively long and narrow crania and broad faces (and strongly positive scores). (Here we judge the size of the "face" from biorbital breadth and minimal frontal breadth; we are considering only its superior aspect.) Modern humans and Neandertals lie at opposite poles along this axis; modern humans have short but broad crania and narrow faces; Neandertals have long but narrow crania and broad faces. Interestingly, Neandertal infants have positive projections on this axis (although

Table 18.12 Correlations of Original Trait Ratios with Individual Scores on the First Two Principal Component Axes for Two Taxon Comparisons (Three Regional Analysis)

		Homo		*Pan*	
Region	*Trait Ratios*	PCA 1	PCA 2	PCA 1	PCA 2
Calotte	Maximum cranial length/CS	0.69	0.64	0.21	−0.98
	Maximum cranial breadth/CS	−0.91	0.41	0.98	0.03
	Minimum frontal breadth/CS	0.80	0.50	0.95	0.19
	Biorbital breadth/CS	0.57	−0.82	−0.99	0.00
Face	Palatal length/FS	−1.00	0.01	−1.00	0.09
	Nasal aperture breadth/FS	0.52	−0.85	0.07	−1.00
	Bi-ectomolare/FS	0.55	0.83	0.99	0.16
Mandible	Bigonial breadth/MS	−0.90	−0.42	0.92	0.03
	Mandibular length/MS	0.86	0.11	−0.94	−0.29
	Symphyseal height/MS	0.82	0.26	−0.78	0.61
	Mandibular corpus height/MS	0.43	−0.79	0.84	0.34
	Mandibular corpus thickness/MS	−0.43	0.80	0.97	−0.11

Note: CS, calotte size; FS, face size; MS, mandible size.

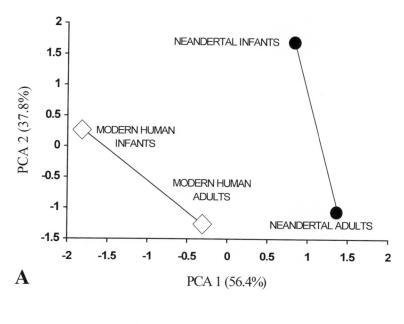

A

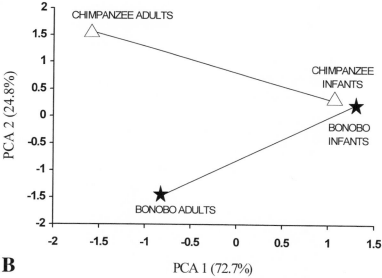

B

Fig. 18.8. PCA of the calotte, showing centroids for infants and adults. *A:* Neandertals versus modern humans. The two-dimensional Euclidean distance (rescaled according to the variance explained by PCA axes 1 and 2) between modern human infant and adult centroids is much smaller than that separating Neandertal infant and adult centroids. *B:* Chimpanzees versus bonobos. The rescaled Euclidean distance between bonobo infant and adult centroids is much smaller than that separating common chimpanzee infant and adult centroids.

they are less strongly positive than those of Neandertal adults), whereas adult modern humans have negative projections (although less strongly negative than those of modern human infants). From infancy to adulthood, both modern humans and Neandertals "travel" along this axis in a positive direction. This indicates a narrowing of the neurocranium relative to frontal breadth and biorbital breadth in both. Although modern humans are more similar in their projections on this axis to Neandertal infants than to Neandertal adults, they do not retain Neandertal infant characteristics into adulthood. Modern humans and Neandertals follow parallel shape changes from different points of "origin" (or infant shapes).

PCA 2 (which explains 37.8% of the variance) separates infants from adults for both modern humans and Neandertals. Infants have positive projections on this axis, and adults have negative projections. Variable loadings show that the major distinction is in relative biorbital breadth. Infants have small biorbital breadths (relative to cranial length, minimum frontal breadth, and maximum cranial breadth). Modern humans change less from infancy to adulthood along this axis than do Neandertals. PCA 2 fails to separate the two taxa.

The PCA of calotte shape in *Pan* differs from that in *Homo* in fundamental ways (Fig. 18.8B). It is the first axis (which explains 72.7% of the variance), not the second, that does the best job of separating infants from adults, and it is the second axis (which explains only 24.8% of the variance), not the first, that best separates taxa. Most of the variance in calotte shape in *Pan* distinguishes infants from adults, whereas most of the variance in calotte shape in *Homo* distinguishes "taxa." Furthermore, the projections of infant bonobos and chimpanzees on PCA 1 and PCA 2 are very close to one another.

Variable loadings demonstrate that the axes separating infants from adults are similar for *Homo* (PCA 2, *Homo* analysis) and for *Pan* (PCA 1, *Pan* analysis). Infants of both *Homo* and *Pan* have small biorbital breadths relative to cranial length, neurocranial breadth, and minimum frontal breadth. Among chimpanzees, cranial length contributes to this distinction much less than does neurocranial breadth and minimum frontal breadth, whereas in humans it contributes more.

The two axes that polarize closely related taxa show little similarity in *Pan* and *Homo*. Bonobo adults can be distinguished from chimpanzee adults almost entirely on the basis of their relatively shorter skulls. Infants of the two species of *Pan* cannot be separated by this criterion. All traits are needed to distinguish modern humans from Neandertals.

Face. The first and second components of the PCA of Neandertal and modern human faces together explain 99.9 percent of the variance; the variance is roughly equally divided between the two (Fig. 18.9*A*). Neandertal and modern human infants are more closely approximated (and, therefore, more similar in facial shape) than are Neandertal and modern human adults. Both Neandertal and modern human infants have positive scores on PCA 1 and PCA 2. Their positive projections on PCA 1 reflect their relatively short but broad palates (Table 18.12). Their positive projections on PCA 2 reflect the fact that infant palates are also broad relative to nasal aperture breadth. In Neandertals, increasingly older individuals have increasingly broader nasal apertures (relative to bi-ectomolare) and therefore increasingly strongly negative projections on PCA 2. There is a very weak trend in the same direction in modern humans. The broad nasal apertures of Neandertal adults are understood as adaptive for life in Europe during the Würm glaciation (Howell 1957; Brose and Wolpoff 1971); alternatively or additionally, the broad Neandertal nasals may have helped to resist excessive biomechanical loads placed on the anterior teeth during certain activities (Mann and Trinkaus 1974; Rak 1986; Trinkaus 1987; Demes 1987).

Our PCA of the face in *Pan* tells a different story (Fig. 18.9*B*). The first axis (PCA 1, explaining 65.8% of the variance) separates infants from adults; the second (PCA 2, explaining 34.2% of the variance) separates the two species. Thus, just as we observed for the calotte in *Pan*, it is the axis separating life cycle stages that explains most of the variance. PCA 1 is a contrast vector separating individuals with relatively wide but short palates (infants) from individuals with relatively narrow but long palates (adults). PCA 2 captures species differences in relative nasal aperture breadth; *P. troglodytes* has a relatively broad nasal aperture throughout postnatal ontogeny and thus strongly negative projections on PCA 2 (see Table 18.12).

Mandible. For *Homo* and *Pan,* the first two axes explain less of the total variance for the mandibles than do the first two axes of our principal component analyses of the calotte and face (Fig. 18.10). This probably reflects the greater number of traits considered in our analysis of mandibular shape. PCA 1 (*Homo* analysis, 51.7% of the variance) largely separates infants (with negative projections) from adults (with positive projections) (Fig. 18.10*A*). PCA 1 is a contrast vector separating individuals with relatively broad and shallow but short mandibles (infants) from individuals with relatively narrow, long, and deep mandibles (adults). This axis also captures the fact that infants have relatively thicker mandibular corpi than do adults. Neandertals travel further along this axis than do modern humans, with increased mandibular length through the development of a retromolar space. It is PCA 2 (explaining

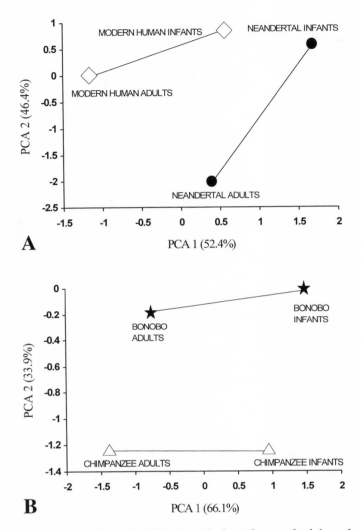

Fig. 18.9. PCA of the face, showing centroids for infants and adults only. *A:* Neandertals versus modern humans. The Euclidean distance (rescaled according to the variance explained by PCA axes 1 and 2) between modern human infant and adult centroids is smaller than that separating Neandertal infant and adult centroids. *B:* Chimpanzees versus bonobos. The rescaled Euclidean distance between bonobo infant and adult centroids is about equal to that separating chimpanzee infant and adult centroids.

30.4% of the variance) that captures the main difference between Neandertal and modern human mandibular shape-change trajectories. Not only do Neandertals have positive projections (throughout ontogeny) on this axis while modern humans have negative projections (again, throughout on-

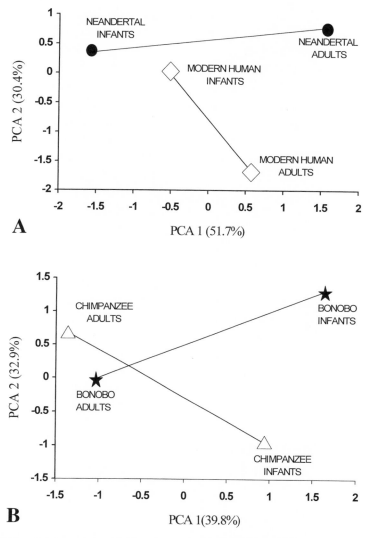

Fig. 18.10. PCA of the mandible, showing centroids for infants and adults only. *A:* Neandertals versus modern humans. The Euclidean distance (rescaled according to the variance explained by PCA axes 1 and 2) between modern human infant and adult centroids is much smaller than that separating Neandertal infant and adult centroids. *B:* Chimpanzees versus bonobos. The rescaled Euclidean distance between bonobo infant and adult centroids is larger than that separating chimpanzee infant and adult centroids.

togeny), but from infancy to adulthood Neandertals change negligibly while modern humans change dramatically along this axis. This axis distinguishes (at the positive end) individuals with relatively thick but shallow mandibular corpi and narrow mandibles (Neandertals) from (at the negative end) individuals with thin but deep mandibular corpi and broad mandibles (modern humans). Modern humans become less and less Neandertal-like (in this respect) throughout ontogeny because of decreasing mandibular corpus thickness relative to corpus height and relative to bigonial breadth.

Figure 18.10*B* shows our principal component analysis of mandibular shape in bonobos and chimpanzees. PCA 1 (capturing 39.75% of the variance) separates infants (with strong positive projections) from adults (with strong negative projections) of both species. As in *Homo*, infants have short but broad mandibles and adults have long but narrow mandibles. Symphyseal height is relatively small and the mandibular corpus is relatively thick in infant *Pan* (again as in humans), but the mandibular corpus is also relatively deep (unlike humans). PCA 2 (explaining 32.9% of the variance) largely captures mandibular shape differences between infant *P. troglodytes* and *P. paniscus*. Common chimpanzee infants have positive projections on this axis, while bonobo infants have negative projections. This is a contrast vector separating individuals with relatively shorter (bonobos) versus those with relatively deeper mandibles (chimpanzees). This distinction is lost in *Pan* from infancy to adulthood.

Discussion and Conclusions

Our results reveal certain consistent signals across different types of analysis. In *Pan*, shape differences between infants and adults are consistently greater than shape differences across taxa at comparable life cycle stages; this is reflected in both Euclidean distance analysis and PCA. This pattern is far less clear-cut for *Homo* (as might be anticipated given the generally weaker growth allometries of all *Homo* compared with *Pan*). Still, Neandertals generally undergo greater (postinfancy) ontogenetic shape changes than do modern humans. This is once again reflected in our principal component analyses, Euclidean distance analyses, and comparisons of allometric coefficients of modern humans and Neandertals. In effect:

—Neandertals generally travel greater distances than do modern humans in two-dimensional PCA space.

—Neandertals generally depart more from their infant morphologies than do modern humans (in n-dimensional shape-space).

—Neandertal growth allometries tend to be stronger than those of modern humans, particularly for the calotte and mandibular regions. Neandertal infants become Neandertal adults via relatively strong positive and negative allometric growth.

Neoteny (*sensu* Gould 1977) carries several theoretical expectations. The first is that the descendant be paedomorphic in form. The second is that this paedomorphosis be achieved through a reduction in the rate of shape change with respect to the rate of growth in size and age at sexual maturation. This generally requires a weakening of growth allometries (i.e., positive growth allometries should become less positive while negative growth allometries become less negative). In other words, growth allometries should converge toward isometry (Shea 1989; Godfrey and Sutherland 1996; Godfrey et al. 1998). This test of neoteny is not reliable, however, when ancestral allometries are nonlinear on logarithmic scales.[5] Other requirements are more easily tested. Under Gould's (1977) "pure" neoteny, maturation should occur at about the same age in ancestor and descendant, and the ancestor and descendant should be roughly equal in adult size. Finally, and most importantly, the descendant should follow its ancestor's shape path. This is the most important prediction of neoteny (and, indeed, of all of Gould's heterochronic processes) because it does not depend on ancestral allometries being simple (linear in logarithmic space). Maintenance of ancestral shape paths is a prediction of neoteny, regardless of whether ancestral allometries are linear or curvilinear on logarithmic scales.

Modern human crania conform to some of the expectations of neoteny with respect to Neandertals. First, they exhibit generally weaker growth allometries. Second, they are more similar in form to those of Neandertal juveniles than Neandertal adults. Assuming that Neandertals matured at the same age as modern humans, then modern humans also exhibit lower rates of craniofacial shape change from infancy to adulthood—a third prediction of neoteny. However, adult Neandertal skulls are generally larger than those of modern humans, and most importantly, *modern humans and Neandertals follow different shape paths.* Many of the differences between modern humans and Neandertals are generated very early in ontogeny (i.e., in prenatal and early postnatal ontogeny). Modern humans do not simply retrace Neandertal developmental steps to a different terminus. By the time they are infants, modern humans and Neandertals are already on different shape paths, particularly for the face and mandible. Indeed, infant Neandertals and infant modern humans are about as distinct in shape (sometimes more, sometimes less, depending on region) from each other as adults of either taxon are from their own infants. Modern human growth allometries are weak because shape changes little from *modern human* (and not Neandertal) infant standards. Modern humans are in the realm of paedomorphosis vis-à-vis Neandertals, but they are by no means perfect paedomorphs.

Ontogenetic craniofacial shape change is stronger in *Pan* than in *Homo*. Indeed, in virtually every analysis, the magnitude of ontogenetic shape change in different regions or for different traits exceeds cross-taxonomic differences at comparable life cycle stages. The general pattern of shape change (as revealed by the directions and magnitudes of allometric coefficients) is more similar in bonobos and chimpanzees than in modern humans and Neandertals. But, like modern humans and Neandertals, bonobos and chimpanzees do not follow the same shape paths, and some shape differences between the taxa are manifested throughout their ontogenies. The Euclidean distances separating infant bonobos and infant chimpanzees are not much smaller than those separating infant modern humans and infant Neandertals. Furthermore, adult pygmy chimpanzees are very similar to adult common chimpanzees; the extent to which they can be described as paedomorphic is *slight*, largely manifested in the shape of the calotte (where adult bonobos are more similar to subadult chimpanzees than to adult chimpanzees). Despite the large ontogenetic shape changes that both bonobos and chimpanzees experience, the adults of these species are more similar in shape than are adult Neandertals and adult modern humans. Finally, the detailed expectations of neoteny are not met (bonobos do not exhibit weakened growth allometries; when compared with common chimpanzees, bonobos are not similar in cranial size but juvenilized in cranial shape).[6]

In summary, our data demonstrate that Neandertals are morphologically distinct from modern humans from early postnatal ontogeny onward and that these differences are at least as great as those manifested by different species of *Pan*. In 1977, Gould proposed that heterochronic shifts played an important role in the evolution of modern humans—namely, that modern humans may be paedomorphic, or juvenilized, in craniofacial shape and that this paedomorphosis may have arisen via neoteny (slower rate of shape change, with little or no change in the overall size or age at maturation of the descendant in comparison to its ancestor). Our developmental data for Neandertals offer little support for this hypothesis.

Acknowledgments

This research was generously supported by a U.S. Fulbright Fellowship (for study in the Netherlands) and a Belgian American Educational Foundation Graduate Fellowship (for study in Belgium) to F.L.W. Funds for F.L.W.'s travel were awarded by the Netherlands-America Commission for Educational Exchange, the Faculteit der Medische Wetenschappen, Rijksuniversiteit Groningen, the Netherlands, as well as by Sigma Xi.

We thank Nancy Minugh-Purvis and Kenneth J. McNamara for inviting us to submit this chapter. Special thanks go to G. N. van Vark, Vakgroep Anatomie en Embryologie, and W. Schaafsma, Wiskunde en Statistiek, Rijksuniversiteit Groningen, the Netherlands; George Maat, Anatomie Museum Rijksuniversiteit Leiden, the Netherlands; John de Vos, Nationaal Natuurhistorisch Museum, Leiden, the Netherlands; and Rosine Orban and Patrick Semal, Institut Royal des Sciences Naturelles de Belgique, Brussels, Belgium, for their help with various aspects of this research. We thank the family of Max Lohest for placing the Spy materials, for the purposes of curation and scientific study, on loan to the Institut Royal des Sciences Naturelles de Belgique. Figures 18.1 and 18.6 were generously provided by Patrick Semal; Figure 18.5 was provided by the Musée National de Préhistoire Les Eyzies de Tayac. Darren Godfrey's assistance with the preparation of Figures 18.4 and 18.7 is greatly appreciated. We are grateful to the heads and curators of the many collections we visited for this research, for their hospitality and for their curation of the materials under their care; they are too numerous to list here. All data used in this study were collected by F.L.W.

Notes

1. Gould (1977, 483) defined *neoteny* as "paedomorphosis (retention of formerly juvenile characters by adult descendants) produced by retardation of somatic development."

2. Although this specimen was discovered in the early part of the nineteenth century, its significance was not understood until much later; see Fraipont (1936), Tillier (1983), and Minugh-Purvis (1988).

3. Analyses of variance (ANOVAs) for nine of the twelve craniofacial traits selected for analysis produced no significant differences among adults of Belgian, Papuan, and mixed populations.

4. Although not every difference is statistically significant, an overall sign test supports the inference that Neandertal allometries are stronger than those of mod-

ern humans. The statistical significance of some differences may be affected by poor sample sizes. The standard errors of the allometric slopes are much higher for Neandertals than for modern humans.

5. Godfrey et al. (1998) discussed the difficulties of using allometric patterns to elucidate heterochronic processes.

6. In fact, even though he labeled pygmy chimpanzees "true neotenes," Shea (1983, 1984) described the craniofacial "paedomorphosis" of pygmy chimpanzees as having arisen through "ontogenetic scaling," which is not a prediction of neoteny (see Shea 1989; Godfrey and Sutherland 1995, 1996; Godfrey et al. 1998; but see Shea, Chap. 4).

References

Anemone, R.L., Mooney, M.P., and Siegel, M.L. 1996. Longitudinal study of dental development in chimpanzees of known chronological age: implications for understanding the age at death of Plio-Pleistocene hominids. *American Journal of Physical Anthropology* 99:119–133.

Boule, M. 1912. *L'Homme Fossile de la Chapelle-aux-Saints*. Paris: Masson and Cie.

Brace, C. L. 1964. The fate of the "classic" Neanderthals: a consideration of hominid catastrophism. *Current Anthropology* 65:3–34.

Bräuer, G., and Smith, F. (eds). 1992. *Continuity or Replacement: Controversies in* Homo sapiens *Evolution*. Rotterdam: Balkema.

Brose, D.S., and Wolpoff, M.N. 1971. Early Upper Paleolithic man and Late Middle Paleolithic tools. *American Anthropologist* 73:1156–1194.

Buikstra, J.E., and Ubelaker, D.H. 1994. *Standards for Data Collection from Human Skeletal Remains*. Fayetteville: Arkansas Archeological Survey.

Corner, B.B., Lele, S., and Richtsmeier, J.T. 1992. Measurement error of three-dimensional landmarks. *Quantitative Anthropology* 3:347–359.

Dean, M.C., Stringer, C.B., and Bromage, T.G. 1986. Age at death of the Neandertal child from Devil's Tower, Gibraltar and the implications for studies of general growth and development in Neandertals. *American Journal of Physical Anthropology* 70:301–309.

Dean, M.C., and Wood, B.A. 1981. Developing pongid dentition and its use for ageing individual crania in comparative cross-sectional growth studies. *Folia Primatologica* 36:111–127.

de Beer, G.R. 1940. *Embryos and Ancestors*. Oxford: Clarendon Press.

Demes, B. 1987. Another look at an old face: biomechanics of the Neandertal facial skeleton reconsidered. *Journal of Human Evolution* 16:297–304.

Fraipont, C. 1936. *Les Hommes Fossiles d'Engis*. Arch. Inst. Pal. Hum. 16. Paris: Masson.

Garn, S.M., Lewis, A.B., and Kerewsky, R.S. 1965. Genetic, nutritional, and maturational correlates of dental development. *Journal of Dental Research* 44:228–242.

Godfrey, L.R., King, S.J., and Sutherland, M.R. 1998. Heterochronic approaches to the study of locomotion. In E. Strasser, J. Fleagle, A. Rosenberger, and H. McHenry (eds.), *Primate Locomotion: Recent Advances*, 277–307. New York: Plenum Press.

Godfrey, L.R., and Sutherland, M.R. 1995. What's growth got to do with it? Process and product in the evolution of ontogeny. *Journal of Human Evolution* 29:405–431.

———. 1996. Paradox of peramorphic paedomorphosis: heterochrony and human evolution. *American Journal of Physical Anthropology* 99:17–42.

Gould, S.J. 1977. *Ontogeny and Phylogeny.* Cambridge: Belknap Press of Harvard University Press.

Howell, F.C. 1957. The evolutionary significance of variation and varieties of "Neanderthal" man. *Quarterly Review of Biology* 32:330–347.

Howells, W.W. 1967. *Mankind in the Making: The Story of Human Evolution.* Garden City, N.Y.: Doubleday.

Hrdlička, A. 1927. The Neanderthal phase of man. *Journal of the Royal Anthropological Institute of Great Britain and Ireland* 57:249–274.

Krovitz, G., Cole, T.M., and Richtsmeier, J.T. 1997. Three-dimensional comparisons of craniofacial growth patterns in Neandertals and modern humans. *American Journal of Physical Anthropology* 24 (suppl): 147.

Kuykendall, K.L., and Conroy, G.C. 1996. Permanent tooth calcification in chimpanzees (*Pan troglodytes*): patterns and polymorphisms. *American Journal of Physical Anthropology* 99:159–174.

Lele, S., and Richtsmeier, J.T. 1991. Euclidean distance matrix analysis: a coordinate free approach for comparing biological shapes using landmark data. *American Journal of Physical Anthropology* 86:415–528.

Lieberman, D.E. 1997. A developmental approach to defining modern humans. *American Journal of Physical Anthropology* 24 (suppl): 155–156.

———. 1998. Sphenoid shortening and the evolution of modern human cranial shape. *Nature* 393:158.

Mann, A.E., and Trinkaus, E. 1974. Neandertal and Neandertal-like fossils from the Upper Pleistocene. *Yearbook of Physical Anthropology* 17:169–193.

McKinney, M.L., and McNamara, K.J. 1991. *Heterochrony: The Evolution of Ontogeny.* New York: Plenum Press.

Minugh-Purvis, N. 1988. Patterns of craniofacial growth and development in Upper Pleistocene hominids. Ph.D. diss., University of Pennsylvannia, Philadelphia.

Montagu, M.F.A. 1989. *Growing Young.* New York: McGraw-Hill.

Nissen, H.W., and Riesen, A.H. 1964. The eruption of the permanent dentition of chimpanzee. *American Journal of Physical Anthropology* 22:285–294.

Privratshy, V. 1981. Neoteny and its role in the process of hominization. *Anthropologie* 19:219–229.

Rak, Y. 1986. The Neanderthal: a new look at an old face. *Journal of Human Evolution* 15:151–164.

Richtsmeier, J.T., Cheverud, J.M., Danahey, S.E., Corner, B.D., and Lele, S. 1993.

Sexual dimorphism of ontogeny in the crab-eating macaque (*Macaca fascicularis*). *Journal of Human Evolution* 25:1–30.

Richtsmeier, J.T., and Walker, A.C. 1993. A morphological study of facial growth. In A.C. Walker and R.E. Leakey (eds.), *The Nariokotome* Homo erectus *Skeleton*, 391–410, 446–447. Cambridge: Harvard University Press.

Shea, B.T. 1983. Allometry and heterochrony in the African apes. *American Journal of Physical Anthropology* 62:275–289.

———. 1984. An allometric perspective on the morphology and evolutionary relationships between pygmy (*Pan paniscus*) and common (*Pan troglodytes*) chimpanzees. In R.L. Susman (ed.), *The Pygmy Chimpanzee: Evolutionary Biology and Behavior*, 89–130. New York: Plenum Press.

———. 1989. Heterochrony in human evolution: the case for neoteny reconsidered. *Yearbook of Physical Anthropology* 32:69–101.

Skinner, M. 1997. Age at death of Gibraltar 2. *Journal of Human Evolution* 32: 469–470.

Smith, B.H., Crummett, T.L., and Brandt, K.L. 1994. Ages of eruption of primate teeth: a compendium for aging individuals and comparing life histories. *Yearbook of Physical Anthropology* 37:177–231.

Smith, F.H., Falsetti, A.B., and Donnelly, S.M. 1989. Modern human origins. *Yearbook of Physical Anthropology* 32:35–68.

Stringer, C.B., and Andrews, P. 1988. Genetic and fossil evidence for the origin of modern humans. *Science* 239:1263–1268.

Stringer, C.B., Dean, M.C., and Martin, R. 1990. A comparative study of cranial and dental development in a recent British population and Neandertals. In C.J. DeRousseau (ed.), *Primate Life History and Evolution*, 115–152. New York: Wiley-Liss.

Stringer, C.B., and Gamble, C. 1993. *In Search of the Neanderthals*. New York: Thames and Hudson.

Tillier, A.-M. 1983. Le crâne d'enfant d'Engis 2: une exemple de distribution des caractères juveniles, primitifs, et néandertaliens. *Bulletin de la Societé Royale Belge d'Anthropologie et de Préhistoire* 94:51–75.

———. 1989. The evolution of modern humans: evidence from young Mousterian individuals. In P. Mellars and C. Stringer (eds.), *The Human Revolution: Behavioural and Biological Perspectives on the Origins of Modern Humans*, 286–297. Princeton: Princeton University Press.

Tompkins, R.L. 1996. Relative dental development of Upper Pleistocene hominids compared to human population variation. *American Journal of Physical Anthropology* 99:103–118.

Trinkaus, E. 1987. The Neandertal face: evolutionary and functional perspectives on a recent hominid face. *Journal of Human Evolution* 16:429–444.

Trinkaus, E., and Shipman, P. 1992. *The Neandertals: Changing the Image of Mankind*. New York: Alfred A. Knopf.

Williams, F.L. 1996. The use of non-linear models to map craniofacial heterochronies in fossil hominids, modern humans and chimpanzees. *American Journal of Physical Anthropology* 22 (suppl): 244.

————. 1997. Mulitvariate analysis of craniofacial heterochrony in Neandertals and modern humans. *American Journal of Physical Anthropology* 24 (suppl): 241.

————. 2001. Heterochronic perturbations in the craniofacial evolution of *Homo* (Neandertals and modern humans) and *Pan* (*P. troglodytes* and *P. paniscus*). Ph.D. diss., University of Massachusetts, Amherst.

Winkler, L.A., Schwartz, J.H., and Swindler, D.R. 1996. Development of the orangutan permanent dentition: assessing patterns and variation in tooth development. *American Journal of Physical Anthropology* 99:205–220.

Wolpoff, M.H. 1979. The Krapina dental remains. *American Journal of Physical Anthropology* 50:67–114.

————. 1986. Describing anatomically modern *Homo sapiens:* a distinction without a definable difference. *Anthropos* 23:41–53.

Wolpoff, M.H., and Caspari, R. 1997. *Race and Human Evolution.* New York: Simon and Schuster.

Chapter 19

Adolescent Postcranial Growth in *Homo neanderthalensis*

ANDREW J. NELSON AND JENNIFER L. THOMPSON

The adult Neandertal form differs from that of typical modern humans in several well-known ways. Among other features, a typical Neandertal has a long, low, and voluminous cranium with pronounced midfacial prognathism, a relatively large rib cage, a radius and tibia that are short relative to their proximal limb segments, robust limb bones, and large joints. These features add up to a description of an individual who was of medium to short stature, stout and barrel-chested, strongly muscled and heavily boned, who would have been well adapted to life as a hunter-gatherer in the circumglacial environment of Upper Pleistocene Europe (Trinkaus 1983a; Stringer and Gamble 1993). The large body core, combined with short stature and distal limb segment shortening, has been interpreted as a response to the selective pressures of life during the Early Glacial Phase of the Würm (Trinkaus 1981; Holliday 1997). The majority of Neandertal fossils come from this period, which can be subdivided into a temperate woodland phase (118,000 to 75,000 years ago) and a cold glacial phase (75,000 to 32,000 years ago) characterized by low temperatures and a drop in sea level (Gamble 1986).

Relatively late arrivals on the European scene, about 40,000 years B.P., are hominids with contrasting morphology: taller yet still rugged individuals without the distal limb shortening characteristic of Neandertals. This morphology of early modern *Homo sapiens* has been ascribed to their adaptation to warmer climes (Trinkaus 1981). This contrast in morphology and late arrival of Upper Paleolithic *H. sapiens* into Europe, coupled with the apparent disappearance of the distinct Neandertal morphotype, has been

442

interpreted within the context of a replacement model of modern human origins (e.g., Stringer and Andrews 1988).

The contrasts between the limb morphology of early modern *H. sapiens* and Neandertals are presumed to represent climatic adaptations following Bergmann's (1847) and Allen's (1877) Rules. According to these ecogeographical rules, species living in colder climates should have greater body mass and relatively shorter limbs than related species living in warmer climates, hence reducing heat loss due to the reduction of body surface area relative to body volume. The limb proportions are traditionally examined by means of the crural index (or tibiofemoral index) (tibia length × 100 / femur length) for the lower limb and the brachial index (or radiohumeral index) (radius length × 100 / humerus length) for the upper limb (after Martin 1928, e.g., Trinkaus 1981). In both cases, low values indicate relatively shortened distal segments, which are generally interpreted as indicating cold adaptation following Allen's Rule (Holliday 1997 and reference therein). The contrasts in limb proportions between Neandertals and early *H. sapiens* adults have been well demonstrated (Trinkaus 1981) and are presumed to reflect the in situ adaptation of Neandertals to the cold climate of Europe versus the adaptation outside Europe of *H. sapiens* to a warmer, possibly African climate (Trinkaus 1986). However, the underlying ontogenetic patterns that produced this contrasting morphology have not been systematically studied in these hominids. One aim of this study is to compare brachial and crural indices to determine whether the ontogeny of these proportions conforms to an ecological/climatic explanation for the morphological contrasts between these two taxa.

Another feature of Neandertal and *H. sapiens* morphology that is not well understood is the ontogeny of postcranial robusticity. Robusticity can be defined as the size of a given skeletal element relative to the whole body (Trinkaus et al. 1991), with bone length generally used as a surrogate for body size. Contrasts in bone robusticity between adults have been systematically studied by several workers (e.g., Trinkaus and Ruff 1989a, 1989b; Churchill 1994; Ruff et al. 1993; Trinkaus et al. 1998) and attributed to differences in mobility, upper limb loading intensity, and behavior patterns (Churchill 1998, 48), although Churchill acknowledges that these contrasts may reflect ecogeographical patterning. Neandertal juveniles demonstrate relatively greater robusticity in their postcranial skeleton than do modern (generally European) humans of equivalent developmental age (Vlček 1973; Heim 1982; Tompkins and Trinkaus 1987; Madre-Dupouy 1992). Explanations for this greater robusticity have included increased activity patterns in Neandertals at an earlier age (Trinkaus et al. 1998) related to the biomechanics of locomotion and/or increased mechanical loading of the skeleton throughout life (Trinkaus 1983b; Tompkins and Trinkaus 1987; Ruff et al. 1993, 1994).

In contrast, Smith (1991) suggested that Neandertal postcranial robusticity and other features are the product of a global acceleration of growth that included accelerated endochondral bone formation. In other words, the musculoskeletal hypertrophy characteristic of Neandertal morphology might be explained as an acceleration of their skeletal growth rates relative to modern humans. Smith also argued, on the basis of work by Minugh-Purvis (1988) and Leigh (1985), that Neandertal subadults are significantly larger at any given age than recent human subadults, as judged by craniofacial and dental development.

Contrary to expectations obtained primarily from craniofacial data, we have demonstrated that at equivalent dental ages Neandertals are absolutely smaller, postcranially, than modern humans (Thompson and Nelson 2000). Thus, the ontogeny of postcranial robusticity needs to be examined in greater detail. Ruff et al. (1994) examined this issue by comparing subadult fossils of similar dental age to recent humans by assessment of cortical area as a percentage of total cross-sectional area to measure medullary stenosis. Their results indicate that *Homo erectus* (as represented by the WT-15000 specimen) followed a developmental trajectory similar to that of modern humans, while Neandertal juveniles followed a different developmental pattern. Churchill (1998) discussed several possible reasons for this apparent phenomenon, and we believe that the examination of postcranial robusticity through ontogeny in Neandertals and early modern Europeans may throw further light on this issue.

The adult form of any species is the termination of a complex ontogenetic program. This program involves both absolute and differential growth of the various parts of the body. The growth program of modern humans has been particularly well studied (see Antón, Chap. 16). The period of growth preceding the final outcome of the adult form is adolescence, with its characteristic growth spurt in linear height. By one definition, the adolescent period ends with the cessation of linear growth (Bogin and Smith 1996). By adolescence, brain growth has finished and most of the teeth are in place. Adolescence is also characterized by the onset of puberty, the attainment of full sexual maturity and secondary sexual characteristics, and the adoption of adult behaviors.

The Le Moustier 1 Neandertal, represented by a cranium as well as by many elements of the postcranium, is a rare example of a Neandertal adolescent (Thompson 1998; Thompson and Bilsborough 1997, 1998a, 1998b; Thompson and Illerhaus 1998; Thompson and Nelson 1999). Previous work (Thompson 1995; Thompson and Illerhaus 2000) has sexed this individual as male, based on its large cranial capacity (see also Illerhaus and Thompson 1999) and large teeth and mandibular ramus flexure. Age estimates obtained on the basis of crown and root formation following Anderson et al. (1976) place Le Moustier 1 at a dental age equivalent to a that of modern adolescent of 15.5 years. However, previous work has found that his femoral length is equivalent to that of a modern human boy of 11–12 years of age (Thompson and Nelson 1997; Nelson and Thompson 1999). Thus, there is discordance between dental age estimates for this specimen and ages suggested by the linear growth of his long bones—or by his stature as reconstructed from long bone lengths. This is important, as it indicates that Le Moustier 1 still had a large proportion of his linear growth to complete if he were to have grown up to be a typical adult male European Neandertal. His stature (138 cm, Thompson and Nelson 2000), estimated using age-appropriate ratios from Feldesman (1992) and Feldesman et al. (1990), is 85.1 percent of the adult male Neandertal mean (Nelson and Thompson 1999). This is equivalent to a modern human boy who is just entering his adolescent growth spurt (Frisancho 1990).

The objective of this study is to examine the evidence for ontogenetic trajectories for limb lengths and limb robusticity in Neandertals and early Upper Paleolithic *H. sapiens* in the context of environmentally diverse modern human populations with the hope of shedding light on how the differences between adults of these groups might arise. In particular, we focus on the adolescent period, as represented by the Le Moustier 1 Neandertal individual. First, we examine Le Moustier 1 in comparison with adult Neandertals and environmentally diverse modern human samples. Then we broaden our focus to include ontogenetic series from two modern samples and various available juvenile Neandertal and early Upper Paleolithic *H. sapiens*. We demonstrate that (1) Neandertals share distal limb shortening with cold-adapted modern populations, a pattern of segmental growth that is set very early in ontogeny, and (2) limb robusticity (defined above after Trinkaus et al. 1991) of Neandertals, also shared with cold-adapted modern populations, does not reach its full expression until during or after the adolescent growth spurt. The latter point has relevance to models of Neandertal robusticity, which posit that high levels of activity must be present at very early ages in these hominids.

Methods and Materials

Our first objective was to assemble a database of modern humans that would allow us to demonstrate aspects of the distinctive Neandertal adult form as compared with those of Upper Paleolithic *H. sapiens* occupants of Europe, who lived within the same circumglacial cold environments (Table 19.1). To clarify which traits could confidently be identified as adaptations to a cold environment, we included in our modern sample a collection of Inuit individuals from Sadlermiut in the Canadian Northwest Territories. This material dates from late prehistoric to historic times (CMOC 1995). To provide a clear contrast between these arctic peoples and populations occupying a warmer environment, we also included a sample of South African Khoisan. The core typical modern sample is geographically and ethnically diverse and includes individuals from pre-European contact California, eighteenth-century Ontario, and England, Japan, and Fiji (Nelson 1995).

Data gathered for this study included lengths of the femur, tibia, humerus, and radius (Martin 1928). In addition, midshaft anteroposterior and medial-lateral diameters were gathered for the femur, tibia, and humerus. These two measurements were combined into a cross-sectional external area, using an elliptical model. The data gathered included maximum diaphyseal lengths for the juvenile individuals and maximum lengths for the adults. All modern individuals were measured by the authors (with supplemental data for radius lengths on the California sample from A. Foley, pers. comm.).

The adult Neandertal sample includes classic European Neandertals only. The *H. sapiens* sample includes individuals from Europe dating to circa 40,000–20,000 B.P. Our two samples both would be experiencing the same periglacial/glacial conditions, which allows comparisons between their ontogenetic patterns. The fossil sample of Neandertal and Upper Paleolithic *H. sapiens* (see Table 19.1) was constructed largely from the literature, supplemented by the authors' own measurements (Nelson 1995; J.L.T. for Le Moustier 1). Individuals were included that had lengths and shaft diameters for any of the major long bones.

Bone robusticity is generally examined using some measure of bone size (diameter, circumference, cross-sectional area) relative to bone length (Martin 1928; Trinkaus et al. 1991). Figures for crural and brachial indices of adult European Neandertals in comparison with a typical non-cold-adapted modern human population and with modern human populations from cold and warm environments are presented in Table 19.2. Midshaft

Table 19.1   Fossil Individuals Used in This Study

European Neandertals	Reference	Upper Paleolithic Homo sapiens	Reference
		Adults	
Fond-de-Forêt 1	Twiesselmann 1961	Abri Pataud 5	Billy 1975
Kiik Koba 1	Nelson 1995	Barma-Grande 2	Verneau 1906
La Chapelle 1	Boule 1912; Churchill 1994	Barma-Grande 5	Verneau 1906
La Ferrassie 1	Churchill 1994; Trinkaus 1980; Lovejoy and Trinkaus 1980	Cro-Magnon 1	Vallois and Billy 1965
La Ferrassie 2	Churchill 1994; Trinkaus 1980; Lovejoy and Trinkaus 1980	Cro-Magnon 2	Vallois and Billy 1965
La Quina 5	Churchill 1994	Cro-Magnon 3	Vallois and Billy 1965
Le Regourdou 1	Churchill 1994	Grotte-des-Enfants 4	Verneau 1906
Neanderthal 1	Twiesselmann 1961	Grotte-des-Enfants 5	Verneau 1906
Spy 2	Churchill 1994; Trinkaus 1976	Grotte-des-Enfants 6	Verneau 1906
		Předmostí 1	Matiegka 1938
		Předmostí 3	Matiegka 1938
		Předmostí 4	Matiegka 1938
		Předmostí 9	Matiegka 1938
		Předmostí 10	Matiegka 1938
		Předmostí 14	Matiegka 1938
		Sungir' 1	Bader et al. 1979
		Juveniles	
Kiik Koba 2	Vlček 1973	Abri Pataud 7	Gambier 1986
La Ferrassie 4	Heim 1982	Cro-Magnon-FD	Gambier 1986
La Ferrassie 4b	Heim 1982	Cro-Magnon-5	Gambier 1986
La Ferrassie 6	Heim 1982	Grotte-des-Enfants 6	Verneau 1906
Le Moustier 1	This study	Předmostí 7	Matiegka 1938
Roc de Marsal 1	Madre-Dupouy 1992	Předmostí 5	Matiegka 1938
Teshik Tash 1	This study	Sungir' 2	Bader et al. 1979
		Sungir' 3	Bader et al. 1979

diameters of the tibia, humerus, and femur were used to calculate a cross-sectional area using an elliptical model [$\pi \times$ (ap diameter / 2) $\times$ (ml diameter / 2)]. Lengths used were maximum (Martin #1) for the femur, condylomalleolar (Martin #1b) for the tibia, maximum (Martin #1) for the humerus, and maximum (Martin #1) for the radius (Martin 1928).

Table 19.2 Crural and Brachial Indices

Sample	Crural Index	Brachial Index
Le Moustier 1	76.1	74.6
Adult European Neandertals	n = 4	n = 5
	mean = 78.8	mean = 69.7
	s = 1.94	s = 1.84
Adult Upper Paleolithic	n = 19	n = 12
Homo sapiens from Europe	mean = 84.5	mean = 74.3
	s = 2.41	s = 2.62
Adult *Homo sapiens*		
Cold-adapted Inuit	n = 9	n = 6
	mean = 77.9	mean = 71.4
	s = 2.87	s = 2.70
Warm-adapted Khoisan	n = 8	n = 8
	mean = 85.2	mean = 76.7
	s = 3.02	s = 2.62
Modern human reference sample	n = 41	n = 21
	mean = 83.1	mean = 76.9
	s = 1.89	s = 3.31

Unfortunately, maximum femoral length is sometimes not reported in the literature for fossil material, where oblique length is reported instead. For these cases, a regression model was calculated on the basis of the core modern sample to estimate maximum length from oblique length for the adults. The resulting predictive equation was $y = 9.992 + 0.986 \times x$, with an r^2 of 0.995 (n = 37). This equation was used to estimate maximum femoral length in the Upper Paleolithic *H. sapiens* sample. The application of this equation to Neandertals led to a consistent overestimation of approximately two millimeters on bones where both measurements were present. Thus, a regression model was calculated based on the seven femora for which both measurements were available. The resulting prediction model was $y = 9.759 \times 0.981$, $r^2 = 0.999$. The relative lengths of the limb segments were examined using the brachial and crural indices in accordance with the extensive literature on these indices and by means of bivariate plots.

Results

The first series of graphics (Figs. 19.1 to 19.5) represent a comparison of adult Neandertals and Le Moustier 1 with the three modern adult samples.

Figure 19.1 illustrates the positions of the Khoisan and Inuit adults with reference to the mixed reference sample for tibial on femoral length. The regression line with confidence intervals is based on the mixed temperate modern comparative sample. The Inuit and Neandertal individuals all lie below, while all but one of the Khoisan individuals lie above the regression line for the reference sample. These positions are consistent with the cold adaptation of the Inuit and the warm climate adaptation of Khoisan relative to the mixed reference sample, all of whom are derived from a temperate climate. The Inuit position results from the relative shortening of the tibia relative to the femur, while the opposite applies for the Khoisan. The positioning of these samples relative to the mixed temperate sample is in keeping with Allen's Rule of relative extremity length in the context of environment. Figure 19.1 also illustrates that Le Moustier 1 possessed the distal shortening characteristic of Neandertal adults.

Figure 19.2 plots radius on humerus length, thus representing elements of the brachial index. Trinkaus (1981) noted that climatic differences seem

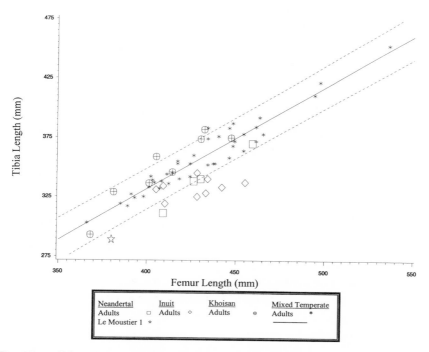

Fig. 19.1. Plot of tibia length on femur length. These are the two elements of the crural index. The least squares regression line and 95 percent confidence intervals are based on the core modern human reference sample.

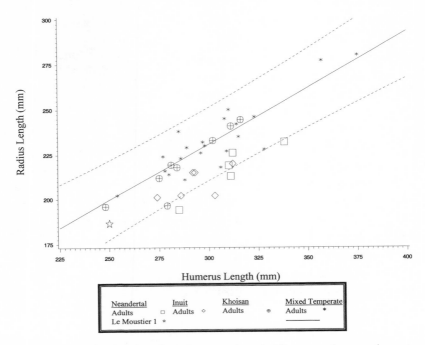

Fig. 19.2. Plot of radius length on humerus length. These are the two elements of the brachial index. The least squares regression line and 95 percent confidence intervals are based on the core modern human reference sample.

to affect the upper limb less than the lower limb. Figure 19.2 indicates the positions of the adult Neandertal individuals relative to the mixed modern reference sample. Here a similar pattern can be seen, with Inuit and Neandertals demonstrating relative shortening of the radii compared with the mixed-temperate modern sample and the Khoisan. Interestingly, the Khoisan brachial indices are not positioned in the upper portion of the temperate range of variation but seem to follow the temperate regression line quite closely. Le Moustier 1 demonstrates some distal shortening but not to the degree seen in adult Neandertals.

Figure 19.3 shows femoral midshaft external cross-sectional area plotted on length; two elements that can be used to describe limb robusticity. Here the adult Neandertals can be seen as more robust than the Khoisan and most of the reference sample. The Inuit are also robust, with some variability visible between the sexes. Le Moustier 1 falls within the range of a typical individual from the reference sample and lies at the lowest end of the adult Neandertal range.

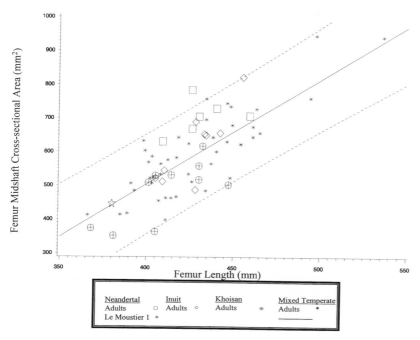

Fig. 19.3.   Plot of femur midshaft cross-sectional area (modeled on an ellipse) on femur length. The least squares regression line and 95 percent confidence intervals are based on the core modern human reference sample.

The robusticity of the Neandertal tibia is well demonstrated in Figure 19.4, in marked contrast to the gracility of the Khoisan tibia. Inuit males share the elevated position of Neandertals, while females do not. Le Moustier 1 has a very robust tibia relative to the reference sample and lies well within the adult Neandertal range of variation for tibial robusticity. In Figure 19.5, the humerus is also seen to be very robust in Neandertals and the Inuit. By contrast, most of the Khoisan individuals lie in the lower half of the range of variation for the reference sample. The Le Moustier 1 humerus is also robust compared with the modern temperate sample.

In summary, these comparisons demonstrate the distal limb segment shortening of the Neandertals, attributed to climatic adaptation, in a manner similar to the Inuit. They also demonstrate distinctive limb robusticity similar to the Inuit. Le Moustier 1 demonstrates clear shortening of the tibia but not the full expression of radius shortening typical of adult Neandertals. Although his femur does not seem to be any more robust than are those of typical modern adults, he demonstrates definite robusticity in

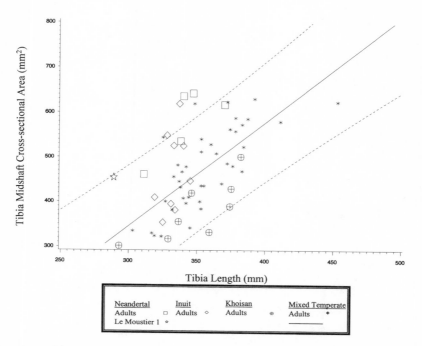

Fig. 19.4. Plot of tibia midshaft cross-sectional area (modeled on an ellipse) on tibia length. The least squares regression line and 95 percent confidence intervals are based on the core modern human reference sample.

the tibia and somewhat less in the humerus. These observations are interesting, in that they suggest possible differences in timing of expression of Neandertal traits. However, a larger ontogenetic context is required to fully evaluate them. For the remainder of this chapter, we will deal only with the Inuit and Khoisan samples, with their full ontogenetic sequences, and a larger sample of Neandertal and early Upper Paleolithic *H. sapiens* juveniles.

Consideration of the tibia on femur plot from this broader perspective reveals that the cold- versus warm-adapted intra–lower limb proportions in adults seem to be the product of an ontogenetic trajectory set very early in life (Fig. 19.6). The two regression lines shown here are those of the entire adult and juvenile Inuit and Khoisan samples. Each regression line is very tight, with r^2 values of 0.99, but the slopes are clearly different. An analysis of covariance and heterogeneity of slopes test demonstrates that the lines are significantly different from each other (Littell et al. 1991). The shallower slope, that of the Inuit, has a value of 0.77 versus 0.86 for the

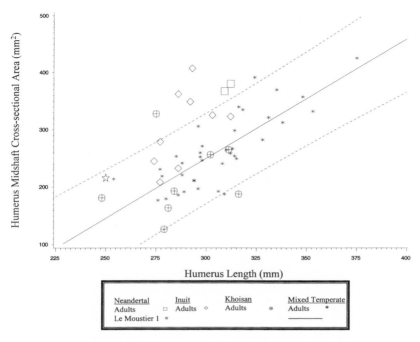

Fig. 19.5. Plot of humerus midshaft cross-sectional area (modeled on an ellipse) on humerus length. The least squares regression line and 95 percent confidence intervals are based on the core modern human reference sample.

Khoisan, indicating that the tibiae of the cold-adapted Inuit grow less per increment of femoral growth throughout their entire ontogenetic sequence than do those of the Khoisan. The Neandertal data points can be seen to follow the Inuit trajectory. In contrast, the early Upper Paleolithic *H. sapiens* individuals fall along the Khoisan trajectory. The youngest specimens of the early Upper Paleolithic *H. sapiens* sample lie just above the Khoisan trajectory.

Figure 19.7 indicates a similar pattern for radius on humerus length. Here the coefficients of determination are also high ($r^2 = 0.99$) and the two regression lines are significantly different from each other. The lower Inuit slope is 0.71, while the steeper Khoisan slope is 0.76. Again, the Neandertals seem to follow the Inuit slope. The early Upper Paleolithic *H. sapiens* sample falls between the two modern slopes, with considerable individual variation. These findings are in keeping with the earlier finding (Fig. 19.2; Trinkaus 1981) that shortening of the radius is not as pronounced as with the leg.

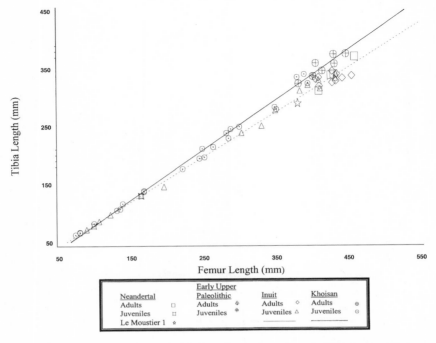

Fig. 19.6. Ontogenetic plot of tibia length on femur length. The least squares regression lines are based on the entire (adults and subadults) Inuit and Khoisan samples.

A consideration of femoral midshaft cross-sectional area on length reveals a nonlinear ontogenetic trajectory (Fig. 19.8). These data illustrate a long, shallow preadolescent slope, which flexes at approximately the time of adolescence, judging by the position of the Le Moustier 1 specimen. It would seem that significant robusticity is produced either at or immediately after the growth spurt, when linear growth ceases but shaft breadths continue to increase. Interestingly, there seems to be a warm climate/cold climate separation in this trajectory, with both the Inuit and Neandertals appearing more robust at any given age than the Khoisan. The early Upper Paleolithic *H. sapiens* individuals broadly follow the Khoisan slope, although variation is considerable: one child is as robust as one of the Inuit children, suggesting that any differences in robusticity between the fossil samples is accentuated after puberty. Within this broader context, the position of Le Moustier 1 suggests that, while he is indeed more robust than most of the equivalent-age Khoisan, he is not more robust than all of the Inuit juveniles.

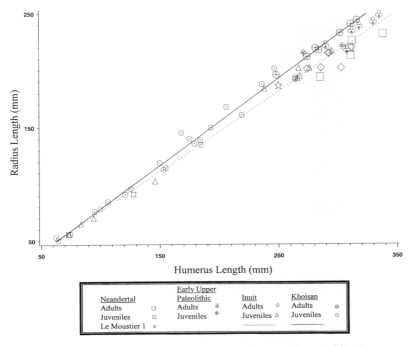

Fig. 19.7. Ontogenetic plot of radius length on humerus length. The least squares regression lines are based on the entire (adults and subadults) Inuit and Khoisan samples.

A similar pattern holds for the tibia in Figure 19.9, except that the warm/cold sample separation is even more clear. Here the flex of the ontogenetic trajectory for most individuals appears at a somewhat longer tibial length (perhaps at a somewhat more advanced position in adolescence) than that achieved by Le Moustier 1. However, Le Moustier 1 is clearly quite robust, as was noted in comparison with the adults (see Fig. 19.4). The position of the early Upper Paleolithic *H. sapiens* individuals is harder to interpret, given the paucity of juvenile specimens. However, the adults are intermediate between the Inuit and Khoisan samples, with somewhat more linear growth trajectory than the cold-adapted Inuit and Neandertals, perhaps related to their longer tibial length.

Finally, the humerus also demonstrates a similar two-stage ontogenetic trajectory for robusticity (Fig. 19.10). Most of the increase in robusticity occurs at a somewhat longer humerus length than that of Le Moustier 1. Again, both Neandertals and Inuit seem more robust than the Khoisan

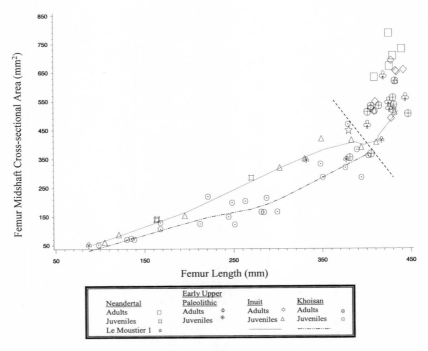

Fig. 19.8. Ontogenetic plot of femur midshaft cross-sectional area (modeled on an ellipse) on femur length. The ontogenetic trajectories for the Inuit and Khoisan subadult samples are indicated by curves fit using a lowess model (Systat version 10). The lowess model was used rather than a least squares model, as the data are clearly nonlinear. The *angled dashed line* shows the approximate location of the flex of the growth trajectory, where predominantly linear growth is replaced by predominantly cross-sectional growth.

throughout growth, while the humerus of the early Upper Paleolithic *H. sapiens* seems less robust, especially after the juvenile stage.

Discussion

These results suggest that the production of distal limb segment shortening and limb robusticity follows two different ontogenetic programs. The relationship between proximal and distal limb segments seems very linear in all samples, indicating that adult intralimb proportions are products of an ontogenetic trajectory set at the earliest stages of growth. The control of the slope of the intralimb trajectory is clearly affected by nat-

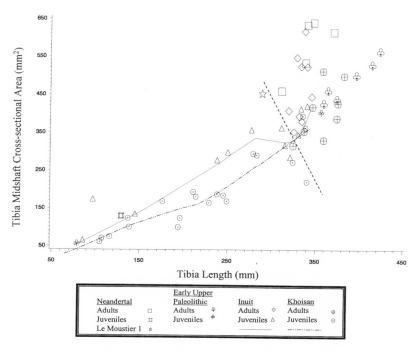

Fig. 19.9. Ontogenetic plot of tibia midshaft cross-sectional area (modeled on an ellipse) on tibia length. The ontogenetic trajectories for the Inuit and Khoisan subadult samples are indicated by curves fit using a lowess model (Systat version 10). The lowess model was used rather than a least squares model, as the data are clearly nonlinear. The *angled dashed line* shows the approximate location of the flex of the growth trajectory, where predominantly linear growth is replaced by predominantly cross-sectional growth.

ural selection acting within the environmental context following Allen's Rule.

The relationship between shaft breadths, represented here as cross-sectional area, and bone length—used together to indicate bone robusticity—follows a different ontogenetic trajectory. Here, the cross-sectional area grows following a relatively shallow (although generally positively allometric) trajectory relative to length until adolescence, whereafter it increases dramatically, especially in the cold-adapted Inuit and Neandertal populations. The exact timing of this change cannot be determined from the present data, but the shift is likely related to the adolescent spurt in linear growth ceasing before somatic growth (Bogin 1988; Tanner 1989), with limb diameters probably following the somatic growth curve.

The position of Le Moustier 1 with respect to the final stature of an av-

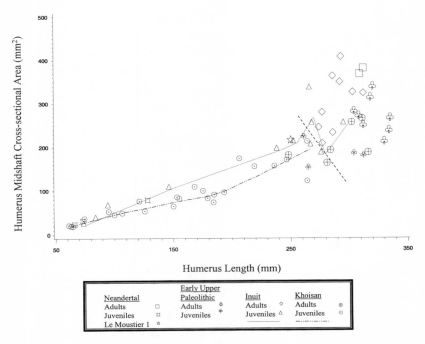

Fig. 19.10. Ontogenetic plot of humerus midshaft cross-sectional area (modeled on an ellipse) on humerus length. The ontogenetic trajectories for the Inuit and Khoisan subadult samples are indicated by curves fit using a lowess model (Systat version 10). The lowess model was used rather than a least squares model, as the data are clearly nonlinear. The *angled dashed line* shows the approximate location of the flex of the growth trajectory, where predominantly linear growth is replaced by predominantly cross-sectional growth.

erage male Neandertal and with respect to the comparative ontogenetic samples suggests that he had not begun or was very early in his growth spurt. An important conclusion offered here is that, while Neandertal juveniles are robust relative to the fairly gracile Khoisan modern humans, they closely follow cold-adapted Inuit juveniles throughout their growth period. Thus, claims of increased robusticity in juvenile Neandertals relative to modern humans in general are not supported by these data. Furthermore, models suggesting that Neandertal robusticity is the product of high activity levels from an early age must demonstrate that (1) Inuit children are highly active from an early age and (2) activity levels in the Inuit, and by extension the Neandertal juveniles, differ significantly from those of Khoisan children throughout ontogeny. Perhaps a more parsimonious explanation, suggested in particular from a consideration of the results from the tibia, is that limb shortening in general, within the selective context of cold climates, affects all limb segments—with the distal segments affected more and the lower limb affected most, so the apparently increased robusticity may in fact represent a relative decrease in length rather than an increase in shaft diameter.

Previous work demonstrated that adult Neandertals share limb proportions with cold-adapted modern human populations, in accordance with Allen's Rule (Trinkaus 1981). As we have demonstrated here, this comparison holds not only for adult limb proportions but also for limb proportions and limb robusticity throughout ontogeny. Moreover, juvenile Inuit seem to represent the most appropriate modern comparative sample for Neandertals, and any discussion of juvenile Neandertal robusticity should consider this context. Finally, these results suggest that Neandertals may have experienced a change in limb robusticity comparable to that experienced by modern humans during adolescence.

Acknowledgments

J. L. Thompson thanks W. Menghin, Director of the Museum für Vor- und Frühgeschichte, Berlin, for permission to examine the original Le Moustier 1 fossil material, and Mrs. Hoffmann for her invaluable assistance during each visit to Berlin. J. L. Thompson is grateful to C. Stringer, British Museum of Natural History; L. Aiello, University College London; and Habil Siegfried Fröhlich, Landesamt für Archäeologie, Halle, for access to comparative cast material and expresses appreciation to R. Kruszynski and J. Grünberg for their assistance. Thanks to Alan Morris, Department of Anatomy, University of Cape Town, for permission to examine human skeletal material in his care. A. J. Nelson acknowledges the support of the Department of Anthropology, University of Western Ontario, and the assistance and generosity of Anne Foley.

Both authors express their thanks to J. Cybulski and C. Merbs for permission to examine the Inuit material at the Museum of Civilization, Hull, with special appreciation for the efforts of Janet Young, who helped make our visit successful. We thank Berndt Hermann for copies of radiographs of the postcrania of Le Moustier 1. B. Bogin provided helpful criticism and encouragement throughout this project. Chris Nelson provided important editorial assistance with the manuscript.

References

Allen, J.A. 1877. The influence of physical conditions in the genesis of species. *Radical Review* 1:108–140.

Anderson, D.L., Thompson, G.W., and Popovich, F. 1976. Age of attainment of mineralization stages of the permanent dentition. *Journal of Forensic Sciences* 21:191–200.

Bader, O.N., Nikitjuk, B.A., and Kharitonov V.M. 1979. The postcranial skeleton of the paleolithic children from the burial at Sungir (preliminary notes). *Voprossy Antropologii* 60:24–37.

Bergmann, C. 1847. Über die Verhltnisse der wäreökonomie der thiere zu ihrer grösse. *Göttinger Studien* 8.

Billy, G. 1975. Étude anthropologique des restes humaines de l'Abri Pataud. In H.L. Movius (ed.), *Excavation of the Abri Pataud*, 201–261. American School of Prehistoric Research, 30. Peabody Museum, Harvard University.

Bogin, B. 1988. *Patterns of Human Growth*. Cambridge: Cambridge University Press.

Bogin, B., and Smith, H. 1996. Evolution of the human life cycle. *American Journal of Human Biology* 8:703–716.

Boule, M. 1912. L'homme fossile de La Chapelle-Aux-Saints (part 4 of 5 articles). *Annales de Paléontologie* 7:65–193.

Churchill, S.E. 1994. Human upper body evolution in the Eurasian Later Pleistocene. Ph.D. diss., University of New Mexico, Albuquerque.

———. 1998. Cold adaptation, heterochrony, and Neandertals. *Evolutionary Anthropology* 7:46–61.

CMOC. 1995. Human Remains Inventory Project Database. Unpublished catalogue on file at the Canadian Museum of Civilization, Hull, Quebec.

Feldesman, M.R. 1992. Femur stature ratio and estimates of stature in children. *American Journal of Physical Anthropology* 87:447–459.

Feldesman, M.R., Kleckner, J.G., and Lundy, J.K. 1990. The femur/stature ratio and estimates of stature in mid- and late Pleistocene fossil hominids. *American Journal of Physical Anthropology* 83:359–372.

Frisancho, A.R. 1990. *Anthropometric Standards for the Assessment of Growth and Nutritional Status*. Ann Arbor: University of Michigan Press.

Gambier, D. 1986. Étude des os d'enfants du gisement aurignacien de Cro-Magnon, Les Eyzies (Dordogne). *Bulletin et Memoir de la Societé d'Anthropologie de Paris* 3, series 14: 13–26.

Gamble, C. 1986. *The Palaeolithic Settlement of Europe*. Cambridge: Cambridge University Press.

Heim, J-L. 1982. Les enfants Néandertaliens de La Ferrassie. Paris: Masson.

Holliday, T.W. 1997. Postcranial evidence of cold adaptation in European Neandertals. *American Journal of Physical Anthropology* 104:245–258.

Illerhaus, B., and Thompson, J.L. 1999. Calculating CT data from matched geometries. *DGZfP Proceedings* BB 67-CD:189–191.

Leigh, S. 1985. Ontogenetic and static allometry of the Neandertal and Modern hominid palate. *American Journal of Physical Anthropology* 66:195.

Littell, R.C., Freund, R.J., and Spector, P.C. 1991. *SAS System for Linear Models*, 3d ed. Cary, N.C.: SAS Institute.

Lovejoy, O., and Trinkaus, E. 1980. Strength and robusticity of Neandertal tibias. *American Journal of Physical Anthropology* 57:465–470.

Madre-Dupouy, M. 1992. *L'Enfant du Roc de Marsal: Cahiers de Paléoanthropologie*. Paris: Éditions du Centre National de la Recherche Scientifique.

Martin, R. 1928. *Lehrbuch der Anthropologie*. Jena: Verlag Von Gustav Fischer.

Matiegka, J. 1938. *Homo předmostensis: L'Homme Fossil di Předmostí en Moravie (Tchécoslovaquie). II. Autres Partes du Squelette*. Prague: Académie Tchèque des Sciences et des Arts.

Minugh-Purvis, N. 1988. Patterns of craniofacial growth and development in Upper Pleistocene hominids. Ph.D. diss., University of Pennsylvania.

Nelson, A.J. 1995. Cortical bone thickness in the primate and hominid postcranium—taxonomy and allometry. Ph.D. diss., University of California, Los Angeles.

Nelson, A.J., and Thompson, J.L. 1999. Growth and development in Neandertals and other fossil hominids: implications for the evolution of hominid ontogeny. In R. Hoppa and C. Fitzgerald (eds.), *Human Growth in the Past*. Cambridge: Cambridge University Press.

Ruff, C.B., Trinkaus, E., Walker, A., and Larsen, C.S. 1993. Postcranial robusticity in *Homo:* I. Temporal trends and mechanical interpretation. *American Journal of Physical Anthropology* 91:21–53.

Ruff, C.B., Walker, A., and Trinkaus, E. 1994. Postcranial robusticity in *Homo:* III. Ontogeny. *American Journal of Physical Anthropology* 93:35–54.

Smith, F.H. 1991. The Neandertals: evolutionary dead ends or ancestors of modern people? *Journal of Anthropological Research* 47:219–238.

Stringer, C.B., and Andrews, P. 1988. Modern human origins. *Science* 24:773–774.

Stringer, C.B., and Gamble, C. 1993. *In Search of the Neanderthals*. New York: Thames and Hudson.

Tanner, J.M. 1989. *Foetus into Man*. Cambridge: Harvard University Press.

Thompson, J.L. 1995. Terrible teens: the use of adolescent morphology in the interpretation of Upper Pleistocene human evolution. *American Journal of Physical Anthropology* 20 (special suppl): 210.

———. 1998. Neanderthal growth and development. In S.J. Ulijaszek, F.E. Johnston, and M.A. Preece (eds.), *The Cambridge Encyclopedia of Human Growth and Development*, 106–107. Cambridge: Cambridge University Press.

Thompson, J.L., and Bilsborough, A. 1997. The current state of the Le Moustier 1 skull. *Acta Praehistorica et Archaeologica* 29:17–38.

———. 1998a. Time for one of the last Neanderthals. Mediterranean Prehistory Online. http://www.med.abaco-mac.it/.

———. 1998b. Time for one of the last Neanderthals. In F. Facchini, A. Palma di Cesnola, M. Piperno, and C. Peretto (eds.), *Proceedings of the XIII Congress of the U.I.S.P.P.-Forli (Italia, 8/14 September)*, 2:289–298. Forli: Abaco.

Thompson, J.L., and Illerhaus, B. 1998. A new reconstruction of the Le Moustier

1 skull and investigation of internal structures using 3-D-CT data. *Journal of Human Evolution* 35:647–665.

———. 2000. CT reconstruction and analysis of the Le Moustier 1 Neanderthal. In C.B. Stringer, N. Barton, and C. Finlayson (eds.), *Proceedings of the Conference on Gibraltar and the Neanderthals, 1848–1998*, 249–255. Oxford: Oxbow Books.

Thompson, J.L., and Nelson, A.J. 1997. Relative postcranial development of Neanderthals. *Journal of Human Evolution* 32:A23–24.

———. 1999. Le Moustier 1, limb proportions, and the ontogeny of the Neandertal form. *Journal of Human Evolution* 36:A22–23.

———. 2000. The place of Neandertals in the evolution of hominid patterns of growth and development. *Journal of Human Evolution* 38:475–495.

Tompkins, R.L., and Trinkaus, E. 1987. La Ferrassie 6 and the development of Neandertal pubic morphology. *American Journal of Physical Anthropology* 73: 233–239.

Trinkaus, E. 1976. The evolution of the hominid femoral diaphysis during the Upper Pleistocene in Europe and the Near East. *Zeitschrift für Morphologie und Anthropologie* 67:291–319.

———. 1980. Sexual difference in Neandertal limb bones. *Journal of Human Evolution* 9:377–397.

———. 1981. Neandertal limb proportions and cold adaptation. In C.B. Stringer (ed.), *Aspects of Human Evolution*, 187–224. London: Taylor and Francis.

———. 1983a. *The Shanidar Neandertals.* New York: Academic Press.

———. 1983b. Neandertal postcrania and the adaptive shift to modern humans. In E. Trinkaus (ed.), *The Mousterian Legacy. BAR International Series* 164: 165–200.

———. 1986. The Neandertals and modern human origins. *Annual Review of Anthropology* 15:193–218.

Trinkaus, E., Churchill, S.E., Villemeur, I., Riley, K.G, Heller, J.A., and Ruff, C.B. 1991. Robusticity versus shape: the functional interpretation of Neandertal appendicular morphology. *Journal of the Anthropological Society of Nippon* 99: 257–278.

Trinkaus, E., and Ruff, C.B. 1989a. Cross-sectional geometry of Neandertal femoral and tibial diaphyses: implications for locomotion. *American Journal of Physical Anthropology* 78:315–316.

———. 1989b. Diaphyseal cross-sectional morphology and biomechanics of Fond-de-Forêt and the Spy 2 femur and tibia. *Anthropology and Prehistory* 100:33–42.

Trinkaus, E., Ruff, C.B., and Churchill, S.E. 1998. Upper limb versus lower limb loading patterns among Near Eastern Middle Paleolithic hominids. In T. Akazawa, K. Aoki, and O. Bar-Yosef (eds.), *Neandertals and Modern Humans in Western Asia*, 391–404. New York: Plenum Press.

Twiesselmann, F. 1961. *Le Fémur Néanderthalien de Fond-de-Forêt (Province de Liège).* Institut Royal des Sciences Naturelle de Belgique, Mémoire 148.

Vallois, H.V., and Billy, G. 1965. Nouvelles recherches sur les hommes fossiles de l'abri de Cro-Magnon. *Anthropologie* 69:47–74.

Verneau, R. 1906. Les Grottes de Grimaldi (Baoussé-Roussé). *Anthropologie* 2 (no. 1). Imprimerie de Monaco.

Vlček, E. 1973. Postcranial skeleton of a Neandertal child from Kiik-Koba, U.S.S.R. *Journal of Human Evolution* 2:537–544.

Chapter 20

Between the Incisive Bone and Premaxilla

From African Apes to Homo sapiens

BRUNO MAUREILLE AND JOSÉ BRAGA

On March 27, 1784, after ten years of research, Johann W. von Goethe, then in Jena, wrote to his friend J. G. Herder: "According to the precepts of the Gospel, I have to inform you, in all haste, of a happiness that engulfs me: I found neither gold nor the silver but that which gives me an inexpressible joy: the premaxillary bone in humans" (Mandelkow 1986). This permitted Goethe to demonstrate that, osteologically, the human skull does not differ from that of any other mammal, leading him to conclude that the distinction between the animal kingdom and humans was no longer acceptable.

Today, most paleoanthropologists regard the absence of the incisive suture (*sutura incisiva*) from the anterior face of the maxilla at birth as a specific diagnostic feature of *Homo* (Wood-Jones 1947; Ashley-Montagu 1935; Friant 1958; Braga 1998). However, despite more than two centuries of attention, variation in the timing of incisive suture closure in different taxa of African apes, as well as in members of the human lineage, remains poorly known. Detailed data on the persistence, timing, and pattern of fusion of this suture might help us to determine when (in terms of ontogeny) the early complete anterior closure of the incisive suture occurred during human evolution and whether the distinctive patterns of incisive suture closure among fossil hominids might be indicative of discrete species and subspecies.

According to the International Code for Nomenclature (Gouazé et al. 1977), one difference between the Hominoidea and the Hominidae is the

464

presence of a *sutura incisiva,* or premaxillary suture, on the anterior aspect of the maxilla.[1] Physical anthropologists interested in the premaxillary suture are mostly concerned with the growth pattern and osteogenic activity of the suture, congenital oral pathologies such as cleft lip/palate, and adaptations of the bony facial skeleton. Premaxillary suture expression is frequently studied only on the palatal aspect of the maxilla, as it supposedly disappears from its anterior aspect before birth.

Here we present a synthetic review of various aspects of the ontogeny of the incisive suture/premaxillary suture in the Hominoidea. In comparing premaxillary suture expression in African apes, extant humans (both juveniles and adults), and most fossil hominids (from *Australopithecus–Paranthropus* to *Homo sapiens*), we have been led to propose new hypotheses concerning the taxonomic position of some Plio-Pleistocene hominids and about the architecture of the Neandertal face.

Comparative Anatomy

The premaxillary or incisive suture represents the articulation between the maxillary and the incisive bone. It is generally assumed, but not known with certainty, that at the beginning of fetal life the timing and extent of expression of the incisive suture is similar in African apes and extant humans (Couly 1991; Mooney and Siegel 1991).

The Anterior Component

Initially, the anterior component of the premaxillary suture extends from the interalveolar septum between the dental laminae for the future deciduous canine and lateral incisor, superiorly to the nasomaxillary suture. At this level, in the African apes, the incisive suture may fail to reach as far superiorly as the nasal bone. This anterior component has never been observed fully patent in a human fetus older than four lunar months (Vallois and Cadenat 1926; Maureille 1994; Williams 1995). Nor have we been able to find any reference in the literature indicating that the incisive suture was fully open for the entire extent of the anterior maxillary face in humans, particularly at the level of the deciduous canine alveolus. In this regard, humans contrast to the African apes because of the embryology of the incisive bone and its position anterior to the maxilla. Apparently, osteogenesis commences in the maxillary ossification center and then spreads anteriorly to reach the incisive bone between twelve and sixteen weeks of fetal age (Woo 1949).

The Palatal Component

In both African apes and humans, the palatal component of the premaxillary suture extends bilaterally from the incisive foramen to the interalveolar septum between the lateral incisor and the canine.

The Nasal Component

Finally, two different nasal components of this suture should be distinguished: from the superior opening of the incisive canal, the suture extends posterolaterally to reach the frontal process of the maxilla. From the nasal face of the frontal process of the maxilla, it extends inferiorly to reach the crista conchalis but is interrupted at this level. Theoretically, the suture also extends above this crest to reach the nasomaxillary suture or the border of the nasal aperture (specifically in African apes).

Materials and Methods

The expression of the premaxillary suture was studied in 215 fetuses and 106 juvenile extant humans between the ages of 0 and 11 years (dental age) and in 160 living adults. All components of the suture were studied: anterior, palatal, and nasal. Data were collected independently and simultaneously by the authors (Braga 1995, 1998, Maureille 1993, 1994).

To obtain a satisfactory picture of the variation in African apes, we studied 543 skulls of three subspecies of the common chimpanzee, *Pan troglodytes;* 169 skulls of the pygmy chimpanzee, *Pan paniscus;* and 396 skulls of the gorilla, *Gorilla gorilla,* of known geographic origin (Table 20.1). In 95 of these specimens, the first molars had not yet erupted. Orangutans were excluded from our sample because we consider *Homo/Pan/Gorilla* as a monophyletic group, as indicated by most molecular studies (Goodman 1964; Ruvolo 1994; Mann and Weiss 1996).

The fossil hominid sample included all available specimens for which one or more components of the incisive suture was observable. A few were taken from literature or casts but, in most cases, we scored these components on the original specimens (Table 20.2) attributed to *Australopithecus*

Table 20.1 Composition of Extant Human and African Ape Samples

	Extant Humans			*African Apes*		
Sample	*No. Juveniles*	*No. Adults*		*Sample*	*No. Juveniles*	*No. Adults*
Unknown	215	0		*Pan troglodytes troglodytes*	84	139
Native Africans	5	32		*Pan troglodytes schweinfurthi*	88	117
Native Ameri. (N)	9	32		*Pan troglodytes verus*	15	100
Native Ameri. (S)	18	0		*Pan paniscus*	102	67
Native Asiatics	0	32		*Gorilla gorilla gorilla*	79	156
Native Australians	5	34		*Gorilla gorilla graueri*	31	75
Native Polynesians	8	0		*Gorilla gorilla beringei*	18	37
European	60	30				

Note: Ameri., Americans; N, North Americans; S, South Americans.

afarensis, Australopithecus africanus, "robust" australopithecines, early *Homo, Homo ergaster/Homo erectus,* archaic *H. sapiens,* and all known middle Paleolithic anatomically modern *H. sapiens* and Neandertals.

To determine whether the premaxillary suture is open only on the surface, we supplemented visual examinations of maxillae with observations from CT scans on the Roc-de-Marsal and Engis 2 Neandertal juveniles. We used a highlight 9008 advantage type CT. CT slices 1.5 mm thick were taken every millimeter and overlapped for the reconstruction. The plane of the CT slices is slightly oblique to the occlusal plane.

Developmental status of the premaxillary suture was scored as patent, intermediate, or absent on its external aspect. This was based solely on visual observations (i.e., to avoid damaging the specimens, we attempted no probing). The status of suture closure was recorded for both its anterior and palatal components in African apes. However, the nasal component was not considered in African apes because of the greater potential risk of intra-observer error due to difficulties in visualizing this portion of the suture, which is often hidden from view by the vomer. However, two other locations, the nasal floor and the nasal face of the maxillary frontal process, were considered for the earliest hominids (*Australopithecus, Paranthropus,* and early *Homo*), *H. erectus,* archaic *H. sapiens,* Neandertals, anatomically modern *H. sapiens* from the Middle Paleolithic, and extant *H. sapiens.*

The anterior component of the suture is considered patent when it is visible from the interalveolar septum of the lateral incisor and canine to the border of the nasal aperture, the nasomaxillary suture, or the frontomaxillary suture. It is considered absent only when there is no trace of it. All other cases

Table 20.2 Analyzed Fossil Samples

Specimen	Site	Specimen	Site
Australopithecus afarensis		*Paranthropus robustus*	
AL 333-105	Hadar	**TM 1517**	Kromdraai
AL 333-86	Hadar	**Sk 12**	Swartkrans
AL 333-1	Hadar	**Sk 13**	Swartkrans
AL 199-1	Hadar	**Sk 46**	Swartkrans
AL-200-1a	Hadar	**Sk 47**	Swartkrans
Australopithecus africanus		**Sk 48**	Swartkrans
Sts 15	Sterkfontein	**Sk 52**	Swartkrans
Sts 17	Sterkfontein	**Sk 79**	Swartkrans
Sts 52a	Sterkfontein	**Sk 83**	Swartkrans
Sts 71	Sterkfontein	**Sk 847**	Swartkrans
Sts 53	Sterkfontein	**Skx 162**	Swartkrans
Stw 73	Sterkfontein	**Skx 265**	Swartkrans
Stw 252a, n, m	Sterkfontein	**Skw 11**	Swartkrans
Stw 391	Sterkfontein	*Homo habilis*	
Taung	Taung	**KNM-ER 1470**	East Turkana
TM (Sts) 1512	Sterkfontein	**KNM-ER 1813**	East Turkana
MLD 6	Makapansgat	*Homo ergaster*	
MLD 9	Makapansgat	**KNM-ER 3883**	East Turkana
MLD 45	Makapansgat	**KNM-WT 15000**	West Turkana
Paranthropus aethiopicus		*Homo erectus*	
KNM-WT 17000	West Turkana	Zkd O1-313	Zhou Kou Dien
KNM-WT 17400	West Turkana	Sangiran 4	Pucangan
Paranthropus boisei			
KNM-ER 732	East Turkana		
KNM-ER 406	East Turkana		

Note: Specimens in bold directly observed from the original.

are intermediate. In apes, the palatal component of the suture was considered patent when visible from the canine/lateral incisor interalveolar septum to the incisive foramen. For extant humans and human fossils, it was considered present when visible from the border of the canine alveolus to the incisive foramen. In this same group, it was scored as intermediate when observed for much of its course and absent when discontinuous or visible for only few a millimeters approaching the incisive foramen. The nasal aspect of the suture was considered patent when observed from the aperture of the incisive canal to the base of the frontal process. As with the palatal component of the suture, intermediate cases are those in which the suture was patent for most of its course; it was regarded as absent if discontinuous or not present. For extant and fossil humans, the suture was recorded as patent on the maxillary frontal process when visible on both sides of the crista conchalis and,

Table 20.2   (*Continued*)

Specimen	Site	Specimen	Site
archaic *Homo sapiens*		**Montmaurin 5**	Montmaurin
Djebel Irhoud	Jebel Irhoud		(Coupe-Gorge)
Bodo	Bodo	**Pech-de-l'Azé**	Pech-de-l'Azé
Kabwe 1	Kabwe	**Pétralona**	Petralona
Kabwe 2	Kabwe	**Roc-de-Marsal**	Roc-de-Marsal
Laetoli H18	Laetoli	**Saccopastore 1**	Saccopastore
Neandertal lineage		**Saccopastore 2**	Saccopastore
Amud 1	Amud	**Saint-Césaire**	Saint-Césaire
Arago XXI	La Caune de l'Arago	Shanidar 1	Shanidar
Atapuerca 405	La Sima de los Huesos	Shanidar 5	Shanidar
Castel di Guido 4	Castel di Guido	**Spy 1**	Spy
Devil's Tower	Gibraltar	Subalyuk 2	Subalyuk
Engis 2	Engis	**Tabun B1**	Mugharet et-Tabun
Forbes' Quarry	Gibraltar	**Tabun C1**	Mugharet et-Tabun
Guattari	Monte Circeo	**Teshik-Tash**	Teshik-Tash
Krapina 46	Krapina	First anatomically	
Krapina 47	Krapina	modern humans	
Krapina 48	Krapina	Klasies River	Klasies River Mouth
Krapina 49	Krapina	AA43/SAS4SHB	
Kůlna	Kůlna	**Skhul IV**	Mugharet-es-Skhul
La Chapelle	La Chapelle-aux-Saints	**Skhul V**	Mugharet-es-Skhul
La Ferrassie 1	La Ferrassie	Qafzeh 4	Djebel Qafzeh
La Ferrassie 2	La Ferrassie	**Qafzeh 5**	Djebel Qafzeh
La Quina H18	La Quina	**Qafzeh 6**	Djebel Qafzeh
La Quina H5	La Quina	**Qafzeh 9**	Djebel Qafzeh
Monsempron	Monsempron	**Qafzeh 11**	Djebel Qafzeh

for the upper part, when it reached the nasomaxillary suture. It was considered intermediate when visible only on one side of the crista (mostly the lower one) and absent when there was no evidence of the suture.

Results

African Apes

We observed clearly distinct patterns of premaxillary suture closure among common and pygmy chimpanzees, gorillas, and humans (Table 20.3 and, for more details, see Braga 1995). First, no completely patent anterior component of the suture was ever seen in the very young common or pygmy

Table 20.3 Incisive/Premaxillary Suture Expression on the Anterior Part of the Face in African Apes and Extant Humans

	Fully Patent Suture			Complete Closure		
Species	% Juveniles	% Adults	Species	% Juveniles	% Adults
Pan troglodytes troglodytes	0	0	Pan troglodytes troglodytes	26.8	86.7
Pan troglodytes schweinfurthi	0	0	Pan troglodytes schweinfurthi	6.3	79.7
Pan troglodytes verus	0	0	Pan troglodytes verus	0	55.6
Pan paniscus	0	0	Pan paniscus	66.8	92.5
Gorilla gorilla gorilla	98.7	24.4	Gorilla gorilla gorilla	0	39.4
Gorilla gorilla graueri	95.2	20	Gorilla gorilla graueri	0	30
Gorilla gorilla beringei	91.7	9.5	Gorilla gorilla beringei	0	54.1
Extant humans	0	0	Extant humans	99.4	100

chimpanzees (Fig. 20.1), whereas it was always found in gorillas of equivalent dental age. In gorillas, the anterior component of the suture starts to close after the eruption of the first permanent molars and is usually, but not always, fused after the eruption of the third molars (see Table 20.3).

Second, in common chimpanzees, anterior closure of the incisive suture also progresses with age. However, it is usually not completed until just after the third molars erupt. In pygmy chimpanzees, complete anterior closure occurs significantly earlier than in common chimpanzees (Braga 1995, 1998). Moreover, complete patency of the palatal aspect of the premaxillary suture occurs in significantly higher proportions in pygmy than in common chimpanzees.

Extant Humans

In human newborns, the premaxillary suture was completely closed on its anterior aspect, except in a few, very rare cases, in which it was present for a few millimeters superiorly on the frontal process of the maxilla. The suture was also found on the palatal aspect of the maxilla, progressively closing from lateral to medial. Less frequently, the suture was found on the floor of the nasal fossa and on the nasal face of the frontal process of the maxilla but only below the crista conchalis or a few millimeters above it. In young specimens between the ages of four and six years, the suture was seen on both its nasal and palatal aspects in fewer than one case out of ten (9.5% of the sample).

In adults, the suture sometimes persists as a scar on the palatal aspect of the maxilla. In such cases we found no relationship between such scars and

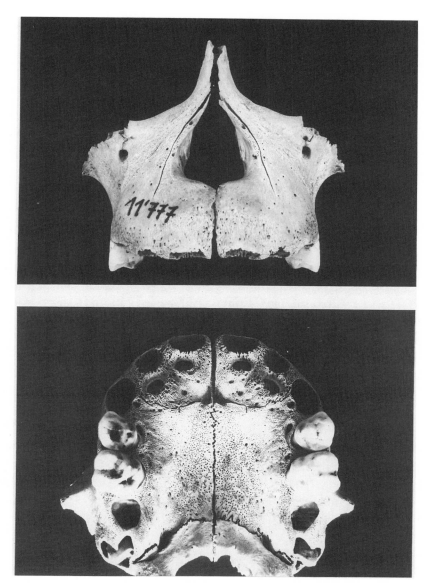

Fig. 20.1. Anterior (*top*) and palatal (*bottom*) components of the incisive suture in a young common chimpanzee. *Source:* Photographs courtesy of J. Braga.

either the geographic origin or sex of the specimen. It was also sometimes possible to see the suture for a few millimeters around the aperture of the incisive canal on the nasal floor. However, in adults it was never found on the nasal face of the frontal process of the maxilla.

These very different expressions of this suture between African apes and extant humans, indicative of differences in growth and development of the maxilla (Schultz 1948; Delaire 1974), suggest that the term *sutura incisiva* should be reserved for specimens presenting a completely or partially patent suture on the anterior aspect of the maxilla postnatally. On the other hand, the term *premaxillary suture* should be reserved for specimens in which postnatal evidence of this suture is limited to the palatal and nasal aspects of the maxilla. However, as the latter is not recognized by the *Nomina Anatomica* (Gouazé et al. 1977), we propose the following Latin term for this structure in extant humans: *sutura premaxillaris*.

Fossil Hominids

The anterior component of the premaxillary suture is clearly seen on such immature specimens attributed to *A. afarensis* as AL 333-86 and AL 333-105. It was also reported by White (1980) on LH 21a. However, in the adult from Hadar (AL 333-1), there is no trace of this feature. Whether it occurs on another adult maxilla from Hadar (AL 417-1d) has not yet been determined. The palatine component of the suture is completely open in WT 17000, a primitive "robust" australopithecine dated from 2.5 million years ago, but completely fused in OMO 323 (right side), the oldest *Paranthropus boisei* specimen known.

An interesting difference is apparent between *A. africanus* and South African robust australopithecines in the closure pattern of the anterior component of the premaxillary suture (Table 20.4). Four *A. africanus* specimens (Taung, MLD 6, MLD 45, and Sts 17) exhibit a partially closed anterior component of the suture. In contrast, all of the South African robust aus-

Table 20.4 Expression of the Premaxillary Suture on the Anterior and Palatine Aspects of the Maxilla in South African Australopithecines and Exant Humans

	Incomplete Closure			Complete Closure	
Sample	% Palatine	% Anterior	Sample	% Palatine	% Anterior
A. africanus	16.6*	41.2	A. africanus	72.2*	58.8
A. robustus	60	0	A. robustus	40	100
Extant humans	27.5	0	Extant humans	82.5	100

* Fully open in Taung, bilaterally.

tralopithecines exhibit a completely closed anterior component, although two of them (SK 47 and SK 13) are developmentally younger than the *A. africanus* specimens exhibiting this feature (for more details, see Braga 1998). We hope that future discoveries of very young South African "robust" australopithecines preserving faces will permit full documentation of the closure pattern in this group. For the moment, the expression of this trait seems closer to that of *Homo* for robust australopithecines than for *A. africanus* specimens. Moreover, *A. africanus* evinces an anterior component of the premaxillary suture very similar to that of some African apes. This observation was proposed some years ago (Wood-Jones 1947; Friant 1958) but has never been recognized by paleoanthropologists.

No trace of either the palatal or the anterior component of the premaxillary suture was seen by us (Table 20.4). These components have not been previously described in any early *Homo* (ER 1813) or *H. ergaster/erectus* (ER 3733, WT 15000) representatives because of the lack of sufficiently preserved faces of very young individuals representing these groups.

Stw 53, an adult specimen from Sterkfontein, is generally considered to belong to the genus *Homo* (Hughes and Tobias 1977). Nevertheless, the expression of the premaxillary suture questions this assignment. On this specimen, the premaxillary suture is clearly patent bilaterally on the anterior face of its maxilla (Fig. 20.2). It is visible for more than one centimeter superiorly on the frontal process, where the suture reaches the nasomaxillary suture. This expression, comparable to that in chimpanzees, *A. afarensis,* and *A. africanus,* strongly suggests that it is preferable to include Stw 53 with australopithecines rather than with the early *Homo* sample. From published photographs (Nieves 1997), we see no trace of the premaxillary suture on the palate of the ten- to eleven-year-old *Homo antecessor* specimen ADT6-69. No more data are available for this specimen or for the three- to four-year-old ADT6-14 from the same site.

No premaxillary suture was seen, even as a scar, in any adult archaic *H. sapiens* representatives, and unfortunately, no juvenile face representing this group is known. Nor was a premaxillary suture seen in adults of the first anatomically modern humans (Vandermeersch 1981), although a possible scar is present on the palatal face of the juvenile Qafzeh 11 right maxilla (Maureille and Bar 1999). According to the present published data, no trace of this suture is reported in any other juvenile specimen from the Qafzeh-Skhul sample.

No premaxillary suture or scar was seen in adult representatives of the Neandertal lineage, including European fossils frequently considered as

Fig. 20.2. Anterior component of the incisive suture in *Australopithecus africanus* (Stw 53). *Source:* Photograph courtesy of J. Braga.

Homo heidelbergensis. Nevertheless, the juvenile sample of the Neandertals is interesting. As shown by one of us (Maureille 1995), most of the specimens between the ages of two and six years evince a widely patent palatal component of the premaxillary suture (Fig. 20.3). In addition, the suture is also seen on the nasal floor and, for the youngest specimens, on the nasal face of the frontal process of the maxilla, below and above the crista conchalis (Table 20.5).

Comparing the expression of palatal and nasal components of the premaxillary suture, using both gross visual examination and CT scans to compare the juvenile Neandertal sample and a sample of fifty-three extant children of equivalent developmental age, we found statistically significant differences in persistence of the premaxillary suture in young Neandertals. Even individuals as old as the three-year-old Roc-de-Marsal (France) Neandertal exhibit a premaxillary suture in the palatal area of the incisor-canine interalveolar septum where there is a real diastema (for more details, see Maureille and Bar 1999).

Finally, from both published data and direct observations on original specimens, we have found no difference in expression of the premaxillary suture between extant humans and the Upper Paleolithic *H. sapiens* (Gambier and Maureille, pers. comm.)

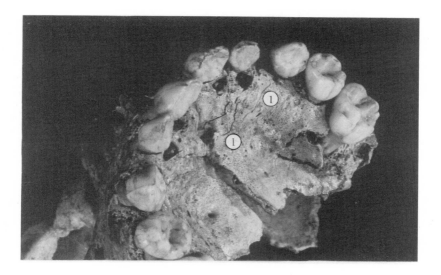

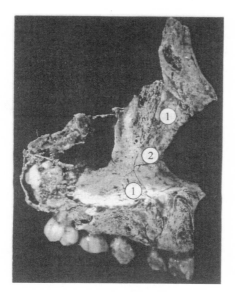

Fig. 20.3. Palatal (*top*) and nasal (*bottom*) components of the premaxillary suture in the three-year-old Neandertal Roc-de-Marsal child. *1*, premaxillary suture; *2*, crista conchalis. *Source:* Photographs courtesy of B. Maureille.

Conclusions

The patterns of closure of the incisive/premaxillary suture in African apes are consistent with the differences revealed by molecular data (Ruvolo et

Table 20.5 Frequencies of the Premaxillary Suture in Extant (n = 52) and Neandertal (n = 2–5) Children Two to Six Years

	Frequency of the Premaxillary Suture (%)						
Sample	A: above the crista conchalis	B: under the crista conchalis	C: on the nasal floor	D: C+B	E: on the palatal floor	F: E+C	G: F+B
Neandertal	100	100	75	66.6	80	75	66.6
Extant	0	9.5	38.1	9.5	28.6	19	0

al. 1994) in that they give the same picture of polytypism as do the molecules. These patterns of closure are completely different from those of extant humans, for which no geographic differences were seen.

Among fossil hominids, *A. afarensis* and *A. africanus* evince a pattern close to that of the pygmy chimpanzees in that the anterior component of the premaxillary suture closes at about the same stage of development. In South African "robust" australopithecines, closure of this suture seems to occur even earlier, although additional young specimens are needed to confirm this suspicion.

The closure of the suture seems to be nearest to the living human condition among the South African "robust" australopithecines and the first members of the *Homo* lineage. Stw 53 does not fit this pattern of premaxillary suture fusion and, therefore, perhaps should no longer be considered as a representative of the early *Homo* lineage.

For *H. ergaster, H. erectus sensu lato,* we have no real information because of the absence of any available young specimen. The Nariokotome youth, KNM-WT 15000, does not show any evidence of the premaxillary suture, nor has a premaxillary suture been reported for any adult *H. erectus* or for the young specimens of *H. antecessor.*

The premaxillary suture is absent on the palate of adult archaic *H. sapiens,* nor is it present in adult anatomically modern humans or members of the *H. heidelbergensis* and Neandertal lineage. Nevertheless, in Neandertals the suture closes later in young individuals than in living humans of the same developmental age. For the moment, we cannot know whether this represents a derived trait of Neandertals or a primitive trait shared with other contemporaneous fossil groups.

Even if data about the timing of the premaxillary suture closure in fossil hominids remain scanty, the present survey provides a fundamental new dataset for those who would like to explore the developmental history of fossil hominids. These data could also be useful in efforts to unravel the phylogeny of the first members of the human lineage, as well as the interpretation of the very peculiar Neandertal facial morphology.

Acknowledgments

We thank all the directors of the different institutions who allowed us to study the original specimens of African apes and human fossils. We also thank N. Minugh-Purvis and K. McNamara, who invited us to participate in this volume.

Note

1. Here we will attempt to demonstrate that *premaxillary suture* is a more appropriate term for extant humans.

References

Ashley-Montagu, M.F. 1935. The premaxilla in the primates. *Quarterly Review of Biology* 10:32–59, 181–208.
Braga, J. 1995. Définition de certains caractères discrets crâniens chez *Pongo, Gorilla* et *Pan:* perspectives taphonomiques et phylogénétiques. Thesis, Université Bordeaux.
———. 1998. Chimpanzee variation facilitates the interpretation of the incisive suture closure in South African Plio-Pleistocene Hominids. *American Journal of Physical Anthropology* 105:121–135.
Couly, G. 1991. *Développement céphalique: embryologie, croissance, pathologie.* Paris: CDP.
Delaire, J. 1974. Considérations sur l'accroissement du prémaxillaire chez l'homme. *Revue de Stomatologie Paris* 75:951–970.
Friant, M. 1958. Sur l'origine de l'homme. *Acta Morphologica Neerlando-Scandinavica* 2:25–27.
Goodman, M. 1964. Man's place in the phylogeny of the primates as reflected in serum proteins. In Washburn, S.L. (ed.), *Classification and Human Evolution,* 204–234. London: Methuen.
Gouazé, A., Baumann, J.A., and Dhem, A. 1977. *Atlas d'Anatomie Humaine:* Vol. 4. *Nomenclature Anatomique Française.* Munich: Urban and Schwarzenberg.

Hughes, A.R., and Tobias, P.V. 1977. A fossil skull probably of the genus *Homo* from Sterkfontein, Transvaal. *Nature* 265:310–312.

Mandelkow, K.R. 1986. *Goethe Briefe: Hamburger Ausgabe,* 1:S-435–436. Munich: Beck.

Mann, A., and Weiss, M. 1996. Hominoid phylogeny and taxonomy: a consideration of the molecular and fossil evidence in an historical perspective. *Molecular Phylogenetics and Evolution* 5:169–181.

Maureille, B. 1993. L'os incisif, particularités chez le Néandertalien immature. *Compte Rendu de l'Academie des Sciences, Paris* 316, série II, no. 6: 831–837.

———. 1994. La face chez *Homo erectus* et *Homo sapiens:* recherche sur la variabilité morphologique et métrique. Thesis, Université Bordeaux.

———. 1995. Un aspect de l'ontogenèse de la face: la sutura incisiva des Néandertaliens. *Anthropologie et Préhistoire* 106:65–74.

Maureille, B., and Bar, D. 1999. The premaxilla in Neandertal and early modern children: ontogeny and morphology. *Journal of Human Evolution* 37:137–152.

Mooney, M.P., and Siegel, M.I. 1991. Premaxillary-maxillary suture fusion and anterior nasal tubercule morphology in the chimpanzee. *American Journal of Physical Anthropology* 85:451–456.

Nieves, J.M. 1997. *Homo antecessor. Blanco y Negro,* June 1, 11–21.

Ruvolo, M. 1994. Molecular evolutionary processes and conflicting gene trees: the hominoid case. *American Journal of Physical Anthropology* 94:89–113.

Ruvolo, M., Pan, D., Zehr, S., Goldberg, T., Disotell, T.R., and von Dornum, M. 1994. Gene trees and hominoid phylogeny. *Proceedings of the National Academy of Science* 91:8900–8904.

Schultz, A.H. 1948. The relation in size between premaxilla diastema and canine. *American Journal of Physical Anthropology* 6:163–179.

Vallois, H.-V., and Cadenat, E. 1926. Le développement du prémaxillaire chez l'homme. *Archives de Biologie* 36:361–425.

Vandermeersch, B. 1981. Les Hommes fossiles de Qafzeh (Israël). Cahiers de Paléontologie. Paris: Éditions du C.N.R.S.

White, T.D. 1980. Additional fossil hominids from Laetoli, Tanzania: 1976–1979 specimens. *American Journal of Physical Anthropology* 53:487–504.

Williams, P.L., Bannister, L.H., Berry, M.M., Collins, P., Dyson, M., Dussek, J.E., and Ferguson, M.W.J. (eds.). 1995. *Gray's Anatomy,* 38th ed. New York: Churchill Livingstone.

Woo, J.-K. 1949. Ossification and growth of the human maxilla, premaxilla and palate bone. *Anatomical Record* 105:737–753.

Wood-Jones, F. 1947. The premaxilla and the ancestry of man. *Nature* 159:439.

Chapter 21

Heterochronic Change in the Neurocranium and the Emergence of Modern Humans

NANCY MINUGH-PURVIS

While all living humans are modern, not every trait or feature we possess is a modern character. This fact, combined with the enormous physical and behavioral variation characterizing living and historically known humanity, makes any attempt to define ourselves from an evolutionary perspective a trying exercise. Without the biological criterion of interfertility and the behavioral criterion of language, both of which are so critical to our recognition of all living humans as members of one single, highly polytypic species, we find that we often lack the necessary tools to recognize ourselves in the fossil record. This inability to recognize early modern humans has rendered paleoanthropologists unable to resolve internal debates regarding modern human origins, as seen in the great proliferation of literature devoted to this topic over the past fifteen years. Clearly, those studying the human paleontological record need to learn what criteria other than interfertility and language are recognizably modern about modern humans and how to recover such evidence of modernity from the fossil record or the debates such as those surrounding the modern human origins controversy since 1980 will continue unresolved (Minugh-Purvis 1996).

Because studies of late Pleistocene hominids tend to look at products rather than processes in data collection and analysis, paleoanthropologists have traditionally looked at morphology in attempting to assess what is and what is not modern using a phenotypic perspective. Unfortunately, we do not yet know to what extent a modern phenotype necessarily reflects a modern genotype, or how these data relate to modern human behavior. One option that offers an intelligent approach to these questions is the ontogenetic perspective. This is because ontogeny is the process that transforms members of living human groups from genotypic transcriptions into functionally, behaviorally modern people.

Each and every genotype includes specific instructions dictating the time at which additions and deletions to an organism's tissue volume (i.e., its growth) should occur. Because each species has a unique core genotype, it follows that each has a unique pattern of growth. However, documenting the complexity of human growth appears at once overwhelming. For example, we must ask at what level do we wish to study growth: proteins? cells? tissues? organs? regions of the body? Moreover, we cannot limit our investigation simply to growth in size. Rather, we also need to examine the patterning of growth, such as its duration, time of initiation, and time of cessation in a given group of structures.

These ontogenetic patterns, and the biological complexes they affect, constitute a unique growth pattern or signature characterizing all normal members of all species, including, of course, ourselves and earlier humans. If a modern human signature is preserved and identifiable in the late Pleistocene hominid paleontological record, we will be well on our way to identifying who is most likely an early modern human. Conversely, deviations from this signature involving not only growth in size but, more importantly, alterations in growth rate and timing, known as heterochronies, will help us identify who might not be an early modern human. Finally, such investigations could potentially lead us to identify the late Pleistocene population(s) that heterochronic changes transformed, ultimately, into modern humans.

Materials and Methods

This study arose as part of a long-term, ongoing investigation of craniofacial ontogeny in Pleistocene hominid evolution in progress since 1978 (see Minugh-Purvis 1988, 1998; Minugh-Purvis et al. 2000). The present investigation used neurocranial remains of thirty-one immature late Pleistocene hominid specimens from Europe, the Middle East, and Uzbekistan (Table 21.1). These included Neandertals (n = 18), early Upper Paleolithic associated Europeans (n = 11), and Levantine children from Skhul and Qafzeh (n = 2). Original and published data from several recent modern

Table 21.1   Late Pleistocene Children Examined (n = 31)

Specimen	Approximate Developmental Age at Death
Neandertals	
Carigüela 2	juvenile, approx. 5–8 yr
Devil's Tower 1	4–5 yr
Engis 2	4–5 yr
Le Fate*	7–9 yr
La Ferrassie 3[†]	juvenile, approx. 10 yr
La Ferrassie 8[†]	1.5–2.5 yr
Krapina 1	juvenile, approx. 6–8 yr[‡]
Krapina 2	juvenile, approx. 9–11 yr
Krapina 17	1–2.5 yr
La Chaise-Delauney	juvenile
Lazaret	8–9 yr
Le Moustier 1[§]	approx. 16 yr
Pech de l'Azé 1	2–3 yr
La Quina H-18	7–8 yr
Roc de Marsal 1	3–4 yr
Shanidar 7[¶]	6–9 mo
Subalyuk 2	2–3 yr
Teshik-Tash 1	10.5–11 yr
Early *Homo sapiens sapiens*	
Kostenki 3	6–7 yr
Kostenki 4	9–11 yr
Mladeč 3	approx. 2–2.5 yr
Parpalló 1	approx. 16 yr
Předmostí 2**	7–8 yr
Předmostí 5**	15–17 yr
Předmostí 6**	2–2.5 yr
Předmostí 7**	11–12 yr
Předmostí 22**	8–9 yr
Sungir' 2	11–12 yr
Sungir' 3	10–11 yr
Late Pleistocene Levantine people	
Skhul 1	4–5 yr
Qafzeh 11[††]	approx. 12 yr

Note: Age determinations are taken from Minugh-Purvis (in preparation) unless otherwise noted.

*Metric data from Giacobini et al. (1984).

[†]Metric data from Heim (1982).

[‡]Age determination taken from Minugh-Purvis et al. (2000).

[§]Metric data from a cast and Weinert (1925).

[¶]Metric data taken from Trinkaus (1934).

**Metric data from Matiegka (1934).

[††]Metric data from Tillier (1992).

skeletal samples were used for comparisons with the late Pleistocene specimens: recent *Homo sapiens sapiens* children from the twenty-sixth to the thirtieth dynasty Gizeh Egyptian E Series (n = 72) (described in Pearson and Davin 1924 and for which I am indebted to Mark Skinner); a late-nineteenth- and early-twentieth-century European sample (n = 20) published by Madre-Dupouy (1992); original data collected by the author from the Libben Amerind Late Woodland site (see Lovejoy et al. 1977 for details on the site and temporal context); the Tepe Hissar Iranian Collection dated between 6,000 and 3,300 B.P. (Dyson 1968a, 1968b); and miscellaneous nineteenth- and twentieth-century osteological preparations from the collections of the Department of Anthropology, University of Pennsylvania. Modern head circumference data for living human populations were taken from studies reported in Meredith (1971).

This study focused on metrical aspects of neurocranial ontogeny because continuous traits are easily measurable. The neurocranium was considered of particular interest and utility for three reasons. First, many neurocranial specimens also include facial remains with teeth, which lend themselves to the most precise available methods for estimating the approximate age at death of a specimen—an important factor in constructing growth curves. Second, the braincase, along with the brain, has evolved in size extremely rapidly, tripling since the earliest known hominids (Tobias 1991). Third, the modern human brain possesses a rather distinctive growth pattern differing considerably from that of other primates (Harvey et al. 1986).

Assuming that estimates of individual specimens' ages at death are reasonably accurate and that the specimens used in reconstructing growth patterns bear some semblence to population samples, we can plot dimensions to reconstruct what we hope might approximate cross-sectional growth curves for the fossil samples. This, in turn, will permit comparisons and elucidate any differences in the growth of these dimensions between the late Pleistocene and modern groups. Measurements obtained for each dimension were compared using plots of raw size by age. The graphs generated in this way provided the basic information needed to delineate a growth signature: growth in absolute size at given developmental ages, information concerning postnatal onset or offset timing of growth for a given dimension, and postnatal maturation rate.

Neurocranial ontogeny was assessed from two perspectives. First, overall neurocranial growth was examined using the single criterion of head circumference. Head circumference is used universally among clinical practitioners and human biologists as a general baseline for brain growth, and a considerable amount of information is available on the growth of this dimension in living modern human populations. Thus, head circumference would seem a logical choice for comparison of neurocranial growth through the late Pleistocene. However, numerous ontogenetic and heterochronic studies support the notion that terminal additions, or growth alterations that occur late in the growth period, most often provide the subtle ontogenetic changes differentiating closely related taxa (McKinney and Gittleman 1995). Work by Thompson and colleagues (Thompson 1998; Thompson and Nelson 2000) and earlier investigators such as Weinert (1925) on the adolescent Neandertal from Le Moustier, France, supports the existence of this phenomenon in hominids. Thus, rather than limit this examination of neurocranial growth and, indirectly, brain growth to the single, conventionally used dimension of cranial circumference, this study also examined constituent regions and bones of the braincase in an attempt to uncover any subtle differences in growth that might have occurred through late Pleistocene hominid evolution.

Results

Cranial Circumference

Examination of overall neurocranial growth as measured by cranial circumference revealed some interesting findings. Meredith (1971) noted a small amount of variation throughout the growth period, by sex and geographic region, in head circumference of six modern human populations (Indians, Bulgarians, U.S. Chinese, U.S. Caucasoids, Europeans, and African Americans). Overshadowing these slight differences, however, is the fact that the overall ontogenetic pattern between modern human groups is remarkably consistent (Fig. 21.1*A*).

When early Upper Paleolithic values are plotted against Meredith's data (Fig. 21.1*B*), they fit fairly well. Neandertals (Fig. 21.1*C*) fit even better. However, the similar curves obtained for these late Pleistocene and modern groups do not necessarily mean that all exhibit the same overall growth pattern for this important human biological complex. First, it is important to acknowledge that the comparison is, in some regards, not very accurate, given that the modern norms are for living children and include scalp thicknesses and hair, whereas the fossil measurements are of bone. An even greater distortion is introduced into comparisons with the fossils by the ac-

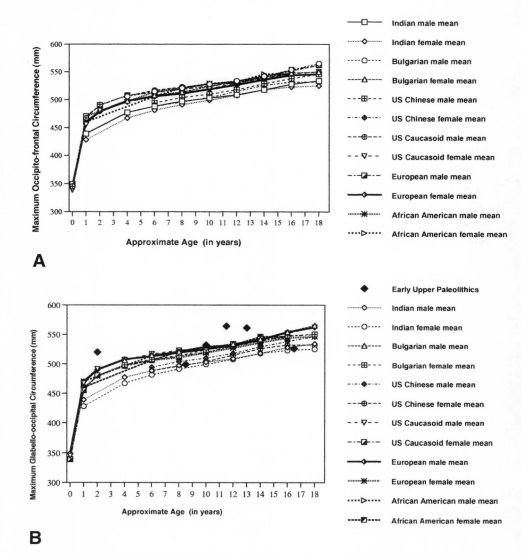

Fig. 21.1. Growth of head circumference in (*A*) six modern human populations; (*B*) living humans compared with early Upper Paleolithic Europeans; (*C*) living humans compared with Neandertals. *Sources:* Modern human data from Meredith 1971; early Upper Paleolithic and Neandertal data from Minugh-Purvis 1988.

quisition, particularly among Neandertals, of robust brow ridges. These prominent cranial superstructures, unrelated to the contents of the cranial cavity, expanded the anteroposterior occipitofrontal diameter of these hominids beginning from at least six to seven years of age. Because of the de-

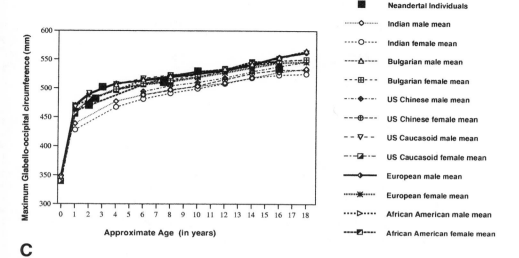

C

Fig. 21.1. (*Continued*)

gree of browridge formation in these hominids, this superstructure is unavoidable in measuring glabello-occipital diameter and thus progressively inflates cranial circumference values through late childhood and into adulthood in Neandertals and, to a lesser extent, the early Upper Paleolithic associated specimens.

Because cranial circumference obviously, then, is a slightly different entity in the groups examined, this measurement may not be appropriate for comparison between the late Pleistocene and modern groups measured. Without additional comparisons, the impression of similarity between the Neandertal, Upper Paleolithic, recent modern, and living human groups— particularly late in the growth period—may not be accurate. This situation further justifies a more detailed examination of the component parts of the neurocranium.

Anterior Vault

Comparison of frontal sagittal length using such dimensions as nasion-bregma and glabella-bregma chord and arc revealed no distinguishable differences between the fossil groups examined, although this measurement, like cranial circumference, becomes increasingly distorted through the growth period by the development of brow ridges among late Pleistocene children. However, in comparisons of frontal breadth, some differences became apparent. Growth patterning in minimum frontal breadth (Fig. 21.2) appears comparable between early Upper Paleolithic and recent modern groups, both of which differ from the Neandertal pattern. In early childhood, both Neandertals and early Upper Paleolithic Europeans are approximately the same size for minimum frontal breadth. However, the Neandertals grow rapidly during this time, predicting the larger adult Neandertal size for this dimension.

Interestingly, offset timing in minimum frontal breadth growth also occurred early among Neandertals, probably in midchildhood (Fig. 21.3A). In contrast, although similar in size to Neandertals at the beginning of early childhood, the early Upper Paleolithic children did not reach their adult size range until after eight years of age, considerably later than the Neandertals but only slightly earlier than the recent modern humans (Fig. 21.3B). These data indicate that Neandertal minimum frontal breadths grew to an absolutely larger size than those of modern humans, despite the fact that, in Neandertals, this dimension grew for a shorter period. The development of a more extended growth period, or slower maturation for

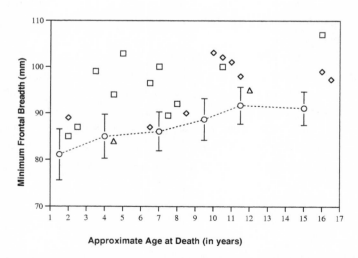

Fig. 21.2.   Growth in minimum frontal breadth in Neandertals, early Upper Paleolithic Europeans, recent humans, and Skhul/Qafzeh. *Squares,* Neandertals; *diamonds,* early Upper Paleolithic associated Europeans; *circles,* recent humans (*bar* indicates ±2 SD); *triangles,* Skhul/Qafzeh. *Source:* Data from Minugh-Purvis 1988 and manuscript in preparation.

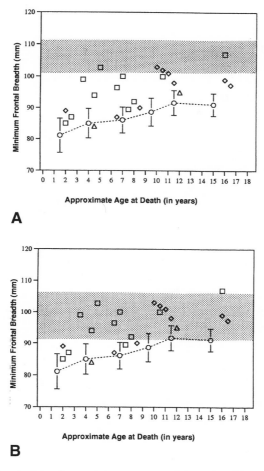

Fig. 21.3. Offset timing of growth in minimum frontal breadth. *Squares,* Neandertals; *diamonds,* early Upper Paleolithic associated Europeans; *circles,* recent humans (*bar* indicates ±2 SD); *triangles,* Skhul/Qafzeh. *A, screened area,* adult Neandertal range; *B, screened area,* adult early Upper Paleolithic associated European range. *Source:* Data from Minugh-Purvis 1988 and manuscript in preparation.

minimum frontal breadth in *H. s. sapiens,* provides evidence of a dissociation between growth in size and rate for this dimension between Neandertals and ourselves. Thus, we find an alteration of growth at the Neandertal/early Upper Paleolithic modern interface, although the early Upper Paleolithic associated Europeans are, in offset timing for growth in minimum frontal breadth, intermediate between these two.

Midvault

Turning to the parietal bone, which comprises the majority of the midvault region, we find that comparisons of parietal length by age in Neandertals and modern humans show marked differences in absolute size throughout the growth period (Fig. 21.4). These parietal length differences in the children are consistent with the well-known large difference in adult mean values for midvault length between Neandertals and early Upper Paleolithic Europeans.

However, in growth *patterning*, both Neandertals and the early Upper Paleolithic associated Europeans reached adult size for bregma-lambda chord (Fig. 21.5) and arc (Fig. 21.6) during early childhood. If correct, this pattern indicates the influence of an ontogenetic shift between Neandertals and Upper Paleolithic people, characterized by greater growth in parietal length per unit of time among the Upper Paleolithic children—there would seem to be no other explanation for the acquisition of the longer midvault segment of the Upper Paleolithic neurocranium. However, the early childhood offset timing among the Upper Paleolithic children contrasts with that seen in the recent human sample, where the growth curve for the parietal sagittal chord does not level off until late childhood or early adolescence. Thus, a second heterochrony occurs for this dimension, this one between the late Pleistocene and more modern *H. s. sapiens,* in which the offset timing, or duration of growth, in this dimension is increased between the fossil specimens (both Neandertals and Upper Paleolithics) and the modern humans.

These patterns suggest that the size changes in midvault length that we see with the emergence of modern humans were accomplished first by an acceleration of growth in early Upper Paleolithic people and then as a prolongation of growth (i.e., a longer duration for midvault growth in recent modern humans). The net result in either case is the longer parietal bones of modern humans as opposed to Neandertals. However, while the early Upper Paleolithic Europeans exhibit a longer midvault than Neandertals, thus *morphologically* sharing the same pattern as modern humans, the Upper Paleolithic European phenotype resulted from a different process than that operating in ourselves.

Analysis of anterior parietal breadths did not provide any clear-cut growth differences between the groups examined. However, some interesting differences in breadth appeared more posteriorly in the neurocranium.

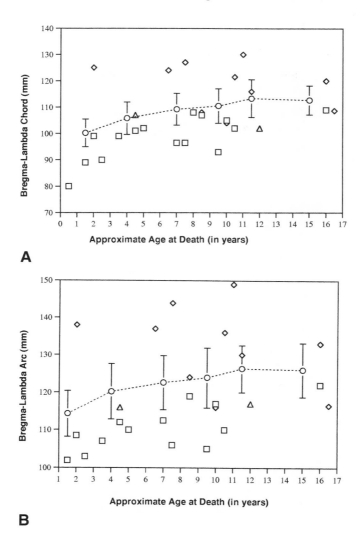

Fig. 21.4. Growth in bregma-lambda length: *squares*, Neandertals; *diamonds*, early Upper Paleolithic associated Europeans; *circles*, recent humans (*bar*, ±2 SD); *triangles*, Skhul/Qafzeh. *A*, chord; *B*, arc. *Source:* Data from Minugh-Purvis 1988 and manuscript in preparation.

Posterior Vault

At the posterior cranial vault, heterochronic trends are reversed. Posterior parietal growth shows a shift toward an earlier offset timing in modern humans compared with Neandertals for lambda-asterion chord (Fig. 21.7*A*)

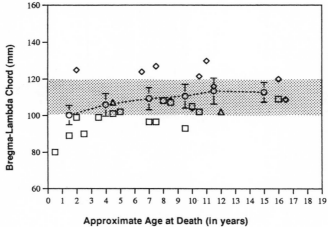

A

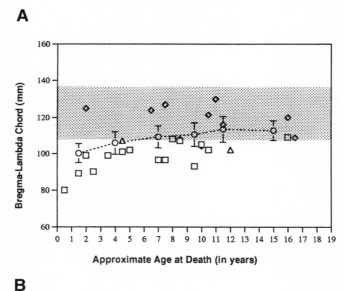

B

Fig. 21.5. Offset timing of growth in bregma-lambda chord. *Squares,* Neandertals; *diamonds,* early Upper Paleolithic associated Europeans; *circles,* recent humans (*bar* indicates ±2 SD); *triangles,* Skhul/Qafzeh. *A, screened area,* adult Neandertal range; *B, screened area,* adult early Upper Paleolithic associated European range. *Source:* Data from Minugh-Purvis 1988 and manuscript in preparation.

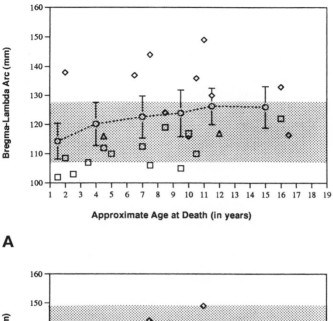

A

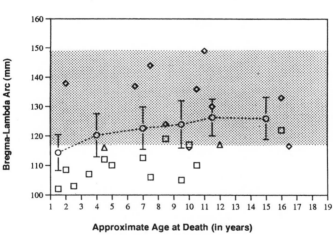

B

Fig. 21.6. Offset timing of growth in bregma-lambda arc. *Squares*, Neandertals; *diamonds*, early Upper Paleolithic associated Europeans; *circles*, recent humans (*bar* indicates ±2 SD); *triangles*, Skhul/Qafzeh. *A, screened area*, adult Neandertal range; *B, screened area*, adult early Upper Paleolithic associated European range. *Source:* Data from Minugh-Purvis 1988 and manuscript in preparation.

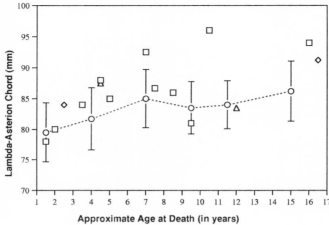

A

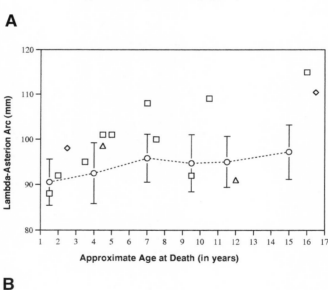

B

Fig. 21.7. Growth in lambdoid chord and arc: *squares,* Neandertals; *diamonds,* early Upper Paleolithic associated Europeans; *circles,* recent humans (*bars,* ±2 SD); *triangles,* Skhul/Qafzeh. *A,* chord; *B,* arc. *Source:* Data from Minugh-Purvis 1988 and manuscript in preparation.

and arc (Fig. 21.7*B*) breadths. In growth patterning, Neandertal posterior parietal breadth shows continued growth through late childhood and into adolescence, while recent modern humans reached adult values earlier—by approximately seven years of age, suggesting an earlier completion of

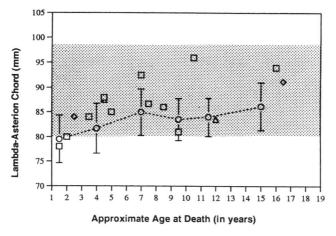

A

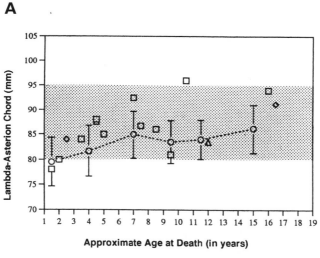

B

Fig. 21.8. Offset timing of growth in the lambda-asterion chord. *Squares*, Neandertals; *diamonds*, early Upper Paleolithic associated Europeans; *circles*, recent humans (*bar* indicates ±2 SD); *triangles*, Skhul/Qafzeh. *A, screened area,* adult Neandertal range; *B, screened area,* adult early Upper Paleolithic associated European range. *Source:* Data from Minugh-Purvis 1988 and manuscript in preparation.

growth in this dimension for the recent modern sample (Fig. 21.8). Unfortunately, only two data points are available for this dimension in the early Upper Paleolithic Europeans, but they seem to approximate the Neandertal pattern, suggesting prolonged growth of this dimension into adoles-

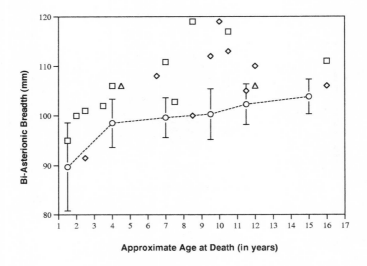

Fig. 21.9. Growth in occipital breadth: *squares*, Neandertals; *diamonds*, early Upper Paleolithic associated Europeans; *circles*, recent humans (*bars*, ±2 SD); *triangles*, Skhul/Qafzeh. *Source:* Data from Minugh-Purvis 1988 and manuscript in preparation.

cence. This hint that early Upper Paleolithic Europeans perhaps followed a pattern of posterior neurocranial vault growth similar to that of Neandertals is supported by the ontogenetic trends seen for another posterior vault dimension, biasterion breadth.

An examination of occipital breadth, using the biasterion chord, shows nearly all the Neandertal children falling above the modern range for occipital breadth throughout ontogeny, whereas the occipital breadth of early Upper Paleolithic Europeans (Fig. 21.9) is generally slightly narrower than that of the Neandertals, a finding consistent with metrical comparisons between the adults of these groups. Growth patterning in the modern children (Fig. 21.10) shows a steady increase in biasterion breadth until approximately four years of age, a much earlier offset than growth in biasterion breadth for both Neandertals (as suggested by Trinkaus and LeMay 1982) and early Upper Paleolithic Europeans, in which occipital breadth seems to have grown until perhaps midchildhood. The fact that no differences are obvious between the Neandertals and the early Upper Paleolithic Europeans for the *pattern* of growth in this dimension suggests that at least some heterochronic changes in the posterior neurocranium have appeared since the emergence of what we have long been calling early modern or anatomically modern humans.

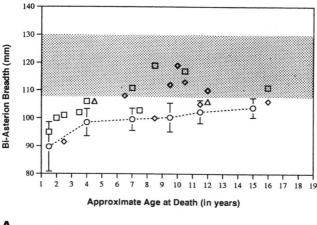

A

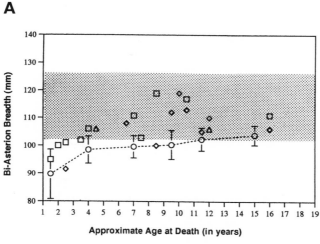

B

Fig. 21.10. Offset timing of growth in biasterion breadth. *Squares,* Neandertals; *diamonds,* early Upper Paleolithic associated modern Europeans; *circles,* recent humans (*bar* indicates ±2 SD); *triangles,* Skhul/Qafzeh. *A, screened area,* adult Neandertal range; *B, screened area,* adult early Upper Paleolithic associated European range. *Source:* Data from Minugh-Purvis 1988 and manuscript in preparation.

Conclusions

These findings clearly indicate that ontogenetic changes have accompanied evolutionary change in the neurocranium of late Pleistocene *Homo,* with some of these changes resulting in phenotypic differences among adults traditionally considered archaic humans and so-called early modern humans. Interestingly, however, other ontogenetic changes led to phenotypic convergences between these groups. Moreover, the fact that such changes are not limited to the archaic–early modern interval but also characterize the early–recent modern interval is significant in that it demonstrates that *neurocranial ontogeny followed a mosaic of heterochronic change describing an evolutionary continuum, which defies any easy division between archaic and modern humans.* Early Upper Paleolithic Europeans, usually regarded as early modern humans based on both behavioral and morphological criteria, were not fully modern in their growth patterning.

These findings also argue strenuously that assessments of ontogenetic change among such closely related groups as late Pleistocene hominids require a fine level of scrutiny. It is striking, for example, that cranial circumference *totally masks* the finer, more subtle differences evident when neurocranial growth is analyzed by component parts. For this reason, we should clearly avoid using gross measurements or single measurements when we attempt to reconstruct or test hypotheses about growth, given that subtle, local heterochronies are more likely to be operating at low levels of phylogeny than are larger, more global patterns of heterochronic change. The present study reaffirms that the samples examined here can only be regarded as a biological continuum. What we have tentatively considered as early Upper Paleolithic associated modern Europeans did not grow entirely like living Europeans; rather, at least in the neurocranium, their growth followed a pattern intermediate between those of Neandertals and ourselves.

Acknowledgments

Thanks to Doug Purvis for his unfailing support, which has made this research possible. Financial support of the LSB Leakey Foundation, Foundation for Research into the Origin of Man, Richard D. Irwin Foundation, Sigma Xi, Department of Neurobiology and Anatomy, MCP Hahnemann University, and Department of Anthropology, University of Pennsylvania, is gratefully acknowledged.

References

Dyson, R. 1968a. Annotations and corrections of the relative chronology of Iran. *American Journal of Archaeology* 72:308–313.

————. 1968b. The archaeological evidence of the second millenium B.C. on the Persian plateau. In *Cambridge Ancient History*, rev. ed., fasc. 66. Cambridge: Cambridge University Press.

Giacobini, G., de Lumley, M.-A., Yokoyama, Y., and Nguyen, H-V. 1984. Neanderthal child and adult remains from a Mousterian deposit in northern Italy (Caverna delle Fate, Finale Ligure). *Journal of Human Evolution* 13:687–707.

Harvey, P.H., Martin, R.D., and Clutton-Brock, T.H. 1986. Life histories in comparative perspective. In B.B. Smuts, D.L. Cheney, R.M. Seyfarth, R.W. Wrangham, and T.T. Struhsaker (eds.), *Primate Societies*, 181–196. Chicago: Chicago University Press.

Heim, J.-L. 1982. *Les enfants Néandertaliens de la Ferrassie*. Paris: Masson.

Lovejoy, C.O., Meindl, R.S., Pryzbeck, T.R., Barton, T.S., Heiple, K.G., and Kotting, D. 1977. Paleodemography of the Libben site, Ottawa County, Ohio. *Science* 198:291–293.

Madre-Dupouy, M. 1992. *L'enfant du Roc de Marsal*. Paris: CNRS.

Matiegka, J. 1934. *Homo Předmostensis: Fosilní člověk z Předmostí na Moravě. I. Lebky*. Prague: Nákladem České Akademie věd a Umění.

McKinney, M., and Gittleman, J.L. 1995. Ontogeny and phylogeny: tinkering with covariation in life history, morphology, and behaviour. In K.J. McNamara (ed.), *Evolutionary Change and Heterochrony*, 21–47. Chichester, England: Wiley.

Meredith, H.V. 1971. Human head circumference from birth to early adulthood: racial, regional, and sex comparisons. *Growth* 35:233–251.

Minugh-Purvis, N. 1988. Patterns of craniofacial growth and development in Upper Pleistocene hominids. Ph.D. diss., University of Pennsylvania.

————. 1996. Recent evolution of the modern human origins controversy, 1984–1994. *Evolutionary Anthropology* 4:140–147.

————. 1998. The search for the earliest modern Europeans: a comparison of the Krapina 1 and es-Skhul 1 juveniles. In T. Akazawa, K. Aoki, and O. Bar-Yosef (eds.), *Neanderthals and Modern Humans in West Asia*, 339–352. New York: Plenum.

————. (in prep.) *Growth of the Neandertal Skull*.

Minugh-Purvis, N., Radovčić, J., and Smith, F.H. 2000. Krapina 1: a juvenile Neandertal from the early Late Pleistocene of Croatia. *American Journal of Physical Anthropology* 111:393–424.

Pearson, K., and Davin, A.G. 1924. On the biometric constants of the human skull. *Biometrika* 16:328–363.

Thompson, J.L. 1998. Neanderthal growth and development. In S.J. Ulijaszek, F.E. Johnston, and M.A. Preece (eds.), *The Cambridge Encyclopedia of Human Growth and Development*, 106–107. Cambridge: Cambridge University Press.

Thompson, J.L., and Nelson, A.J. 2000. The place of Neandertals in the evolution of hominid patterns of growth and development. *Journal of Human Evolution* 38:475–495.

Tillier, A.-M. 1992. The origins of modern humans in southwest Asia: ontogenetic aspects. In T. Akazawa, A. Kenichi, and T. Kimura (eds.), *The Evolution and Dispersal of Modern Humans in Asia*, 15–28. Tokyo: Hokusen-Sha.

Tobias, P.V. 1991. *Olduvai Gorge:* Vol. 4. *The Skulls, Endocasts, and Teeth of* Homo habilis. Cambridge: Cambridge University Press.

Trinkaus, E. 1983. *The Shanidar Neandertals.* New York: Academic Press.

Trinkaus, E., and LeMay, M. 1982. Occipital bunning among later Pleistocene hominids. *American Journal of Physical Anthropology* 57:27–35.

Weinert, H. 1925. *Der Schädel des eiszeitlichen Menschen von Le Moustier in neuer zusammensetzung.* Berlin: Julius Springer.

Glossary

Acceleration. Heterochronic process involving faster rate of development in the descendant; produces a peramorphic trait in the adult phenotype.

Allometry. The study of size and shape; the change in size and shape observed. Complex allometry occurs when the ratios of the specific growth rates of the traits compared are not constant and a log-log plot comparing the traits does not yield a straight line (k in the allometric formula not remaining constant). When there is no change in shape with size increase, isometry is said to occur. In other words, a log-log plot yields a straight line with a slope of k equal to 1. Positive allometry occurs when trait x is increasing more slowly than trait y (slope, $k > 1$). Negative allometry is the reverse (slope, $k < 1$).

Deceleration (syn. neoteny). Slower rate of developmental events in the descendant; produces paedomorphic traits when expressed in the adult phenotype.

Dissociated heterochrony. An ontogenetic change in rate or timing of a trait that does not occur in some other trait; consequently, some traits may be paedomorphic, others peramorphic.

Epigenetics. The sum of the genetic and nongenetic factors acting upon cells to control gene expression selectively to produce increasing phenotypic complexity during development and evolution.

Global heterochrony. Ontogenetic change in rate or timing that affects the entire individual.

Heterochrony. Change in timing or rate of developmental events relative to the same event in the ancestor.

Heterotopy. The displacement of the development of an organ or tissue in space rather than in time (heterochrony).

Hypermorphosis. Delayed cessation (offset) of developmental events in the descendant; produces peramorphic traits when expressed in the adult phenotype. Late sexual maturation can produce global hypermorphosis; this has also been termed *terminal hypermorphosis*. Delayed cessation in local growth fields can also produce hypermorphosis.

Isomorphosis. The unusual situation of paedomorphosis being followed by peramorphosis, or vice versa, resulting in no effective morphological change in the descendant.

Neoteny (syn. deceleration). Slower rate of developmental events in the descendant; produces paedomorphic traits when expressed in the adult phenotype.

499

Ontogeny.   Growth (usually size increase) and development (both size increase and differentiation of traits) of the individual.

Paedomorphosis.   The retention of subadult ancestral traits in the descendant adult.

Peramorphosis.   Development of traits beyond that of the ancestral adult.

Postdisplacement.   Late initiation ("onset") of developmental events in the descendant; produces paedomorphosis when expressed in the adult phenotype.

Postformation.   Paedomorphosis arising from inital shape underdevelopment.

Predisplacement.   Early initiation ("onset") of developmental events in the descendant; produces peramorphosis when expressed in the adult phenotype.

Preformation.   Peramorphosis arising from initial shape overdevelopment.

Progenesis (syn. time hypomorphosis).   Early cessation (offset) of developmental events in the descendant; produces paedomorphic traits when expressed in the adult phenotype. Early sexual maturation will produce global progenesis; this has also been termed **terminal progenesis.** Early cessation in local growth fields can also produce progenesis.

Sequential heterochrony.   Prolongation or contraction of ontogenetic growth or life-history stages in the descendant relative to the ancestor.

Sequential hypermorphosis (syn. proportional growth prolongation).   Prolongation of ontogenetic growth or life-history stages in the descendant relative to the ancestor.

Sequential progenesis.   Contraction of ontogenetic growth or life-history stages in the descendant relative to the ancestor.

Time hypomorphosis (syn. progenesis).   Early cessation (offset) of developmental events in the descendant; produces paedomorphic traits when expressed in the adult phenotype. Early sexual maturation will produce global time hypomorphosis, but early cessation in local growth fields can also produce time hypomorphosis.

Index